FUZZY MULTICRITERIA DECISION-MAKING

FUZZY MULTICRITERIA DECISION-MAKING

MODELS, METHODS AND APPLICATIONS

Witold Pedrycz

University of Alberta, Canada
Systems Research Institute,
Polish Academy of Sciences, Poland

Petr Ekel and Roberta Parreiras

Pontifical Catholic University of Minas Gerais,
Belo Horizonte, Brazil

WILEY

A John Wiley and Sons, Ltd., Publication

This edition first published 2011

Registered office
John Wiley & Sons Ltd, The Atrium, Southern Gate, Chichester, West Sussex, PO19 8SQ, United Kingdom

For details of our global editorial offices, for customer services and for information about how to apply for permission to reuse the copyright material in this book please see our website at www.wiley.com.

Library of Congress Cataloguing-in-Publication Data

Pedrycz, Witold, 1953-
Models and algorithms of fuzzy multicriteria decision-making and their applications/Witold Pedrycz, Petr Ekel, Roberta Parreiras. – 1st ed.
p. cm.
Includes index.
ISBN 978-0-470-68225-8 (cloth)
1. Decision-making. 2. Decision-making–Mathematical models. 3. Fuzzy decision-making. I. Ekel, Petr. II. Parreiras, Roberta. III. Title.
T57.95.P42 2011
003′.56–dc22

2010025490

A catalogue record for this book is available from the British Library.

Print ISBN: 9780470682258
ePDF ISBN: 9780470974049
oBook ISBN: 9780470974032

Typeset in 10/12pt Times by Aptara Inc., New Delhi, India
Printed and bound in Singapore by Markono Print Media Pte Ltd.

Contents

About the Authors **xi**

Foreword **xiii**

Preface **xvii**

1 Decision-Making in System Project, Planning, Operation, and Control: Motivation, Objectives, and Basic Concepts **1**
1.1 Decision-Making and its Support 1
1.2 Optimization and Decision-Making Problems 6
1.3 Multicriteria Decision-Making 9
1.4 Group Decision-Making 11
1.5 Fuzzy Sets and their Role in Decision-Making Processes 14
1.6 Conclusions 17
References 18

2 Notions and Concepts of Fuzzy Sets: An Introduction **21**
2.1 Sets and Fuzzy Sets: A Fundamental Departure from the Principle of Dichotomy 21
2.2 Interpretation of Fuzzy Sets 24
2.3 Membership Functions and Classes of Fuzzy Sets 26
2.4 Fuzzy Numbers and Intervals 28
2.5 Linguistic Variables 29
2.6 A Generic Characterization of Fuzzy Sets: Some Fundamental Descriptors 31
2.7 Geometric Interpretation of Sets and Fuzzy Sets 37
2.8 Fuzzy Sets and the Family of α-cuts 38
2.9 Operations on Fuzzy Sets 41
2.9.1 Triangular Norms and Triangular Conorms as Models of Operations on Fuzzy Sets 43
2.9.2 Negations 46
2.10 Fuzzy Relations 47
2.10.1 The Concept of Relations 47
2.10.2 Fuzzy Relations 49
2.10.3 Properties of the Fuzzy Relations 50

	2.10.4	*Domain and Codomain of Fuzzy Relations*	50
	2.10.5	*Representation of Fuzzy Relations*	51
	2.10.6	*Equality of Fuzzy Relations*	51
	2.10.7	*Inclusion of Fuzzy Relations*	51
	2.10.8	*Operations on Fuzzy Relations*	51
	2.10.9	*Union of Fuzzy Relations*	52
	2.10.10	*Intersection of Fuzzy Relations*	52
	2.10.11	*Complement of Fuzzy Relations*	52
	2.10.12	*Transposition of Fuzzy Relations*	52
	2.10.13	*Cartesian Product of Fuzzy Relations*	52
	2.10.14	*Projection of Fuzzy Relations*	53
	2.10.15	*Cylindrical Extension*	54
	2.10.16	*Reconstruction of Fuzzy Relations*	55
	2.10.17	*Binary Fuzzy Relations*	55
	2.10.18	*Transitive Closure*	56
	2.10.19	*Equivalence and Similarity Relations*	57
	2.10.20	*Compatibility and Proximity Relations*	58
2.11	Conclusions		59
	References		61
3	**Selected Design and Processing Aspects of Fuzzy Sets**		**63**
3.1	The Development of Fuzzy Sets: Elicitation of Membership Functions		63
	3.1.1	*Semantics of Fuzzy Sets: Some General Observations*	63
	3.1.2	*Fuzzy Set as a Descriptor of Feasible Solutions*	65
	3.1.3	*Fuzzy Set as a Descriptor of the Notion of Typicality*	66
	3.1.4	*Vertical and Horizontal Schemes of Membership Function Estimation*	68
	3.1.5	*Saaty's Priority Approach of Pairwise Membership Function Estimation*	70
	3.1.6	*Fuzzy Sets as Granular Representatives of Numeric Data – A Principle of Justifiable Granularity*	73
	3.1.7	*Design of Fuzzy Sets through Fuzzy Clustering: From Data to their Granular Abstraction*	76
	3.1.8	*Fuzzy Equalization as a Way of Building Fuzzy Sets Supported by Experimental Evidence*	80
	3.1.9	*Several Design Guidelines for the Formation of Fuzzy Sets*	81
3.2	Aggregation Operations		82
	3.2.1	*Averaging Operations*	83
3.3	Transformations of Fuzzy Sets		84
	3.3.1	*The Extension Principle*	84
	3.3.2	*Fuzzy Numbers and Fuzzy Arithmetic*	87
	3.3.3	*Interval Arithmetic and α-cuts*	88
	3.3.4	*Fuzzy Arithmetic and the Extension Principle*	90
	3.3.5	*Computing with Triangular Fuzzy Numbers*	93
	3.3.6	*Addition*	93
	3.3.7	*Multiplication*	95
	3.3.8	*Division*	96

3.4	Conclusions	97
	References	101
4	**Continuous Models of Multicriteria Decision-Making and their Analysis**	**103**
4.1	Continuous Models ($\langle \mathbf{X}, \mathbf{M} \rangle$ Models) of Multicriteria Decision-Making	103
4.2	Pareto-Optimal Solutions	104
4.3	Approaches to the Use of DM Information	105
4.4	Methods of Multiobjective Decision-Making	107
4.5	Bellman–Zadeh Approach and its Application to Multicriteria Decision-Making	113
4.6	Multicriteria Resource Allocation	115
4.7	Adaptive Interactive Decision-Making System for Multicriteria Resource Allocation	118
4.8	Application of the Bellman–Zadeh Approach to Multicriteria Problems	120
4.9	Conclusions	132
	References	134
5	**Introduction to Preference Modeling with Binary Fuzzy Relations**	**137**
5.1	Binary Fuzzy Relations and their Fundamental Properties	137
5.2	Preference Modeling with Binary Fuzzy Relations	141
5.3	Preference Structure of Binary Fuzzy Preference Relations	145
5.4	A Method for Constructing a Fuzzy Preference Structure	146
5.5	Consistency of Fuzzy Preference Relations	148
5.6	Conclusions	151
	References	152
6	**Construction of Fuzzy Preference Relations**	**155**
6.1	Preference Formats	155
	6.1.1 Ordering of Alternatives	156
	6.1.2 Utility Values	157
	6.1.3 Fuzzy Estimates	160
	6.1.4 Multiplicative Preference Relations	161
	6.1.5 Fuzzy Preference Relations	163
6.2	Ordering of Fuzzy Quantities and the Construction of Fuzzy Preference Relations	166
6.3	Transformation Functions and their Use for Converting Different Preference Formats into Fuzzy Preference Relations	172
	6.3.1 Transformation Functions for ARFPR	172
	6.3.2 Transformation Functions for NRFPRs	180
6.4	A Method for Repairing Inconsistent Judgments	185
	6.4.1 Method for Repairing an Inconsistent ARFPR to Satisfy Weak Transitivity	186
6.5	Conclusions	188
	References	189

7	**Discrete Models of Multicriteria Decision-Making and their Analysis**	**193**
7.1	Optimization Problems with Fuzzy Coefficients and their Analysis	193
7.2	Discrete Models ($\langle \mathbf{X}, \boldsymbol{R} \rangle$ Models) of Multiattribute Decision-Making	202
7.3	Basic Techniques of Analysis of $\langle \mathbf{X}, \boldsymbol{R} \rangle$ Models	203
7.4	Interactive Decision-Making System for Multicriteria Analysis of Alternatives in a Fuzzy Environment	210
7.5	Multicriteria Analysis of Alternatives with Fuzzy Ordering of Criteria	219
7.6	Multicriteria Analysis of Alternatives with the Concept of Fuzzy Majority	221
7.7	Multicriteria Analysis of Alternatives Based on an Outranking Approach (Fuzzy Promethee)	226
7.8	Application Examples	232
7.9	Conclusions	243
	References	245
8	**Generalization of a Classic Approach to Dealing with Uncertainty of Information for Multicriteria Decision Problems**	**247**
8.1	Classic Approach to Dealing with Uncertainty of Information	247
8.2	Choice Criteria	250
8.3	Generalization of the Classic Approach	251
8.4	Modification of the Choice Criteria	252
8.5	General Scheme of Multicriteria Decision-Making under Uncertainty	253
8.6	Application Example	254
8.7	Conclusions	260
	References	261
9	**Group Decision-Making: Fuzzy Models**	**263**
9.1	Group Decision-Making Problem and its Characteristics	263
9.2	Strategies for the Analysis of Group Decision-Making Problems: Multiperson and Multiattribute Aggregation Modes	265
9.3	The Different Levels of Influence of Each Expert in the Construction of the Collective Opinion	268
9.4	Aggregation Operators for Constructing Collective Opinions on the Basis of Fuzzy Models and their Properties	269
9.5	Consistency of Pairwise Judgments in Group Decision-Making	275
9.6	Fuzzy Group Decision-Making Methods	278
9.7	Conclusions	288
	References	291
10	**Use of Consensus Schemes in Group Decision-Making**	**293**
10.1	Consensus in Group Decision-Making	293
10.2	Consensus Schemes: Definition and Motivation	295
10.3	Fuzzy Concordance and Fuzzy Consensus Measures	295
10.4	Moderator Interventions	304
10.5	Optimal Consensus in a Fuzzy Environment	305

10.6 Consensus Schemes in Fuzzy Environment 309
10.6.1 Guidance Procedure of CSFE 311
10.6.2 Guidance Procedure of CSFPR 312
10.6.3 Guidance Procedure of CSRA 318
10.7 An Application Related to the Balanced Scorecard Methodology 324
10.8 Conclusions 330
References 332

Index **335**

About the Authors

Witold Pedrycz is a Professor and Canada Research Chair (CRC – Computational Intelligence) in the Department of Electrical and Computer Engineering, University of Alberta, Edmonton, Canada. He is also with the Systems Research Institute of the Polish Academy of Sciences, Warsaw, Poland. He also holds an appointment of special professorship in the School of Computer Science, University of Nottingham, UK. In 2009 Dr. Pedrycz was elected a foreign member of the Polish Academy of Sciences. His main research directions involve Computational Intelligence, fuzzy modeling and Granular Computing, knowledge discovery and data mining, fuzzy control, pattern recognition, knowledge-based neural networks, relational computing, and Software Engineering. He has published numerous papers in this area. He is also an author of 14 research monographs covering various aspects of Computational Intelligence and Software Engineering. Witold Pedrycz has been a member of numerous program committees for IEEE conferences in the area of fuzzy sets and neurocomputing. Dr. Pedrycz is intensively involved in editorial activities; he is an Editor-in-Chief of ***Information Sciences*** and Editor-in-Chief of ***IEEE Transactions on Systems, Man, and Cybernetics – part A***. He currently serves as an Associate Editor of ***IEEE Transactions on Fuzzy Systems*** and a number of other international journals. He has edited a number of volumes; the most recent one is entitled "Handbook of Granular Computing" (J. Wiley, 2008). In 2007 he received a prestigious Norbert Wiener award from the IEEE Systems, Man, and Cybernetics Council. He is a Fellow of the IEEE and a recipient of the IEEE Canada Computer Engineering Medal 2008. In 2009 he has received a Cajastur Prize for Soft Computing from the European Centre for Soft Computing for "*pioneering and multifaceted contributions to Granular Computing*".

Petr Ekel is a Professor in the Graduate Program in Electrical Engineering, Pontifical Catholic University of Minas Gerais and a Supervisor of Ph.D. Studies in the Graduate Program in Electrical Engineering, Federal University of Minas Gerais, Belo Horizonte, Brazil. He is also a President and Principal Consultant in ASOTECH – Advanced System Optimization Technologies, Belo Horizonte, Brazil. He has experience in research, teaching, consulting, and public service. His research activities in the field of system modeling, optimization, and control are concerned with discrete and fuzzy programming, fuzzy multicriteria decision-making, fuzzy preference modeling, fuzzy identification and control, construction of rational mathematical models and their analysis on the basis of experimental design, planning, operation, and control of power systems and subsystems. The results of his investigations have been reflected in more than 260 publications, including pioneering research related to applications of fuzzy set theory in power engineering. More than 45 projects with industry, including projects based on applying techniques of decision-making in a fuzzy environment and fuzzy identification and control, have been realized under his leadership or with his participation. Petr Ekel has been a chair and a member of scientific, advisory, and program committees of more than 80 conferences related to system modeling, optimization, and control, including conferences in the areas of computational intelligence and fuzzy sets. He has been a member of the editorial boards of eight international journals. He is an academician of the Ukrainian Academy of Engineering Sciences.

Roberta Parreiras received the PhD degree in Electrical Engineering from the Federal University of Minas Gerais, Belo Horizonte, Brazil, in 2006. In 2008, she concluded her post-doctoral activities in the Pontifical Catholic University of Minas Gerais, Belo Horizonte, Brazil, where she is currently as an Assistant Professor in the Department of Electrical Engineering. She is also a Consultant in ASOTECH – Advanced System Optimization Technologies, Belo Horizonte, Brazil. Her research has been reported in papers and book chapters in the areas of fuzzy multicriteria decision-making, group decision-making, consensus schemes, multiobjective optimization, evolutionary search algorithms, and robust optimization. Her activities as a consultant in ASOTECH have been mainly associated with planning, including strategic planning, in power engineering.

Foreword

We live in an age which is widely referred to as the age of information. Information has a position of centrality in modern society. Today, it would be hard to live without mobile phones, computers, fax machines, copiers, TV, radio and other artifacts of information. However, in our preoccupation with information we tend to lose sight of the fact that information is not the final destination. What lies beyond information is decision-making. In the final analysis, information is merely a basis for making rational decisions. From the time we wake up in the morning until the time we go to bed, we make a multitude of decisions. Most everyday decisions are made on a subconscious level, but some involve a conscious analysis of consequences of decisions—an analysis aimed at making a decision which in some sense is better than others.

In large measure, the information age began with the pioneering work of Shannon on information theory. Shannon's first presentation of his theory took place in New York in 1946. As a student at Columbia University, I attended his presentation. I was deeply impressed by Shannon's ideas. He opened the door to a new world—the world of information and digital information processing.

Two years before the debut of information theory, von Neumann and Morgenstern published a path-breaking book, "Theory of Games and Economic Behavior". Decision analysis, as we know it today, is the brain-child of two great minds—von Neumann and Morgenstern—and other great minds who followed them. The von Neumann-Morgenstern theory was driven by a quest for a theory which is rigorous, precise and prescriptive. The degree to which von Neumann and Morgenstern have achieved their objective is still a matter of discussion and debate.

Information has many attributes. Among them there are two that stand out in importance—uncertainty and imprecision. Roughly speaking, uncertainty relates to randomness of information while imprecision relates to fuzziness, that is, to unsharpness of class boundaries. Randomness and fuzziness are distinct phenomena, but more often than not what we observe is a mixture. The classical, Aristotelian, bivalent logic is the logic of classes with crisp, (sharp) boundaries. By contrast, fuzzy logic may be viewed as the logic of classes with unsharp boundaries. In fuzzy logic, everything is or is allowed to be a matter of degree.

Like most theories in science, the von Neumann-Morgenstern theory is based on bivalent logic. Based as it is on bivalent logic, the von Neumann-Morgenstern theory addresses the issue of randomness but not that of fuzziness. To deal with randomness, von Neumann and Morgenstern developed the Expected Utility Theory (EUT). The price of being based on bivalent logic is proclivity of EUT to counter-intuitive conclusions. The paradoxes of Allais, Ellsberg and others brought to light serious shortcomings of EUT. In my view, no bivalent-logic-based theory of decision analysis can be paradox-free.

The first attempt to address the issue of fuzziness in decision analysis was made in the 1972 paper by Bellman and I, "Decision-making in a fuzzy environment". A key idea in this paper involves aggregation of a collection of fuzzy goals (criteria) and fuzzy constraints through conjunction. The issue of uncertainty was not substantively addressed. In a later, 1976 paper, "The linguistic approach and its application to decision analysis", I introduced the concepts of fuzzy Pareto-optimality and linguistic preference relations. Since then, a number of papers in the literature have addressed the issue of fuzziness in decision analysis.

The book "Fuzzy Multicriteria Decision-Making: Models, Methods and Applications", or FMD for short, co-authored by Witold Pedrycz, Petr Ekel and Roberta Parreiras, is the first comprehensive treatment of both fuzziness and randomness in decision analysis. As such, it opens the door to construction of far more realistic models of decision problems that can be constructed through the use of bivalent-logic-based theories. In FMD, employment of fuzzy set theory reflects a fact of life: the closer one gets to reality the fuzzier it looks.

What one finds in FMD goes far beyond what has appeared in the literature of decision analysis. Since fuzzy set theory is used extensively in FMD, FMD includes a very thorough and skillfully-organized exposition of those parts of fuzzy set theory which are of relevance to decision analysis. An exposition of applications of fuzzy set theory to decision analysis begins in Chapter 4. In Chapters 4 and 5, principal models of preference relations, among them models based on the ideas of Wald, Laplace, Savage and Hurwicz, are discussed with insight and attention to detail. A realistic example involving resource allocation is analyzed. Chapters 6 and 7 offer a broad panorama of concepts and techniques which have a position of centrality in the fuzzy-set-theory-based approach to decision analysis. In the main, what these chapters offer are important generalizations of existing approaches, but there is much that is new and original. The last two chapters deal with issues which are rarely discussed in textbooks on decision analysis—group decision-making and consensus formation. Based on fuzzy set theory, the authors have succeeded in developing significantly more realistic models of group decision-making and consensus formation that can be found in the literature. What is presented in these chapters is of relevance to policy-making and societal decision processes. What should be underscored is that a concept which has a position of centrality in FMD is that of a fuzzy preference relation. This concept has been an object of discussion in the literature of decision-analysis, but in FMD it plays a far more important role.

A final comment. There is an important issue in decision-making under uncertainty which has not received as much attention in the literature as it deserves. The issue is that of decision-making under second-order uncertainty, that is, uncertainty about uncertainty. Imprecise (uncertain) probabilities fall into this category. The importance of decision-making with imprecise probabilities derives from the fact that most real-world probabilities are not known precisely. In early attempts to deal with imprecise probabilities it was assumed that an imprecise probability distribution is an element of a convex set. Then, a minimax criterion was employed to find an optimal solution. A problem with the minimax criterion is that it is much too conservative. Many variations on the minimax criterion have been suggested, but none has gained wide acceptance. At this juncture, the problem of decision-making under second-order uncertainty is far from solution. I presume that this is the reason why the problem of decision-making under second-order uncertainty is not addressed in Chapter 8.

As a test of a theory to deal with first-order and second-order uncertainties, I should like to suggest a quartet of problems.

Assume that I am given two boxes, each containing twenty black and white balls.

Problem 1. A ball is picked at random from box 1. If I pick a white ball, I get a_1 dollars. If I pick a black ball, I lose b_1 dollars. If I pick a ball at random from box 2, I get a_2 dollars if the ball is white and I lose b_2 dollars if the ball is black. I can count the number of white balls and black balls in each box. Which box should I choose?

In Problem 2, I am shown the boxes for a few seconds, not long enough to count the balls. I form a perception of the number of white and black balls in each box. These perceptions lead to perception-based (fuzzy) imprecise probabilities. The question is the same: Which box should I choose?

In Problem 3, I am given enough time to be able to count the number of white and black balls, but it is the gains and losses that are perception-based (fuzzy). The question remains the same.

In Problem 4, probabilities, gains and losses are perception-based (fuzzy). The question remains the same.

These four problems are representative of problems which decision-makers encounter in the real world.

The work of Pedrycz, Ekel and Parreiras is a role model of exposition. Carefully worked out examples are crafted to facilitate understanding. Definitions are carefully formulated and make reference to earlier work. Exercises at the end of chapters make the book very useful as a textbook. FMD is highly informative and highly reader-friendly.

In sum, the importance of the work of Pedrycz, Ekel and Parreiras is hard to exaggerate. However, their work is not intended to be read during lunch hour. FMD requires careful study. FMD's wealth of new, original and applicable results makes it a must reading for all who are concerned with decision analysis. The authors and the publisher, John Wiley, deserve a loud applause.

Lotfi A. Zadeh
August 23, 2010
Berkeley, CA

Preface

This book presents a comprehensive, constructive, well-balanced, fuzzy set modeling framework for a timely, challenging, and important area of multicriteria decision-making. It focuses on ways of representing and handling diverse manifestations of uncertainty and the remarkably multicriteria nature of problems encountered in system projects, planning, operation, and control. The focus of the book is on multiobjective and multiattribute individual and group decision-making. We stress the hands-on nature of the exposition of the overall material and the book comes with a wealth of detailed appealing examples and carefully selected real-world case studies.

We stress the existence of alternative methods for the solution of the most complicated decision-making problems. Especially, diverse techniques for multicriteria analysis of alternatives on the basis of fuzzy preference modeling are presented. The choice of a specific technique is a prerogative of a decision-maker or of a group of decision-makers; it is based on the specificity of the problem and possible sources of available information and its uncertainty.

There have been a number of comprehensive publications in the area of fuzzy decision-making, each of them adhering to some pedagogy and highlighting a certain perspective on the decision-making process. The key features of this book, which determine its focus, can be highlighted as follows:

- It describes a complete set of models and methods based on the direct application of fuzzy sets or their combination with other approaches to uncertainty representation and handling for multicriteria decision-making, including multiobjective, multiattribute, and group decision-making. We aim at providing constructive answers to the fundamental decision questions "*what* should we do?" and "*how* should we do it?" which emerge in the planning, design, operation, and control of complex systems.
- Taking into account that different experts involved in a decision-making process as well as different criteria taken into consideration can demand the use of different ways to represent preferences, the book includes the description of several preference formats, which cover a majority of real situations encountered when preparing information for decision-making. The book presents transformation functions for converting different preference formats into fuzzy preference relations. It bridges an acute gap between decision-making in a fuzzy environment and classical, widely applied decision-making technologies, such as utility theory and an analytic hierarchy process (AHP) approach.
- It describes different aggregation strategies and procedures for constructing collective opinions in group decision-making. The main differences between these strategies are associated

with: the points in the process of the multicriteria analysis in which aggregation of the opinion of experts is carried out; the way the experts are considered (mutually dependent or independent); and the character of estimates being aggregated (fuzzy or linguistic estimates, fuzzy preference relations, or fuzzy nondominance degrees).
- It presents different consensus schemes which allow different ways of organizing the meetings among the experts involved in a decision-making process. We show how a level of consensus among the experts and a level of concordance among pairs of opinions can be assessed and monitored.
- It describes ways of evaluating the consequences of decision-making, including the quantification of particular risks or regrets (monocriteria estimates) and aggregated risks or regrets (multicriteria estimates), which are based on a generalization of the classic approach to dealing with uncertainty in decision-making problems.

Due to the coverage of the material, the book will appeal to those active in various areas in which decision-making becomes of paramount relevance: operational research, systems analysis, engineering, management, and economics. Given the way in which the material is structured, the book can serve as a useful reference source for graduate and senior undergraduate students in courses related to the areas indicated above, as well as for courses on decision-making, risk management, numerical methods, and knowledge-based systems. The book will be of interest to system analysts and researchers in areas where decision-making technologies are paramount.

The book is organized into 10 chapters. In Chapter 1, which is of an introductory nature, we offer the reader a broad perspective on the fundamentals of decision-making problems and discuss generic notions of decision-making problems such as criteria, objectives, and attributes. Diverse manifestations of the uncertainty factor, its relevance, and visibility in decision-making problems are stressed. We also discuss fundamental differences between optimization and decision-making problems. The main objectives, concepts, and characteristics of group decision-making are presented. The role of fuzzy sets is stressed in the general framework of decision-making processes along with their advantages in application to individual and group decision-making problems. The chapter also presents all the required notation and terminology used throughout the book.

The basic concepts of fuzzy sets are introduced in Chapter 2. The fundamental idea of partial memberships, which are conveniently quantified through membership functions and individual membership degrees, is discussed. We present the underlying rationale behind fuzzy sets regarded as information granules and then move on to a detailed description of fuzzy sets by considering the most commonly encountered classes of membership functions and directly relating these classes to the semantics of fuzzy sets. The basic operations on fuzzy sets are further elaborated. The fundamental concepts of fuzzy relations and their main properties, which are of direct relevance to decision-making problems, are discussed. In Chapter 3, which is an immediate continuation of Chapter 2, we present the development aspects of fuzzy set ideas by focusing on the main issues related to the design of fuzzy sets, logic operations, and aggregation of fuzzy sets, and their transformations (mappings).

In Chapter 4, the questions of the construction, analysis, and application of continuous multicriteria decision-making models (multiobjective or $\langle \mathbf{X}, \mathbf{M} \rangle$ models) are considered. The basic definitions related to multicriteria decision-making as well as the commonly utilized approaches to multiobjective decision making are discussed. Particular attention is given to the classic and well-established Bellman–Zadeh approach to decision-making in a fuzzy environment and its application to multicriteria problems. We show that this approach is a convincing means to develop harmonious solutions to multiobjective problems. We illustrate its direct use by solving problems on the multicriteria allocation of resources (or their shortages) as well as some important power engineering problems.

Chapter 5 provides an introduction to preference modeling realized in terms of binary fuzzy relations and addresses certain difficulties that arise in the extension of the classical or Boolean preference structures of binary relations to the fuzzy environment. To alleviate these difficulties, we recall some concepts related to binary fuzzy relations and specific t-norms, t-conorms, and negation operators. We introduce fuzzy preference structures of binary fuzzy relations as well as develop a method for constructing these fuzzy structures, without losing important characteristics that are present in the classical preference structures of binary relations.

Chapter 6 is dedicated to an important problem of forming fuzzy preference relations to analyze multiattribute decision-making models ($\langle \mathbf{X}, \boldsymbol{R} \rangle$ models). Techniques based on the direct and indirect construction of preference relations are considered. Experts involved in an individual or group decision-making process may present their preferences in heterogeneous forms. Different criteria can also demand the use of different preference forms. Taking this into account, the chapter considers five preference formats which cover a significant part of real situations and which arise in preparing preference information. Considering this as well as the rationality of utilizing fuzzy preference relations for a uniform preference representation, the chapter studies diverse transformation functions required to convert different preference formats into fuzzy preference relations. Some aspects of eliminating inconsistencies in the judgments provided by experts are also tackled here.

In Chapter 7, the essence and key features of problems of multicriteria evaluation, comparison, choice, prioritization, and/or ordering of alternatives in a fuzzy environment, based on the analysis of $\langle \mathbf{X}, \boldsymbol{R} \rangle$ models, are discussed. There exist two types of situations which generate these models. The first type is associated with a direct statement of multiattribute decision-making problems when the consequences of the problems' solution cannot be estimated with a single criterion. The second type, illustrated in the chapter by analyzing continuous as well as discrete optimization models with fuzzy coefficients, is related to problems that may be solved on the basis of a single criterion; however, if the uncertainty of information does not permit one to obtain unique solutions, it is possible to reduce these problems to multiattribute decision-making by applying additional criteria. Diverse techniques of the multicriteria analysis of alternatives in a fuzzy environment developed on the basis of fuzzy preference modeling are considered. These techniques are directly aimed at individual decision-making. However, they can be and are applied to decision-making in a group environment. We stress that although the presented techniques can lead to different solutions, this situation is quite natural and should not be treated as an impediment of the underlying methods. On the contrary, given several methods, the most adequate technique can be selected by taking into account the essence of the problem, possible sources of information, and associated uncertainty.

In Chapter 8, the generalization of the classic approach to dealing with uncertainty of information (based on constructing and analyzing payoff matrices) in monocriteria decision-making for multicriteria problems is discussed. The ways of constructing aggregated payoff matrices, modifying the choice criteria, and evaluating particular (monocriteria) and aggregated (multicriteria) risks or regrets in decision-making are studied. We propose a general scheme of multicriteria decision-making, based on a unified application of the generalization of the classic approach and the use of the analysis of $\langle \mathbf{X}, \mathbf{M} \rangle$ and $\langle \mathbf{X}, \boldsymbol{R} \rangle$ models. The special feature of this scheme is the utilization of all available quantitative information to the highest extent in order to reduce the decision uncertainty regions; if a resolving capacity of the processing of formal information does not lead to unique solutions, the scheme resorts to the application of qualitative information based on the knowledge, experience, and intuition of experts involved in a decision-making process.

The last two chapters are dedicated to different approaches for solving decision-making problems in a group environment. In particular, Chapter 9 is concerned with a certain approach which consists of using aggregation procedures regarded as the exclusive arbitration scheme to arrive at an evaluation, comparison, choice, prioritization, and/or ordering of alternatives for the group. This type of dictatorial arbitration scheme does not require achieving a consensus within a group of decision-makers. Three strategies, based on different aggregation mechanisms, are considered. In each of them, the experts involved in the decision process are seen in a different way: either as mutually dependent individuals who act synergistically in the process of decision-making; or as independent individuals who are capable of solving the decision-making problem independently of the other members. We include some examples to illustrate how these strategies are utilized to solve group decision problems by means of different techniques for multiattribute decision-making.

Finally, Chapter 10 presents a suite of procedures for achieving a consensus in the analysis of discrete multicriteria decision-making problems, which involves the evaluation, comparison, choice, prioritization, and/or ordering of alternatives, in a group environment. The chapter presents two different approaches for constructing collective opinions under a rubric of satisfactory consensus: the consensus schemes and the procedures for constructing an optimized consensus. Whereas the former approach requires the experts to review and update their respective opinions in an iterative discussion process, the latter approach represents an attempt to automate the process of constructing and improving the collective opinion, in such a way that the level of consensus in the group is elevated. Each approach has its own advantages and drawbacks. The selection of the most suitable method for a specific application depends mostly on the available time and on the cost of facilitating meetings among the members of the group.

As has been noted, the book can be used in a variety of senior undergraduate and graduate courses. While, in general, one can adhere to the linear flow of coverage of the main topics presented in the consecutive chapters, depending upon the prerequisites, some chapters can be briefly reviewed. For instance, assuming familiarity with the concepts of fuzzy sets, Chapters 2 and 3 could be briefly reviewed with more focus on the design of fuzzy sets and their operational framework.

We would like to take this opportunity to acknowledge support from the National Council for Scientific and Technological Development of Brazil (CNPq) – the research presented in this book was partially supported under CNPq grants 307406/2008-3 and 307574/2008-9. Support from the Natural Sciences and Engineering Research Council of Canada (NSERC) and Canada Research Chair (CRC) Program is highly acknowledged.

We would like to express our thanks to colleagues and friends, namely R. C. Berredo, A. F. Bondarenko, E. A. Galperin, I. V. Kokshenev, O. Machado Jr. J. S. C. Martini, C. A. P. S. Martins, R. M. Palhares, V. A. Popov, A. V. Prakhovnik, J. C. B. Queiroz, F. H. Schuffner Neto, G. L. Soares, R. Schinzinger (in memoriam), L. D. B. Terra, J. A. Vasconcelos, and

V. V. Zorin, for their encouragement and support. We would also like to thank our graduate students W. J. Araujo, M. F. D. Junges, B. Mendonça Neta, J. G. Pereira Jr. (software development), M. R. Silva (software development), and V. V. Tkachenko for their dedication and hard work.

We are very grateful to the editorial team at John Wiley & Sons, Ltd, especially Debbie Cox and Nicky Skinner, for providing truly professional assistance, expert advice, and continuous encouragement during the realization of this project.

Witold Pedrycz
Petr Ekel
Roberta Parreiras
Edmonton and Belo Horizonte, April 2010

1

Decision-Making in System Project, Planning, Operation, and Control: Motivation, Objectives, and Basic Concepts

The intent of this introductory chapter is to offer the reader a broad perspective on the fundamentals of decision-making problems, provide their general taxonomy in terms of criteria, objectives, and attributes involved, stress the relevance and omnipresence of the uncertainty factor, and highlight the aspects of rationality of decision-making processes. We also highlight the fundamental differences between optimization and decision-making problems. The main objectives, concepts, and characteristics of group decision-making are presented. The role of fuzzy sets is stressed in the general framework of decision-making processes. The main advantages of their application to individual and group decision-making processes are briefly discussed. The chapter also clarifies necessary notations and terminology (such as $\langle \mathbf{X}, \mathbf{M} \rangle$ models and $\langle \mathbf{X}, \boldsymbol{R} \rangle$ models) used throughout the book.

1.1 Decision-Making and its Support

The life of each person is filled with alternatives. From the moment of conscious thought to a venerable age, from morning awakening to nightly sleeping, a person meets the need to make a decision of some sort. This necessity is associated with the fact that any situation may have two or more mutually exclusive alternatives and it is necessary to choose one among them. The process of decision-making, in the majority of cases, consists of the evaluation of alternatives and the choice of the most preferable from them.

Making the "correct" decision means choosing such an alternative from a possible set of alternatives, in which, by considering all the diversified factors and contradictory requirements, an overall value will be optimized (Pospelov and Pushkin, 1972); that is, it will be favorable to achieving the goal sought to the maximal degree possible.

Fuzzy Multicriteria Decision-Making: Models, Methods and Applications Witold Pedrycz, Petr Ekel and Roberta Parreiras

If the diverse alternatives, met by a person, are considered as some set, then this set usually includes at least three intersecting subsets of alternatives related to personal life, social life, and professional life. The examples include, for instance, deciding where to study, where to work, how to spend time on a weekend, who to elect, and so on.

At the same time, if we speak about any organization, it encounters a number of goals and achieves these goals through the use of diverse types of resources (material, energy, financial, human, etc.) and the performance of managerial functions such as organizing, planning, operating, controlling, and so on (Lu *et al.*, 2007). To carry out these functions, managers engage in a continuous decision-making process. Since each decision implies a reasonable and justifiable choice made among diverse alternatives, the manager can be called a decision-maker (DM). DMs can be managers at various levels, from a technological process manager to a chief executive officer of a large company, and their decision problems can vary in nature. Furthermore, decisions can be made by individuals or groups (individual decisions are usually made at lower managerial levels and in small organizations, and group decisions are usually made at high managerial levels and in large organizations). The examples include, for instance, deciding what to buy, when to buy, when to visit a place, who to employ, and so on. These problems can concern logistics management, customer relationship management, marketing, and production planning.

A person makes simple, habitual decisions easily, frequently in an automatic and subconscious way, not leaving much to intensive thinking. However, in many cases, alternatives are related to complex situations which are characterized by a discrepancy of requirements and multiple criteria, ambiguity in evaluating situations, errors in the choice of priorities, and others. All these factors substantially complicate the process of taking decisions.

Furthermore, various facets of uncertainty are commonly encountered in a wide range of decision-making problems, which are inherently present in the project, planning, operation, and control of complex systems (engineering, economical, ecological, etc.). In particular, diverse manifestations of the uncertainty factor are associated, for instance, with:

- the impossibility or inexpediency of obtaining sufficient amounts of reliable information;
- the lack of reliable predictions of the characteristics, properties, and behavior of complex systems that reflect their response to external (the surroundings) and internal actions;
- poorly defined goals and constraints in the project, planning, operation, and control tasks;
- the impossibility of formalizing a number of factors and criteria.

This situation should be considered as being natural and unavoidable in the context of complex systems. It is not difficult to understand that it is impossible, in principle, to reduce these problems to exact and well-formulated mathematical problems; to do this, it is necessary, in one way or another, "to take away" the uncertainty and position some hypothesis. However, the construction of a hypothesis is a prerogative of the substantial analysis; this is the formalization of informal situations. One of the ways to address the problem is the formation of subjective estimates carried out by experts, managers, and DMs in general, and the definition of the corresponding preferences.

Thus DMs are forced to rely on their own subjective ideas of the efficiency of possible alternatives and importance of diverse criteria. Sometimes, this subjective estimation is the only possible basis for combining the heterogeneous physical parameters of a problem to be solved into a unique model, which permits decision alternatives to be evaluated (Larichev,

1987). At the same time, there is nothing unusual and unacceptable in the subjectivity itself. For instance, experienced managers perceive, in a broad and well-informed manner, how many personal and subjective considerations they have to bring into the decision-making process. On the other hand, successes and failures of the majority of decisions can be judged by people on the basis of their subjective preferences.

However, the most complicated aspect is associated with the fact that a realm of problems solved by humans in diverse areas has been changed (Trachtengerts, 1998). New, more complicated, and unusual problems have emerged. For many centuries, people made decisions by considering one or two main factors, while ignoring others that were perceived to be marginal to the essence of the problem. They lived in a world where changes in the surroundings were few and new phenomena arose "in turn" but not simultaneously.

At the present time, this situation has changed. A considerable number of problems, or probably the majority of them, are multicriteria in nature, where it is necessary to take into account many factors. In these problems, a DM has to evaluate a set of influences, interests, and consequences which characterizes decision alternatives. For example, in decision-making dealing with the formation of an enterprise, it becomes necessary to consider not only the expected profits and necessary investment, but also market dynamics, the actions of competitors, and ecological, political, and social factors, etc.

Taking into account all the aspects listed above, it is necessary to stress that recognition of the factor of subjectivity of a DM in the process of decision-making conflicts with the fundamental methodological principle of operational research: the search for an *objectively* optimal solution. Recognition of the ultimate right of a DM in the subjectivity of decisions is a sign of the appearance of a new paradigm of multicriteria decision-making (Kuhn, 1962). However, in decision-making with multiple criteria, an objective component always exists. Usually, this component includes diverse types of constraints imposed by the environment on possible decisions (availability of resources, temporal constraints, ecological requirements, social situations, etc.).

A large number of psychological investigations demonstrate that DMs, not being provided with additional analytical support, use simplified and, sometimes, contradictory decision rules (Slovic, Fischhoff, and Lichtenstein, 1977).

Further, Lu *et al.* (2007) share the opinion given above (Trachtengerts, 1998) and indicate that decision-making in the activities of organizations is more complicated and difficult because the number of available alternatives is much larger today than ever before. Due to the availability of information technology and communication systems, especially the Internet and its search engines, we can find more information quickly and therefore more alternatives can be generated. Second, the cost of making errors can be great because of the complexity of operations, automation, and the chain reaction that an error can cause in many parts, in both the vertical and horizontal levels, of the organization. Third, there are continuous changes in the fluctuating environment and more uncertainties in the impacting elements, including information sources and information itself. More importantly, the rapid change of the decision environment requires decisions to be made quickly. These reasons cause organizational DMs to require increasing technical support to help make high-quality decisions. A high-quality decision, such as in bank management, is expected to bring greater profitability, lower costs, shorter distribution times, and increased shareholder value, attracting more new customers, or resulting in a certain percentage of customers responding positively to a direct mail campaign.

Decision support consists of assisting a DM in the process of decision-making. For instance, this support may include (Trachtengerts, 1998):

- assisting a DM in the analysis of an objective component, that is, in the understanding and evaluation of the existing situation and constraints imposed by the surroundings;
- revealing DM preferences, that is, revealing and ranking priorities, considering the uncertainty in DM estimates, and shaping the corresponding preferences;
- generating possible solutions, that is, shaping a list of available alternatives;
- evaluating possible alternatives, considering DM preferences and constraints imposed by the environment;
- analyzing the consequences of decision-making;
- choosing the best alternative, from the DM's point of view.

Computerized decision support, in any case, is based on the formalization of methods for obtaining initial and intermediate estimates given by a DM and on the algorithm for a proper decision process.

The formalization of methods for generating alternatives, their evaluation, comparison, choice, prioritization, and/or ordering, and, if necessary, concordance is a very complicated processes. One of the main complexities and challenges is associated with the fact that a DM, as a rule, is not ready to provide quantitative estimates in the decision process, is not accustomed to the evaluation of proper decisions on the basis of applying formal mathematical methods, and analyzes the consequences of decisions with difficulty.

As a matter of fact, decision support systems have existed for a long time, for example, councils of war, ministry boards, various meetings, analytical centers, and so on (Trachtengerts, 1998). Although they were never called decision support systems, they executed the functions of such systems, at least partially.

The term "decision support system" appeared at the beginning of the 1970s (Eom, 1995). There are several definitions of this concept, such as that given in Larichev and Moshkovich (1996): "Decision support systems are man-machine objects, which permit a DM to use data, knowledge, objective and subjective models for the analysis and solution of semi-structured or unstructured problems".

Taking into account this definition, it is necessary to indicate that one of the important features of decision-making problems is associated with their structures. In particular, it is possible to distinguish structured, semi-structured, and unstructured problems of decision-making (Simon, 1977; Larichev and Moshkovich, 1996; Lu *et al.*, 2007). The latter two types of decision-making problems are also called *ill-structured*.

In *structured problems* (quantitatively formulated problems), essential relationships are established so convincingly that they can be expressed in numbers or symbols which receive, ultimately, numerical estimates. Such problems can be described by existing "traditional" mathematical models. Their analysis becomes possible by applying standard methods leading to the solution.

Unstructured problems (qualitatively expressed problems) include only a description of the most important resources, indicators, and characteristics. Quantitative relationships between them are not known. These problems cannot be described by existing traditional mathematical models and cannot be analyzed by applying standard methods.

Finally, *semi-structured problems* (or mixed problems) include quantitative as well as qualitative elements. As these are examined, qualitative, little-known, poorly explored, uncertain parameters have a tendency to dominate. These problems fall between structured and unstructured problems, having both structured and unstructured elements. The solutions to these problems involves a combination of both standard solution procedures and active DM participation.

According to the classification given above, typical problems in operational research can be called structured. This class of problems is widely used in the project planning, operation, and control of engineering systems. For example, it is possible to talk about the design of forms of an aircraft hull, planning of water supply systems, control of power systems, and so on.

The distinctive characteristics of unstructured problems are as follows (Larichev and Moshkovich, 1996):

- uniqueness of choice in the sense that, at any time, the problem is a new one for a DM or it has new properties in comparison to a similar problem solved in the past;
- uncertainty in the evaluation of alternative solutions;
- the qualitative character of the evaluations of problem solutions, most often formulated in verbal form;
- the evaluation of alternatives obtained only on the basis of the subjective preferences of a DM;
- the estimates of criteria obtained only from experts.

Typical unstructured problems are associated, for example, with planning new services, hiring executives, selecting a locale for a new branch, choosing a set of research and development projects, and alike.

If we speak about *semi-structured problems*, their solutions are based on applying traditional analytical models as well as models based on DM preferences. As an example, one can look at the problem (Trachtengerts, 1998) related to liquidation of the consequences of extraordinary situations associated with radioactive contamination. In the solution of this problem, analytical models can be applied to define the degree and character of radioactive contamination for given temporal intervals. At the same time, models based on DM preferences can be applied in the choice of measures for liquidation of the consequences of radioactive contamination. It is possible to qualify many problems associated with economical and political decisions, medical diagnostics, and so on, as semi-structured problems.

Returning to the issue of computerized decision support, we should note that, due to the large number of components (variables, functions, and parameters) involved in many decisions, this has become a basic requirement to assist DMs in considering and examining the implications of various courses of decision-making (Lu *et al.*, 2007). Furthermore, the impact of computer technologies, particularly the Internet, on organizational management is increasing rapidly. Interaction and cooperation between users and computers are growing to cover more and more aspects of organizational decision-making activities. Internet- or intranet-based computerized information systems have now become vital to all kinds of organizations.

Thus, computer applications in organizations are moving from transaction processing and monitoring activities to problem analysis and finding solutions (Lu *et al.*, 2007). Internet- or intranet-based online analytical processing and real-time decision support are becoming the cornerstones of modern management, in particular within the elaboration of e-commerce, e-business, and e-government. There is a trend toward providing managers with information systems that can assist them directly with their most important task, that is, making decisions.

A detailed description of the advantages generated by applying computerized decision support systems for individual as well group decision-making is given, for instance, in Lu *et al.* (2007). At the same time, these authors indicate that the important issue is that, with computerized decision support technologies, many complex decision-making problems can now be handled effectively. However, these technologies can be better used in analyzing structured problems rather than semi-structured and unstructured problems. In an unstructured problem, only part of the problem can be supported by advanced tools such as intelligent decision support systems. For semi-structured problems, the computerized decision support technologies can improve the quality of information on which the decision is based by providing not just a single, unique solution, but a range of alternative solutions from the decision uncertainty regions. Their occurrence and essence will be discussed in the next section.

1.2 Optimization and Decision-Making Problems

Is there any difference between the notions of "optimization" and "decision-making"? Are these notions synonymous or not? Partial answers to these questions have been given in the previous section. However, deeper and more detailed considerations are beneficial here.

A traditional optimization problem is associated with the search for an extremum (minimum or maximum, according to the essence of the problem) of a certain objective function, which reflects our interests, when observing diverse types of constraints (imposed on allowable resources, physical laws, standards, industrial norms, etc.). Formally, it is possible to represent an optimization problem as follows:

$$F(x) \to \operatorname*{extr}_{x \in \mathbf{L}} \tag{1.1}$$

where $\mathbf{L}$ is a set of feasible solutions in $\mathbf{R}^n$ defined by the constraints indicated above.

To solve the problem (1.1) we should find a vector $\boldsymbol{x}^0$ such that

$$\boldsymbol{x}^0 = \arg \operatorname*{extr}_{x \in \mathbf{L}} F(x) \tag{1.2}$$

If numerical details of the problem (1.1) have been provided and we can obtain a unique solution without any guidance or assistance from a DM, then we are concerned with an optimization problem.

Generally, an optimization problem may be complicated from the mathematical point of view, and we need a large amount of computing time to generate a solution. Can human participation in the search for a solution be useful? Definitely, such participation could be useful, because, for instance, the introduction of heuristics or a change of initial points for a search can reduce the time necessary to obtain an optimal solution. However, in principle, a unique solution to the problem can be obtained without human participation.

At the same time, the presence of any type of uncertainty can call for human participation in order to arrive at a unique solution to the problem.

For instance, the uncertainty of information gives rise to some decision uncertainty regions. As shown in Figure 1.1, the uncertainty of information $\delta F(x)$ in the estimation of an objective function $F(x)$ leads to a situation where formally the solutions coming from a region δx

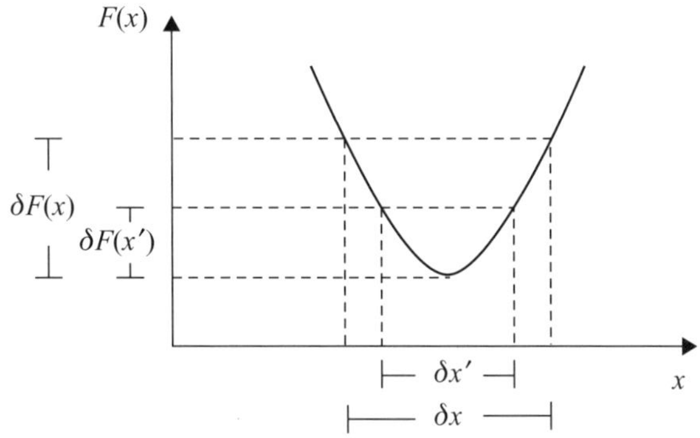

Figure 1.1 Decision uncertainty region and its reduction through the reduction of the level of uncertainty of information.

cannot be distinguished, thus giving rise to a decision uncertainty region. Taking this into consideration, the formal formulation (1.1) can be transformed to the following:

$$F(x, \theta) \rightarrow \underset{x \in \mathbf{L}(\theta)}{\text{extr}} \tag{1.3}$$

where θ is a vector of uncertain parameters, whose existence changes the essence of (1.1). In particular, we can say that the solution (1.2) is an optimal solution for a concrete realization of θ (a concrete hypothesis); however, for some other realization (another hypothesis), it is no longer optimal.

What are the ways to reduce this uncertainty region? The first way is to "buy" information (let us not forget that any information has some cost associated with it), for example, by acquiring additional measurements or examining experts to reduce the level of uncertainty. As shown in Figure 1.1, the reduction of the uncertainty $\delta F(x)$ to $\delta F(x')$ permits one to obtain a reduced decision uncertainty region with $\delta x' < \delta x$.

However, if there is no possibility of reducing the uncertainty of information, we can resort to some alternative approach. This way is associated with introducing additional criteria to try to reduce the decision uncertainty regions. As demonstrated in Figure 1.2, introduction of the objective function $F'(x)$ allows us to reduce the decision uncertainty region as well, arriving at $\delta x' < \delta x$.

On the other hand, the existence of more than one objective function may be considered as uncertainty as well. This comes in as the uncertainty of goals. Although the nature of this type of uncertainty is not the same as the uncertainty of available information, it also leads to the generation of decision uncertainty regions.

To focus our attention, let us consider the simple problem of minimizing two objective functions $F_1(x) = F_1(x_1, x_2)$ and $F_2(x) = F_2(x_1, x_2)$, considering a set of feasible solutions $\mathbf{L}$. We can transform $\mathbf{L}$ from the decision space to some region $\mathbf{L}_F$ of the space of objective functions $F_1(x)$ and $F_2(x)$ (or, simply, the objective space). In Figure 1.3, we can see that point a corresponds to the best solution ($\min_{x \in \mathbf{L}} F_1(x)$) from the point of view of the first objective

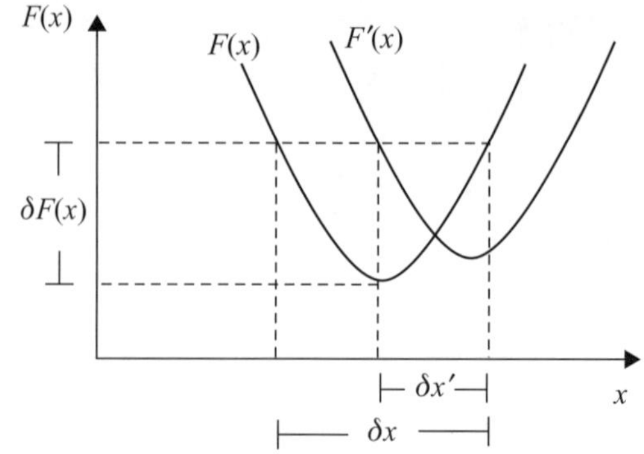

Figure 1.2 Decision uncertainty region and its reduction through the introduction of additional criteria.

function. On the other hand, point b corresponds to the best solution ($\min_{x\in\mathbf{L}} F_2(x)$) when considered from the viewpoint of the second objective function.

Is point c a solution to the problem? Yes, it is. Can we improve this solution? Yes, we can do that by passing to point d. Can we improve this solution? Yes, this is possible by passing to point e. Can we improve this solution? This is possible by passing to point f. Can we improve this solution? We cannot advance here. It is possible to pass to point g, but this step does not make the resulting solution any better: we can improve it from the point of view of $F_1(x)$ but deteriorate its quality from the point of view of $F_2(x)$. In a similar way, by passing to point h, we can improve the solution from the point of view of $F_2(x)$ but deteriorate it from the point of view of $F_1(x)$.

Thus, formally, the solution to the problem presented in the objective space is a boundary Ω_F^P of $\mathbf{L}_F$ located between points a and b. The set $\Omega^P \subseteq \mathbf{L}$ corresponding to Ω_F^P is the problem solution, which is called a Pareto-optimal solution set. This concept of optimality was proposed by Edgeworth (1881) and was further generalized by Pareto (1886). Although we say that Ω^P is the problem solution, from a formal point of view this is not a solution that can be implemented. In reality, it is the decision uncertainty region. The choice of a particular Pareto-optimal solution is based on the DM's involvement.

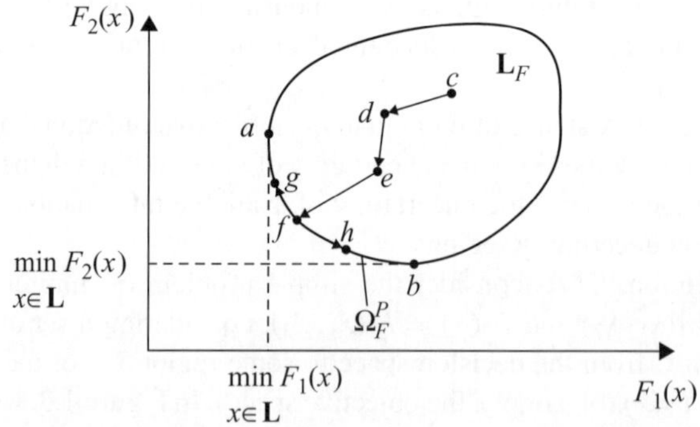

Figure 1.3 The concept of Pareto-optimal solutions.

The more difficult situations are associated with problems where there exists an uncertainty of information as well as an uncertainty of goals.

The problems of an optimization character, which include the uncertainty of information and/or the uncertainty of goals and demand human participation in their solution, are inherent problems in decision-making. Taking this into consideration, it is necessary to make some additional observations.

One of the most important criteria (Larichev, 1984) for classifying decision-making problems is the existence or lack of an objective model for the corresponding problem. Note here that it is not uncommon to encounter situations where it is impossible to talk about the existence of objective functions in decision-making problems. The models which can be used for analyzing these problems reflect "a point of view" and, in a more general sense, the "world outlook" of a DM. In these cases, an obvious question is how to choose actions which correspond, in the best way, to the preferences of a DM (Keeney and Raifa, 1976) and are based on his/her knowledge, experience, and intuition. Taking this into account, semi-structured and unstructured problems, classified in the previous section, are subjects of decision-making.

In conclusion, the following general tendency is visible. If we solve an optimization problem, we generally look for the *best* solution. If we talk about a decision-making problem, the methodology used to solve it is quite distinct: we do not look for the best solution, but apply information arriving from different sources and try to eliminate some alternatives, which are dominated by other alternatives, in order to reduce the decision uncertainty regions.

1.3 Multicriteria Decision-Making

The uncertainty of goals in decision-making is an important manifestation of uncertainty that relates to the multicriteria character of many problems encountered in the project, planning, operation, and control of complex systems of different nature. Some professionals in the field of decision-making and systems analysis (for example, Lyapunov, 1972) agree that, from the general point of view, this type of uncertainty is the most difficult to treat and overcome because "we simply do not know what we want". In reality, this type of uncertainty cannot be effectively captured on the basis of applying formal models and methods, because sometimes the unique information sources are the individuals who make decisions.

Multicriteria decision-making is related to making decisions in the presence of multiple and conflicting criteria. Multicriteria decision-making problems may range from everyday decision problems, such as the purchase of a car, to those affecting entire nations, as in the judicious use of money to preserve national security (Lu *et al.*, 2007).

However, even with this existing diversity, all multicriteria decision-making problems share the following common characteristics (Hwang and Yoon, 1981):

- **multiple criteria**: each problem has multiple criteria, which can be objectives or attributes;
- **conflicting criteria**: multiple criteria conflict with each other;
- **incommensurable units**: criteria may have different units of measurement;
- **design/selection**: solutions to multicriteria decision-making problems are either to design the best alternative(s) or to select the best one among previously specified finite alternatives.

Taking the above into account, we distinguish two types of *criteria*: *objectives* and *attributes*. In such a manner, multicriteria decision-making problems can be classified into two wide classes:

- *multiobjective* decision-making;
- *multiattribute* decision-making.

The main difference between these two classes is that the first concentrates on continuous decision spaces and the second focuses on problems with discrete decision spaces.

To elaborate further, some basic concepts and terminology are given below. They are in line with the notation presented in the literature (Hwang and Masud, 1979; Hwang and Yoon, 1981; Lu *et al.*, 2007).

Criteria form the standard of judgment or rules to test acceptability. In the multicriteria decision-making literature, they indicate objectives and/or attributes.

Objectives are the reflection of the desire of DMs and indicate the direction on which DMs want to concentrate. Multiobjective decision-making problems, as a result, involve the design of alternatives that optimize or at least satisfy the objectives of DMs.

Goals are entities desired by DMs and expressed in terms of a specific state in space and time. Thus, while objectives give the desired direction, goals give a desired (or target) level to achieve.

Attributes are the characteristics, qualities, or performance parameters of alternatives. Multiattribute decision-making problems involve the selection of the "best" alternative from a pool of preselected alternatives described in terms of their attributes.

Multiobjective decision-making is known as the continuous type of multicriteria decision-making and its main characteristics are that DMs need to achieve multiple objectives while these objectives are noncommensurable and conflict with each other. A multiobjective decision-making model includes a vector of decision variables, objective functions that describe the objectives, and constraints. DMs attempt to maximize or minimize the objective functions.

Multiattribute decision-making is related to making a preference decision (that is, comparison, choice, prioritization, and/or ordering) over the available alternatives that are characterized by multiple, usually conflicting, attributes. The main peculiarity of multiattribute decision-making problems is that there are usually a limited number of predetermined alternatives, which are associated with a level of achieving the attributes. Based on the attributes, the final decision is made.

Finally, we should discuss in detail the concept of *alternatives*. How to generate alternatives is a significant part of the process of multiobjective and multiattribute decision-making model building (Lu *et al.*, 2007). In almost all multiobjective decision-making models, the alternatives can be generated automatically by the models. In the case of multiattribute decision-making models, however, it is necessary to generate alternatives manually. Sometime, the essence of the problem defines the number of alternatives. However, in general, how and when to stop generating alternatives becomes a very important issue. Generating alternatives significantly depends on the availability and cost of information, and also requires reliance on expertise in the problem area. Alternatives can be generated with the use of heuristics as well, and they could come from either individuals or groups.

The issues related to the necessity of setting up and solving multicriteria problems as well as the classification of decision-making situations, which need the application of the multicriteria approach, have been discussed in many works (for instance, Larichev, 1984; Gomes, Gomes, and Almeida, 2002). It is possible to identify two major types of situations, which call for the application of a multicriteria approach:

- Problems whose solution consequences cannot be estimated with a single criterion: these problems are associated with the analysis of models including economic as well as physical indices (when alternatives cannot be reduced to comparable form) and also by the need to consider indices whose cost estimation is hampered (for example, many power engineering problems are considered on the basis of technological, economical, ecological, and social nature criteria).
- Problems that may be solved on the basis of a single criterion (or several criteria). However, if the uncertainty of information does not permit the derivation of unique solutions, it is possible to reduce these problems to multicriteria decision-making by applying additional criteria, including those of a qualitative character (for example, "flexibility of development", "complexity of maintenance", "attractiveness of investments", and so on, whose utilization is based on the knowledge, experience, and intuition of involved experts). This can serve as a convincing means to contract the corresponding decision uncertainty region. It could be regarded as an intuitively appealing approach exercised in the practice of decision-making.

According to the major types of situations outlined above, two classes of models, so-called $\langle \mathbf{X}, \mathbf{M} \rangle$ models and $\langle \mathbf{X}, \boldsymbol{R} \rangle$ models (Ekel, 2001; Ekel, 2002) can be constructed. Both of these classes of models are comprehensively discussed in the book. The $\langle \mathbf{X}, \mathbf{M} \rangle$ models correspond to multiobjective decision-making problems. In the book, their analysis is illustrated by considering the problems of multicriteria allocation of resources or their shortages (with the presentation of an adaptive interactive decision-making system (AIDMS1), which is dedicated to their solution) as well as important classes of power engineering problems (multiobjective power and energy shortage allocation as applied to load management, multiobjective power system operation, multiobjective optimization of network configurations in distribution systems, and energetically effective (bicriteria) voltage control in distribution systems). At the same time, the $\langle \mathbf{X}, \boldsymbol{R} \rangle$ models correspond to the multiattribute decision-making problems and include a vector of fuzzy preference relations (Orlovsky, 1981; Fodor and Roubens, 1994), which play the role of attributes. We will present an interactive system for multicriteria decision-making (MDMS) dedicated to the analysis of the $\langle \mathbf{X}, \boldsymbol{R} \rangle$ models. In the book, the construction and application of the $\langle \mathbf{X}, \boldsymbol{R} \rangle$ models is illustrated by considering the problems of substation planning in power systems, reactive power source choice at a power system bus, energy planning (selection of the most appropriate technology in a renewable energy diffusion plan) as well as managerial activities.

Finally, the $\langle \mathbf{X}, \boldsymbol{R} \rangle$ models are also used in the present book in problems of group decision-making, which are briefly discussed in the next section.

1.4 Group Decision-Making

Group decision-making is defined as a decision situation in which there is more than one individual involved. The group members have their own attitudes and motivations, recognize the existence of a common problem, and attempt to reach a collective decision (Lu *et al.*, 2007). The necessity of applying procedures of group decision-making is associated with the following considerations.

There are many situations, for instance at the high managerial levels of organizations, when the decision problems involve wide domains of knowledge which are beyond a single individual (this is particularly true when the decision environment becomes more complex and multifaceted). As a consequence, it is usually necessary to allocate more than one professional to the decision process. This is particularly valid in environments with a diverse workforce, where decisions require multiple perspectives and different areas of expertise of the individuals represented in the group. The following are among the advantages of group over individual decision-making (Tan, Teo, and Wei, 1995):

- Group decision-making allows more intellectual resources to be gathered to support the decision. The resources available to the group include the individual competencies, intuition, and knowledge.
- With the participation of multiple experts, it becomes possible to distribute among them the labor related to acquiring and processing the vast amount of information pertaining to the decision.
- If the group members exhibit divergent interests, the final decision tends to be more representative of the needs of the organization.

It is possible to indicate some important characteristics of group decision-making as follows (Lu *et al.*, 2007):

- the group performs a decision-making task;
- group decision-making may cover the whole process of transfer from generating ideas for solving a problem to implementing solutions;
- group members may be located at the same place or at different places;
- group members may work at the same or different times;
- group members may work for the same or different departments or organizations;
- the group can be at any managerial level;
- there may be conflicting opinions in the group decision process among group members;
- the decision might have to be accomplished in a short time;
- group members might not have complete information for the decision;
- some required data, information, or knowledge for a decision may be located in many sources and some may be external to the organization.

Quite often, the group members may be at different locations and may be working at different times. Thus, they need to communicate, collaborate on, and access a diverse set of information sources, which can be met by the development of the Internet and its derivates (intranets and extranets). The questions of constructing and utilizing Web-based group decision support systems are discussed, for instance, in Lu *et al.* (2007).

With regard to the common goals and interests of the experts in group decision-making, it is worthy distinguishing two different environments, namely *cooperative* and *noncooperative* work. In cooperative decision-making, all the experts are supposed to work together, in order to achieve a decision for which they will share the responsibility. In noncooperative decision-making, the experts play the role of antagonists or disputants over some common interest for which they must negotiate (Lu *et al.*, 2007). It should be made clear that this book addresses problems of group decision-making in the cooperative environment.

As in cooperative work the experts share responsibility for the decision (and, as indicated above, they also may participate in the implementation of the selected solution), it is important to guarantee that each member is satisfied with the selected solution. Clearly the commitment of the group to the implementation of the outcomes depends upon the level of consensus achieved by the group. Under this perspective, a group decision constructed by means of domination and enforced concessions should be considered inferior to an individual decision, because it will probably face more difficulties in its implementation. Therefore, achieving a genuine consensus on the solution is an important task for the group. However, it should be indicated that achieving perfect concordance among the experts is extremely rare. Although, ideally, the condition for terminating the decision-making process under group settings should be the achievement of a unanimous solution, in reality, because that unanimous solution hardly ever exists, it is sufficient to meet the alternative that is the most satisfactory for the group as a whole. Otherwise, the decision will probably take longer than is admissible or affordable.

Among the reasons for the occurrence of discordance among the group, we can identify the following:

- Although group members are supposed to share the primary goal, which obviously is to meet the solution which most benefits the organization, their secondary goals may be just partially shared. For instance, when each expert is representative of a different department, it is natural that they would have specific interests associated with the priorities and needs of their respective departments.
- Each expert usually has a distinct perception of the problem and intuition which may be difficult to formalize and communicate to the other members.
- In general, no single expert knows the entire domain of the decision problem. Each expert usually has access to different profiles of information. In particular, certain members of the group may have privileged access to secure information.

In general, these factors can be diminished by promoting discussions among the experts, in an attempt to pool all relevant information pertaining to the decision. Indeed, by pooling the undistributed information, it is possible to increase the chances of achieving better decisions than each member could obtain without help. However, the existence of abundant intellectual resources is not sufficient to guarantee high-quality decisions, as the group may fail to wisely consider, evaluate, and integrate the profiles of information and perspectives held by the other members of the group (Bonner, Baumannb, and Dalal, 2002; van Ginkel and van Knippenberg, 2009). The current literature identifies some factors that can adversely affect the decision process, leading to low-quality decisions. We can distinguish the following:

- The pressure for early consensus that is due to the need to obtain a solution rapidly.
- The pressure of concordant majorities on the other experts, which is reflected by the group's tendency to prematurely converge on a single solution, once a majority supports a position (even if such solution is not good).
- The problem of critical pooling of nondistributed (centralized) information, which can be described as follows: the information supporting the best alternative is not shared among

all experts, whereas all experts have information supporting the inferior alternatives. Under these circumstances, the group may prematurely achieve a consensus on a bad solution that is apparently good, as the information shared among most experts has more chance of being recalled than the information that is available to just a few experts. One way of reducing this specific problem is to stimulate each member to focus on information related to their respective areas of expertise during the discussion (Stasser and Vaughan, 2000).

In this context, it is important to stress the importance of the moderator (or facilitator) in the discussion among the experts. As indicated in Wong and Aiken (2003), the participation of a moderator, which may be human or automated, in the decision process, always results in better outcomes. The moderator is supposed to act as an arbiter responsible for controlling the information flow across the group. In this way, the moderator does not participate directly in the decision, but is supposed to enhance the ability of the group to make decisions (Griffith, Fuller, and Northcraft, 1998).

Among the tasks of a moderator we can identify the following: (1) define the rules for the group decision process and the tasks of each member, select the appropriate group technology, support the group in formulating the problems, and define the outcomes to be achieved; (2) develop the schedule to be accomplished, identify controversial opinions across the group, identify conflicting topics that should be focused on in the discussion, and verify if the current level of concordance among experts is acceptable (Ngwenyama, Bryson, and Mobolurin, 1996).

It is important to indicate that, in real-world applications, sometimes it is impossible to promote the consensus and thereby the exchange of information among the experts, due to logistic, timing, or monetary constraints. In this case, the invited professionals may give their opinions individually and then the group decision is dictatorially constructed with the use of an aggregation rule, despite the existence of substantial discordances among the experts. We can distinguish the following most common approaches for dealing with this situation:

- the use of a majority rule, according to which the group decision is constructed in concordance with the opinion of the majority in the group (Lu *et al.*, 2007);
- the use of a rule determined by a member of the group with authority to make the ultimate decision for the group (Lu *et al.*, 2007);
- the search for a collective opinion that minimizes the major discordance in the group, in such a way that no expert is extremely dissatisfied with the group outcomes (Parreiras *et al.*, 2010).

1.5 Fuzzy Sets and their Role in Decision-Making Processes

As elaborated in Section 1.1, various types of uncertainty are commonly met in a wide range of decision-making problems, which are inherently encountered in the project, planning, operation, and control of complex systems. Taking these types of uncertainty into account when constructing mathematical models serves as a vehicle for increasing the adequacy of the models and, as a result, the credibility and factual efficiency of decisions based on their analysis. Considering this, it is necessary to note that the starting point in the formation of mathematical models is the requirement of a strict correspondence of these models to the level of uncertainty of information used for their construction. Observing just this correspondence, we can talk about the adequacy of the presentation of the object, system, or process and the possibility of obtaining a real effect as a result of solving the corresponding problems of an optimization character. Any simplification of reality or its idealization, undertaken with the purpose of using rigorous mathematical models, distorts the nature of many problems and diminishes the practical value of results obtained on the basis of analyzing these models. Following this line of thought, researchers (for instance, Belyaev and Krumm, 1983; Rommelfanger, 2004), for a number of reasons, have doubts about the validity or, at least, the expediency of taking into account the uncertainty factor within the framework of traditional approaches (first of all, approaches based on probability theory, for instance, Dantzig, 1955; Grassman, 1981; Wagner, 1982). In particular, Belyaev and Krumm (1983) indicate that, similar to the solution of problems on the basis of deterministic methods, when we assume exact knowledge of the information, which usually does not correspond to reality, the application of probabilistic methods also supposes exact knowledge of the distribution laws and their parameters, which does not always correspond to the real possibilities of obtaining the entire spectrum of the probabilistic description.

In general, the approaches highlighted above do not ensure an adequate or sufficiently rational consideration of the uncertainty factor along with an entire spectrum of its manifestations.

Giving up the traditional approaches to the construction of mathematical models and, the application of the fuzzy set theory (Dubois and Prade, 1980; Zimmermann, 1996; Pedrycz and Gomide, 1998), established by Zadeh (1965), may play and plays a significant positive role in overcoming the difficulties that are present. The utilization of this theory opens an interesting avenue of giving up "excessive" precision, which is inherent in the traditional modeling approaches, while preserving reasonable rigor. The principle of incompatibility coined by Zadeh (1973) offers an interesting view of the tradeoffs between precision and relevance of the models: "As the complexity of a system increases, our ability to make precise and yet significant statements about its behavior diminishes until a threshold is reached beyond which precision and significance (or relevance) become almost mutually exclusive characteristics". Furthermore, operating in a fuzzy parameter space allows one not only to be oriented toward the contextual or intuitive aspect of the qualitative analysis as a fully substantial aspect, but, by means of fuzzy set theory, to use this aspect as a sufficiently reliable source for obtaining quantitative information. Finally, fuzzy sets allow one to reflect in an adequate way on the essence of the decision-making process. In particular, since the "human factor" has a noticeable effect and occupies a very visible position in making decisions in many real-world problems, we can capitalize on the way in which fuzzy sets help quantify the linguistic facet of available data and preferences (Dubois and Prade, 1980; Zimmermann, 1996; Pedrycz and Gomide, 1998).

We also have to bear in mind that the quest for attaining the maximum effectiveness in decision-making in the presence of uncertainty requires, first of all, that a significant effort be directed toward finding ways to remove or, at least, partially overcome the uncertainty factor (Popov and Ekel, 1987). In particular, this can be attained by aggregating information that arrives from different sources, being both formal and informal in nature. This aggregation allows one (Ekel and Popov, 1985) to supplement the characteristics of the uncertain initial information by justified assumptions about the differentiated confidence (reliability) of its various values which could be reflected by choosing appropriate membership functions (Dubois and Prade, 1980; Zimmermann, 1996; Pedrycz and Gomide, 1998).

However, taking the above into account, it is necessary to indicate that the issues related to the relationships between probability theory and fuzzy set theory, as well as an interpretation of membership functions, have been the subject of intensive discussions of methodological and philosophical character over the years. Thus, it should be emphasized that the decision-making approaches based on fuzzy set theory do not compete with probabilistic methods, but these two approaches are orthogonal in nature. Furthermore, we can witness some hybrid approaches in which fuzzy sets and probability are used in a synergistic way. Likewise, recent years have seen intensive investigations which have applied fuzzy set theory in combination with other approaches to deal with diverse facets of uncertainty in problems of an optimization character. These developments offer the advantages of both a fundamental nature (as we exercise the possibility of obtaining more effective, less "cautious" solutions as well as the ability to consider simultaneously different manifestations of the uncertainty factor) and a computational character.

Finally, it is possible to distinguish two principal ways of solving problems under conditions of uncertainty. In applying the first way, one obtains (at least, theoretically) an exact solution for fixed values of the uncertain parameters, and then estimates its stability for variations of such parameters (for example, by performing multivariant computations). The second way presupposes the tracking of the effect of the uncertainty at all stages along the path toward the final decision. This approach can be implemented on the basis of fuzzy set theory. It is much more complicated than the first one, but is also much more fruitful and highly promising.

As mentioned above, in many real-world problems we have to take into account the criteria, constraints, indices, and so on, of a qualitative character. Thus, it should be emphasized that this type of information was taken into account in the past. However, it was used only after obtaining solutions on the basis of the use of formal models, with the disruption of these solutions (to consider information of a qualitative character) and without any sufficient justification. As such approaches reduce the essential value of the obtained solutions, it remains necessary to develop ways of introducing this type of information directly into the decision-making processes. Fuzzy sets can be considered here as a sound way of proceeding along this path.

Returning to the considerations of Section 1.1, it is necessary to highlight that one of the most important criteria for classifying decision-making problems (Larichev, 1987) is the existence or lack of an objective model for the problem. Taking this into consideration, it should be noted that it is not uncommon to encounter situations, as mentioned in Section 1.1, where it is next to impossible to speak about the existence of objective functions in decision-making problems. The corresponding models reflect the "world outlook" of a DM. In these cases, an obvious question is how to choose actions which correspond, in the best way, to the preferences of the individual (Keeney and Raifa, 1976). Considering that the manner of human thinking, including the perception of preferences, is vague and subjective, fuzzy set theory can play an important role in individual and group preference modeling (Fedrizzi and Kacprzyk, 1990; Fodor and Roubens, 1994).

The application of fuzzy sets to preference modeling and analysis of the corresponding decision-making problems provides a flexible environment which permits us to deal with the inherent fuzziness of perception and, in this manner, to incorporate more human consistency into preference models. Besides, a stimulus for utilizing fuzzy set theory stems, as indicated above, from one of its most important facets that concerns the linguistic aspect commonly applied to different decision-making problems and different preference structures (Herrera and Viedma, 2000; Xu, 2005). In particular, it is possible to distinguish among several directions in decision-making by applying the linguistic aspect of fuzzy set theory, such as multicriteria decision-making (Buckley, 1995; Rasmy *et al.*, 2002), group decision-making (Yager, 1993; Herrera, Herrera-Viedma, and Verdegay, 1995), diverse consensus schemes (Herrera, Herrera-Viedma, and Verdegay, 1995; Bordogna, Fedrizzi, and Passi, 1997), decision-making on the basis of information granularity (Borisov *et al.*, 1989; Herrera, Herrera-Viedma, and Martínez, 2000), and so on. In principle, all these directions are associated with analyzing the $\langle \mathbf{X}, \boldsymbol{R} \rangle$ models mentioned above. Taking into account the rationality of analyzing $\langle \mathbf{X}, \mathbf{M} \rangle$ models on the basis of fuzzy sets as well, it is possible to assert that their utilization in the statement and solution of decision-making problems provides answers to the fundamental questions "What should we do?" ($\langle \mathbf{X}, \boldsymbol{R} \rangle$ models) and "How should we do?" ($\langle \mathbf{X}, \mathbf{M} \rangle$ models) arising in the project, planning, operation, and control of complex systems of diverse nature.

Finally, we should be aware that the development and application of diverse types of uncertainty expressed in the language of fuzzy sets not only serves as a vehicle for improving the adequacy of the constructed models and, consequently, enhancing the credibility and factual efficiency of decisions based on their analysis, but also becomes highly beneficial to the formation of convincing and effective human-oriented (in contrast to machine-oriented) interfaces between a DM and a computer. This aspect becomes crucial given the important and general trend of computerized "intellectualization" of decision-making pursuits.

Although the themes related to fuzzy decision-making have been widely and deeply studied, this area brings about a number of open questions associated not only with methods of decision-making in a fuzzy environment, but also with their combination with other branches of the mathematics of uncertainty (Wang, 2007). From the practical point of view, only some theoretical results have been translated to concrete algorithms and their implementation. In this context, one of the essential objectives of this book is to fill certain theoretical and practical gaps when considering the uncertainty and multicriteria factors in system projects, planning, operation, and control.

1.6 Conclusions

We have discussed the fundamental questions of the appearance and essence of decision-making problems arising in the project, planning, operation, and control of complex systems of diverse nature. The relevance and omnipresence of the uncertainty factor and its influence on the character of the analyzed decision-making models have been considered. The structured, unstructured, and semi-structured problems of decision-making have been classified with a distinct focus on unstructured problems. The main functions of decision support frameworks have been briefly discussed. The fundamental differences between optimization and decision-making problems have also been considered. The models of multicriteria decision-making have been characterized and classified with the split into two main categories of so-called $\langle \mathbf{X}, \mathbf{M} \rangle$ and $\langle \mathbf{X}, \boldsymbol{R} \rangle$ models, which are the subject of comprehensive considerations in this book. The essence, main concepts, and characteristics of group decision-making have been discussed. Finally, the role of fuzzy set theory in decision-making processes has been discussed, including consideration of its advantages. First of all we stressed the fundamental benefit stemming from the use of fuzzy sets that is the possibility of obtaining more effective, less "cautious" solutions to the decision-making problems, as well as the abilities of incorporating different facets and manifestations of the uncertainty factor.

References

Belyaev, L.S. and Krumm, L.A. (1983) Applicability of probabilistic methods in energy calculations. *Power Engineering*, **21** (2), 3–10.

Bonner, B.L., Baumannb, M.R., and Dalal, R.S. (2002) The effects of member expertise on group decision-making and performance. *Organizational Behavior and Human Decision Processes*, **88** (2), 719–736.

Bordogna, G., Fedrizzi, M., and Passi, G. (1997) A linguistic modeling of consensus in group decision-making based on OWA operators. *IEEE Transactions on Systems, Man, and Cybernetics*, **A-27** (1), 126–132.

Borisov, A.N., Alekseev, A.V., Merkur'eva, G.V., *et al.* (1989) *Fuzzy Information Processing in Decision-Making Systems*, Radio i Svyaz, Moscow (in Russian).

Buckley, J.J. (1995) The multiple judge, multiple criteria ranking problem: a fuzzy set approach. *Fuzzy Sets and Systems*, **13** (1), 23–37.

Dantzig, G.B. (1955) Linear programming under uncertainty. *Management Science*, **1** (2), 197–207.

Dubois, D. and Prade, H. (1980) *Fuzzy Sets and Systems: Theory and Applications*, Academic Press, New York.

Edgeworth, F.Y. (1881) *Mathematical Physics*, P. Keegan, London.

Ekel, P.Ya. (2001) Methods of decision-making in fuzzy environment and their applications. *Nonlinear Analysis: Theory, Methods and Applications*, **47** (5), 979–990.

Ekel, P.Ya. (2002) Fuzzy sets and models of decision-making. *Computers and Mathematics with Applications*, **44** (7), 863–875.

Ekel, P.Ya. and Popov, V.A. (1985) Consideration of the uncertainty factor in problems of modelling and optimizing electrical networks. *Power Engineering*, **23** (2), 45–52.

Eom, S.B. (1995) Decision support systems research: reference disciplines and a cumulative tradition. *Omega*, **23** (5), 511–523.

Fedrizzi, M. and Kacprzyk, J. (eds) (1990) *Multiperson Decision-Making Models Using Fuzzy Sets and Possibility Theory*, Kluwer, Boston.

Fodor, J. and Roubens, M. (1994) *Fuzzy Preference Modelling and Multicriteria Decision Support*, Kluwer, Boston, MA.

Gomes, L.F.A.M., Gomes, C.F.S., and Almeida, A.T. (2002) *Managerial Decision-Making: Multicriteria Approach*, Atlas, Sao Paulo (in Portuguese).

Grassman, W.K. (1981) *Stochastic Systems for Management*, North-Holland, New York.

Griffith, T.L., Fuller, M.A., and Northcraft, G.B. (1998) Facilitator influence in group support systems: intended and unintended effects. *Information Systems Research*, **9** (1), 20–36.

Herrera, F. and Viedma, E.H. (2000) Linguistic decision analysis: steps for solving decision problems under linguistic information. *Fuzzy Sets and Systems*, **115** (1), 67–82.

Herrera, F., Herrera-Viedma, E., and Martínez, L. (2000) A fusion approach for managing multi-granularity linguistic term sets in decision-making. *Fuzzy Sets and Systems*, **114**, (1), 43–58.

Herrera, F., Herrera-Viedma, E., and Verdegay, J.L. (1995) A sequential selection process in group decision-making with linguistic assessment. *Information Sciences*, **85** (2), 223–239.

Hwang, C.L. and Masud, A.S. (1979) *Multiple Objective Decision-Making: Methods and Applications*, Springer-Verlag, Berlin.

Hwang, C.L. and Yoon, K. (1981) *Multiple Attribute Decision-Making: Methods and Applications – A State-of-the-Art Survey*, Springer-Verlag, Berlin.

Keeney, R. and Raifa, H. (1976) *Decisions with Multiple Objectives: Preferences and Value Trade-offs*, John Wiley & Sons, Inc., New York.

Kuhn, T.S. (1962) *The Structure of Scientific Revolutions*, University of Chicago Press, Chicago.

Larichev, O.I. (1984) Psychological validation of decision methods. *Journal of Applied Systems Analysis*, **11** (1), 37–46.

Larichev, O.I. (1987) *Objective Models and Subjective Decisions*, Nauka, Moscow (in Russian).

Larichev, O.I. and Moshkovich, E.M. (1996) *Qualitative Methods of Decision-Making*, Nauka, Moscow (in Russian).

Lu, J., Zhang, G., Ruan, D., and Wu, F. (2007) *Multi-Objective Group Decision-Making: Methods, Software and Applications with Fuzzy Set Techniques*, Imperial College Press, London.

Lyapunov, A.A. (ed.) (1972) *Operational Research: Methodological Aspects*, Nauka, Moscow (in Russian).

Ngwenyama, O.K., Bryson, N., and Mobolurin, A. (1996) Supporting facilitation in group support systems: techniques for analyzing consensus relevant data. *Decision Support Systems*, **16** (2), 155–168.

Orlovsky, S.A. (1981) *Problems of Decision-Making with Fuzzy Information*, Nauka, Moscow (in Russian).

Pareto, V. (1886) *Cours d'Economie Politique*, Lousanne Rouge, Lousanne.

Parreiras, R.O., Ekel, P.Ya., Martini, J.S.C., and Palhares, R.M. (2010) A flexible consensus scheme for multicriteria group decision-making under linguistic assessments. *Information Sciences*, **180** (7), 1075–1089.

Pedrycz, W. and Gomide, F. (1998) *An Introduction to Fuzzy Sets: Analysis and Design*, MIT Press, Cambridge, MA.

Popov, V.A. and Ekel, P.Ya. (1987) Fuzzy set theory and problems of controlling the design and operation of electric power systems. *Soviet Journal of Computer and System Sciences*, **25** (4), 92–99.

Pospelov, D.A. and Pushkin, V.M. (1972) *Thinking and Machines*, Sovetskoe Radio, Moscow (in Russian).

Rasmy, M.H., Lee, S.M., Abd El-Wahed, W.F., *et al.* (2002) An expert system for multiobjective decision-making: application of fuzzy linguistic preferences and goal programming. *Fuzzy Sets and Systems*, **127** (2), 209–220.

Rommelfanger, H. (2004) The advantages of fuzzy optimization models in practical use. *Fuzzy Optimization and Decision-Making*, **3** (4), 295–309.

Simon, H.A. (1977) *The New Science of Management Decision*, Prentice Hall, Englewood Cliffs, NJ.

Slovic, P., Fischhoff, B., and Lichtenstein, S. (1977) Behavioral decision theory. *Annual Review of Psychology*, **28** (1), 1–39.

Stasser, G. and Vaughan, S.I. (2000) Pooling unshared information: the benefits of knowing how access to information is distributed among group members. *Organizational Behavior and Human Decision Processes*, **82** (1), 102–116.

Tan, B. C. Y., Teo, H.H., and Wei, K.K. (1995) Promoting consensus in small decision-making groups. *Information & Management*, **28** (4), 251–259.

Trachtengerts, E.A. (1998) *Computer Support of Decision-Making*, SINTEG, Moscow (in Russian).

van Ginkel, W.P. and van Knippenberg, D. (2009) Knowledge about the distribution of information and group decision-making: when and why does it work? *Organizational Behavior and Human Decision Processes*, **108** (2), 218–229.

Wagner, H.M. (1982) *Operations Research: An Introduction*, Macmillan, New York.

Wang, P.P. (2007) Guest editorial: mathematics of uncertainty. *Information Sciences*, **177** (23), 5141–5142.

Wong, Z. and Aiken, M. (2003) Automated facilitation of electronic meetings. *Information & Management*, **41** (2), 125–134.

Xu, Z. (2005) On method for uncertain multiple attribute decision-making problems with uncertain multiplicative preference information on alternatives. *Fuzzy Optimization and Decision-Making*, **4** (2), 131–139.

Yager, R.R. (1993) Nonnumeric multi-criteria multi-person decision-making. *Group Decision and Negotiation*, **2** (11), 81–93.

Zadeh, L.A. (1965) Fuzzy sets. *Information and Control*, **8** (3), 338–353.

Zadeh, L.A. (1973) Outline of a new approach to the analysis of complex systems and decision processes. *IEEE Transactions on Systems, Man, and Cybernetics*, **3** (1), 28–44.

Zimmermann, H.J. (1996) *Fuzzy Set Theory and Its Application*, Kluwer, Boston.

2

Notions and Concepts of Fuzzy Sets: An Introduction

In this chapter, we introduce the essential concepts of fuzzy sets. We focus on the fundamental idea of partial membership, which is conveniently quantified through membership functions and membership degrees. We present the underlying rationale and then move on to a detailed description of fuzzy sets by discussing the most commonly encountered classes of membership functions and relating these classes to the semantics of fuzzy sets. We elaborate on the basic operations on fuzzy sets (intersection, union, complement, negation) and discuss the concepts of fuzzy relations and their main properties, which are of direct relevance in the context of decision-making.

2.1 Sets and Fuzzy Sets: A Fundamental Departure from the Principle of Dichotomy

Conceptually and algorithmically, fuzzy sets constitute one of the most fundamental and influential notions in science and engineering. The notion of a fuzzy set is highly intuitive and transparent since it captures what really becomes the essence of a way in which the real world is being perceived and described in our everyday activities. We are faced with categories of objects whose "belongingness" to a given category (concept) is always a matter of degree. There are numerous examples in which we encounter elements whose allocation to the concept we want to define can be satisfied to some extent. One may eventually claim that continuity of transition from full belongingness to full exclusion is the major and ultimate feature of the physical world and natural systems. For instance, we may qualify an indoor environment as *comfortable* when its temperature is kept *around* 20 °C. If we observe a value of 19.5 °C it is very likely we still feel quite *comfortable*. The same holds if we encounter 20.5 °C – humans usually do not discriminate between changes in temperature within the range of one degree Celsius. A value of 20 °C would be fully compatible with the concept of *comfortable* temperature, yet 0 °C or 30 °C would not. In these two cases, as well for temperatures close to these two values, we would describe them as being *cold* and *warm*, respectively. We could question whether the temperature of 25 °C is viewed as *warm* or *comfortable* or, similarly, if 15 °C is *comfortable* or *cold*. Intuitively, we know that 25 °C is somehow between *comfortable* and *warm* while 15 °C is between *comfortable* and *cold*. The value 25 °C is partially compatible with the term *comfortable* and *warm*, and somewhat compatible or, depending on the observer's perception, incompatible with the term of *cold* temperature. Similarly, we may say that 15 °C is partially compatible with the comfortable and cold temperature, and slightly compatible or incompatible with the warm temperature. In spite of this highly intuitive and apparent categorization of environment temperatures into the three classes, namely *cold*, *comfortable*, and *warm*, we note that the transition between the classes is not instantaneous and sharp. Simply, when moving across the range of temperatures, these values become gradually perceived as *cold*, *comfortable*, or *warm*. A similar phenomenon happens when we are dealing with the concept of the height of people. An individual of height 1 meter is *short*, whereas a person of 1.90 m is perceived to be *tall*. Again the question is, what is the range of height values that could qualify a person to be *tall*? Does a height of 1.85 m discriminate between *tall* and *short* individuals? Or maybe 1.86 m would be the right choice? In asking these questions, we know that they do not make too much sense. We realize that the nature of these concepts is such that we cannot use a single number – a transition between the notion of *tall* and *short* is not abrupt in any way. Hence we cannot assign a single number that does a good job. This sends a clear message: the concept of dichotomy does not apply when defining even simple concepts. An illustration of the concept of dichotomy is included in Figure 2.1(a). In contrast, defining a concept where we do not confine ourselves to the dichotomy is illustrated in Figure 2.1(b).

Fuzzy sets and the corresponding membership functions form a viable and mathematically sound framework to formalize these concepts. When talking about heights of Europeans we may refer to real numbers within the interval [0,3] to represent a universe of heights that range between 0 and 3 m. This universe of discourse (space) is suitable for describing the concept of *tall* people.

Let us denote by **X** a universe of discourse (space) of all elements. The universe can be either continuous or discrete. For instance, the closed interval [0,3] constitutes a continuous and bounded universe whereas the set $\mathbf{N} = \{0, 1, 2, \ldots\}$ of natural numbers is discrete and countable but there are no bounds.

Consider the universe of discourse $\mathbf{X} = [0,3]$ and the collection S of values in **X** that are less than a threshold value τ in **X**, for example, $\tau = 1.8$. Consider the sets $S = \{x \in \mathbf{X} \mid 0 < x < 1.8\}$ and $T = \{x \in \mathbf{X} \mid 1.8 \leq x \leq 3.0\}$, Figure 2.2. Each set is a class whose members

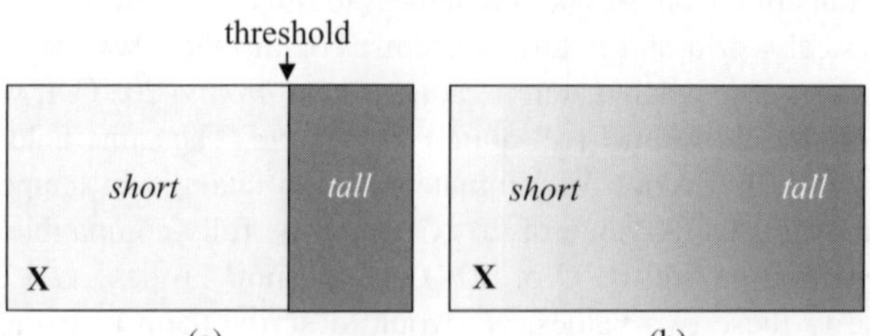

Figure 2.1 Contrasting the concept of a set and the principle of dichotomy itself versus a relaxation of the concept of complete inclusion and exclusion.

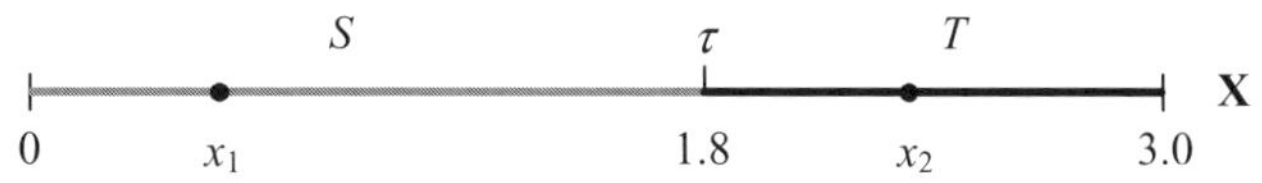

Figure 2.2 A set as a collection of numerical values located in the corresponding intervals.

are elements of the universe that satisfy the same property. This set is equivalent to a list of elements of the universe that are members of the set.

Given a certain value in **X**, the process of dichotomization (binarization) imposes a binary, all or none, classification decision: either accept or reject the value as belonging to a given collection. For instance, consider the set S shown in Figure 2.2. Clearly, the point x_1 belongs to S whereas x_2 does not, that is, $x_1 \in S$ and $x_2 \notin S$. Similarly for the set T we have $x_1 \notin T$ and $x_2 \in T$. If we denote the accept decision by 1 and the reject decision by 0, for short, we may express the classification (assignment) decision of $x \in \mathbf{X}$ through a characteristic function as follows:

$$S(x) = \begin{cases} 1 & \text{if} \quad x \in S \\ 0 & \text{if} \quad x \notin S \end{cases} \qquad T(x) = \begin{cases} 1 & \text{if} \quad x \in T \\ 0 & \text{if} \quad x \notin T \end{cases} \tag{2.1}$$

In general, a characteristic function of set A defined in **X** assumes the following form:

$$A(x) = \begin{cases} 1 & \text{if} \quad x \in A \\ 0 & \text{if} \quad x \notin A \end{cases} \tag{2.2}$$

The empty set $\oslash$ has a characteristic function that is identically equal to zero, $\oslash(x) = 0$ for all x in **X**. The universe **X** itself comes with the characteristic function that is identically equal to one, that is, $\mathbf{X}(x) = 1$ for all x in **X**. Also, a singleton $A = \{a\}$, a set with only a single element, has a characteristic function such that $A(x) = 1$ if $x = a$ and $A(x) = 0$ otherwise.

Characteristic functions $A : \mathbf{X} \rightarrow \{0, 1\}$ induce a constraint with well-defined boundaries on the elements of the universe **X** that can be assigned to a set A. The fundamental idea of a fuzzy set is to relax this requirement by admitting intermediate values of class membership. Therefore we may assign intermediate values between 0 and 1 to quantify our perception on how compatible these values are with the class, with 0 meaning incompatibility (complete exclusion) and 1 compatibility (complete membership). Membership values thus express the degrees to which each element of the universe is compatible with the properties distinctive to the class. Intermediate membership values mean that no "natural" threshold exists and that elements of a universe can be members of a class and at the same time belong to other classes with different degrees. Allowing for gradual, hence less strict, membership degrees is the crux of fuzzy sets.

Formally, a fuzzy set A is described by a membership function mapping the elements of a universe **X** to the unit interval [0,1] (Zadeh, 1965; Zadeh, 1975):

$$A{:}\mathbf{X} \rightarrow [0, 1] \tag{2.3}$$

The membership functions are therefore synonymous of fuzzy sets. In a nutshell, membership functions generalize characteristic functions in the same way as fuzzy sets generalize sets.

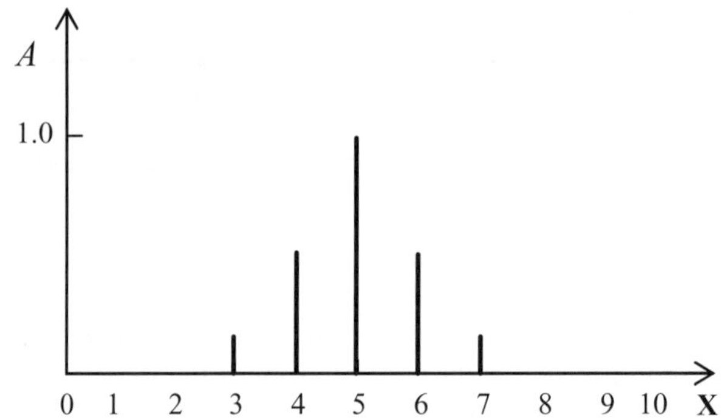

Figure 2.3 Fuzzy set A defined in a discrete universe **X**.

Fuzzy set can also be viewed as a set of ordered pairs of the form $\{x, A(x)\}$ where x is an element of **X** and $A(x)$ denotes its corresponding degree of membership. For a finite universe of discourse $\mathbf{X} = \{x_1, x_2, \ldots, x_n\}$, A can be represented by an n-dimensional vector $\boldsymbol{A} = [a_1, a_2, \ldots, a_n]$ with the elements $a_i = A(x_i)$. Figure 2.3 illustrates a fuzzy set whose membership function captures the concept of an integer *around* 5. Here $n = 10$. The fuzzy set linguistically expressing the quantity *around 5* in a finite universe of 10 integers $\boldsymbol{A} = [0, 0, 0, 0.2, 0.5, 1.0, 0.5, 0.2, 0, 0, 0]$. An equivalent notation of A can be read as $A = \{0/1, 0/2, 0/3, 0.2/4, 0.5/4, \ldots, 0/10\}$.

The choice of the unit interval for the values of membership degrees is usually a matter of convenience. The choice of the very detailed membership values (up to several decimal digits), say $A(4) = 0.9865$, is not crucial or even counter-productive. We should stress here that in describing membership grades, the ultimate objective is to reflect an order of the elements in A in terms of their belongingness to the fuzzy set (Dubois and Prade, 1979).

Being more descriptive, we may view fuzzy sets as elastic constraints imposed on the elements of a universe. As emphasized previously, fuzzy sets deal primarily with the concept of elasticity, graduality, or absence of sharply defined boundaries. In contrast, when dealing with sets we are concerned with rigid boundaries, lack of graded belongingness, and sharp, binary boundaries. Gradual membership means that no natural boundary exists and that some elements of the universe of discourse can, contrary to sets, coexist (belong) to different fuzzy sets with different degrees of membership.

2.2 Interpretation of Fuzzy Sets

In fuzzy sets, the concept of fuzziness comes with a precise meaning. Fuzziness primarily means lack of precise boundaries of a collection of objects and, as such, is an evident manifestation of imprecision and a particular type of uncertainty. Let us make some observations in this regard.

First, it is worth indicating that fuzziness is both conceptually and formally different from the fundamental concept of probability. In general, it is difficult to foresee the result of tossing a fair coin once it is impossible to know if either a head or tail will occur for certain. We may, at most, say that there is a 50% chance for a head or tail to occur, but as soon as the coin falls, uncertainty vanishes. But, in the case of a person's height, imprecision remains. Formally, fuzzy sets are membership functions that are treated as mappings from a given universe of discourse to the unit interval as presented in (2.3). In contrast, probability is a set function, a mapping whose universe is a set of subsets of a domain.

Second, there are differences between fuzziness, generality, and ambiguity. A notion is general when it applies to a multiplicity of objects and keeps only a common essential property. An ambiguous notion stands for several unrelated objects. Therefore, from this point of view fuzziness does not mean either generality or ambiguity and applications of fuzzy sets exclude these categories. Fuzzy set theory assumes that the universe is well defined and has its elements assigned to classes by means of a numerical scale.

Applications of fuzzy sets to areas such as data analysis, reasoning under uncertainty, and decision-making suggest different interpretations of membership grades in terms of similarity, uncertainty, and preference (Dubois and Prade, 1997; Dubois and Prade, 1998). From the similarity point of view, $A(x)$ means the degree of compatibility of an element $x \in \mathbf{X}$ with representative elements of A. This is the primary and most intuitive interpretation of a fuzzy set, one that is particularly suitable for data analysis. An example is the case where we question how to qualify an environment as *comfortable* when we know that the current temperature is 25 °C. As discussed at the beginning of this chapter, such quantification is a matter of degree. For instance, assuming a universe of discourse of $\mathbf{X} = [0,40]$ and choosing 20 °C as representative of *comfortable* temperature, we note, Figure 2.4, that the degree at which 25 °C is comfortable is 0.2. In the example, we have adopted piecewise linear decreasing functions of the distance between temperature values and the representative value 20 °C to determine the corresponding membership degree.

Now let us assume that values of a variable "x" are located within the support of a fuzzy set A. Then, given a value "v" of $\mathbf{X}$, $A(v)$ expresses a possibility that $x = v$ given that "x" is in A is all that is known about it. In this situation, the membership degree of a given tentative value "v" to the class A reflects the degree of plausibility that this value is the same as "x". This idea reflects a type of uncertainty because, if the membership degree is high, our confidence about the value of "x" may still be low, but if the degree is low, then the tentative value may be rejected as an implausible candidate. The variable labeled by the class A is uncontrollable. This allows the assignment of fuzzy sets to possibility distributions as presented in possibility theory (Zadeh, 1978; Zadeh, 1999). For instance, suppose someone said they felt comfortable

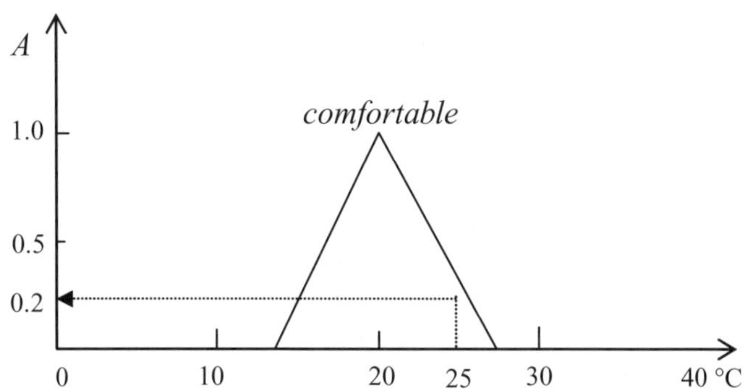

Figure 2.4 Membership function of a fuzzy set of *comfortable* temperature.

when watching a soccer game. In this situation the membership degree of a given tentative temperature value, say 25 °C, reflects the degree of plausibility that this value of temperature is the same as the one when the person felt comfortable. Note that the temperature value felt is unknown, but there is no question if it did occur or not. Possibility is whether an event may occur and with what degree. On the contrary, probability is about whether an event will occur.

Finally, assume that A reflects a preference on the values of a variable "x" in $\mathbf{X}$. For instance, "x" can be a decision variables and fuzzy set A is an elastic constraint characterizing feasible values and decision-maker preferences. In this case $A(v)$ denotes the grade of preference in favor of "v" as the value of "x". This interpretation prevails in fuzzy optimization and decision analysis. For instance, we may be interested in finding a *comfortable* value of temperature. The membership degree of a candidate temperature value "v" reflects our degree of satisfaction with the particular temperature value chosen. In this situation, the choice of the value is controllable in the sense that the value being adopted depends on our choice.

2.3 Membership Functions and Classes of Fuzzy Sets

Formally speaking, any function $A : \mathbf{X} \to [0,1]$ could be qualified to serve as a membership function describing the corresponding fuzzy set. In practice, the form of the membership functions should be reflective of the problem at hand for which we construct fuzzy sets. They should reflect our perception (semantics) of the concept to be represented and further used in problem solving, the level of detail we intend to capture, and a context in which the fuzzy sets are going to be used. It is also essential to assess the type of fuzzy set from the standpoint of its suitability when handling the ensuing optimization procedures. With these criteria in mind, we elaborate on the most commonly used categories of membership functions. All of them are defined in the universe of real numbers, that is, $\mathbf{X} = \mathbf{R}$.

Triangular membership functions. These are described by their piecewise linear segments described in the form

$$A(x, a, m, b) = \begin{cases} 0 & \text{if} \quad x \leq a \\ \dfrac{x-a}{m-a} & \text{if} \quad x \in [a, m] \\ \dfrac{b-x}{b-m} & \text{if} \quad x \in [m, b] \\ 0 & \text{if} \quad x \geq b \end{cases} \tag{2.4}$$

Using more concise notation, the above expression can be written in the form $A(x, a, m, b) = \max\{\min[(x-a)/(m-a), (b-x)/(b-m)], 0\}$. The meaning of the parameters is straightforward: "m" denotes a modal (typical) value of the fuzzy set while "a" and "b" are the lower and upper bounds, respectively. They could be sought as the extreme elements of the universe of discourse that delineate the elements belonging to A with nonzero membership degrees.

Triangular fuzzy sets (membership functions) are the simplest possible models of grades of membership as they are fully defined by only three parameters. As mentioned, the semantics is evident as the fuzzy sets are expressed on a basis of knowledge of the spreads of the concepts and their typical values. The linear change in the membership grades is the simplest possible model of membership one could think of. Taking the absolute value of the derivative of the triangular membership function, which could be sought as a measure of sensitivity of A, $|dA/dx|$, we conclude that its sensitivity is constant for each of the linear segments of the fuzzy set.

Trapezoidal membership functions. These are piecewise linear functions characterized by four parameters, "a", "m", "n", and "b", each of which defines one of the four linear parts of the membership function. They assume the following form:

$$A(x) = \begin{cases} 0 & \text{if} \quad x < a \\ \dfrac{x-a}{m-a} & \text{if} \quad x \in [a, m] \\ 1 & \text{if} \quad x \in [m, n] \\ \dfrac{b-x}{b-n} & \text{if} \quad x \in [n, b] \\ 0 & \text{if} \quad x > b \end{cases} \tag{2.5}$$

Using an equivalent notation, we can rewrite A as follows:

$$A(x, a, m, n, b) = \max\{\min[(x-a)/(m-a), 1, (b-x)/(b-n)], 0\}$$

Γ-membership functions. These are expressed in the following form:

$$A(x) = \begin{cases} 0 & \text{if} \quad x \le a \\ 1 - \exp[-k(x-a)^2] & \text{if} \quad x > a \end{cases} \quad \text{or} \quad A(x) = \begin{cases} 0 & \text{if} \quad x \le a \\ \dfrac{k(x-a)^2}{1+k(x-a)^2} & \text{if} \quad x > a \end{cases} \tag{2.6}$$

where $k > 0$.

***S*-membership functions**. These functions are of the form

$$A(x) = \begin{cases} 0 & \text{if} \quad x \le a \\ 2\left(\dfrac{x-a}{b-a}\right)^2 & \text{if} \quad x \in [a, m] \\ 1 - 2\left(\dfrac{x-b}{b-a}\right)^2 & \text{if} \quad x \in [m, b] \\ 1 & \text{if} \quad x > b \end{cases} \tag{2.7}$$

The point $m = (a + b)/2$ is the crossover point of the S-function.

Gaussian membership functions. These membership functions are described by the following relationship:

$$A(x, m, \sigma) = \exp\left(-\frac{(x-m)^2}{\sigma^2}\right) \tag{2.8}$$

Gaussian membership functions are described by two important parameters. The modal value (m) represents the typical element of A while s denotes a spread of A. Higher values of s correspond to larger spreads of the fuzzy sets.

Exponential-like membership functions. These membership functions are described in the form

$$A(x) = \frac{1}{1 + k(x - m)^2} \qquad k > 0 \tag{2.9}$$

The spread of the exponential-like membership function increases as the values of "k" get lower.

2.4 Fuzzy Numbers and Intervals

In practice, exact values of parameters of models are not so common. Normally uncertainty and imprecision arise due to lack of knowledge and incomplete information reflected in system structure, parameters, inputs, and possible bounds.

Fuzzy numbers and intervals model imprecise quantities and capture our innate concept of approximate numbers such as about 5, around 10, and intervals such as below 100, around 2 and 3, above 10. Fuzzy quantities are intended to model our intuitive notions of approximate numbers and intervals as a generalization of numbers and intervals, as Figure 2.5 suggests. In general, fuzzy quantities summarize numerical data by means of linguistically labeled fuzzy sets whose universe is **R**, the set of real numbers. For instance, if a value of a real variable is certain, say $x = 2.5$, then we can represent it as a certain quantity, a singleton whose

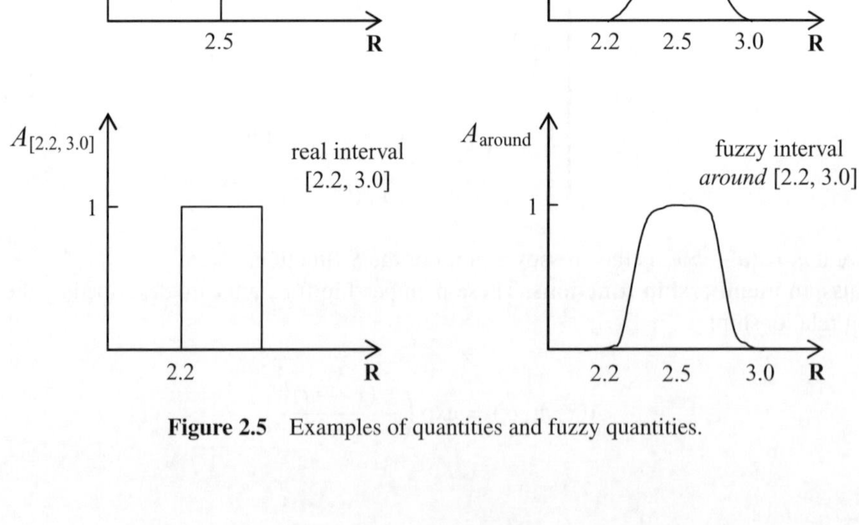

Figure 2.5 Examples of quantities and fuzzy quantities.

characteristic function is $A_{2.5}(x) = 1$ if $x = 2.5$ and $A_{2.5}$ (0) = 0 otherwise, as shown in Figure 2.5. In this situation, the quantity has both a precise value and precise meaning. If we are uncertain of the value of the variable, but certain about its bounds, then the quantity is uncertain and can be represented, for instance, by the closed interval [2.2,3.0], a set whose characteristic function is $A_{[2.2,3.0]}(x) = 1$ if $x \in [2.2,3.0]$, and $A_{[2.2,3.0]}(x) = 0$ otherwise. Here the variable is characterized by an imprecise value, but its meaning is precise. When bounds also are not sharply defined, the quantities become fuzzy numbers or intervals, respectively, as Figure 2.5 illustrates. In these cases both fuzzy numbers and intervals also are quantities with precise meaning, but with imprecise values.

2.5 Linguistic Variables

One can often deal with variables describing phenomena of physical or human systems assuming a finite, quite small number of descriptors.

We often describe observations about a phenomenon by characterizing its states which we naturally translate in terms of the idea of a variable. For instance, we may refer to an environment through words such as comfortable, sunny, and nice. In particular, we can qualify the environmental condition through the variable temperature with values chosen in a range such as the interval $X = [0,40]$. Alternatively, temperature could be qualified using labels such as cold, comfortable, and warm. A precise numerical value such as 20 °C seems simpler to characterize the environment than the ill-defined term comfortable. But the linguistic label comfortable is a choice of one out of three values, whereas 20 °C is a choice of one out of many. The statement could be strengthened if the underlying meaning of comfortable is conceived as about 20 °C. While the numerical quantity 20 °C can be visualized as a point in a set, the linguistic temperature value comfortable can be viewed as a collection of temperature values in a bounded region centered at 20 °C. The label comfortable can, therefore, be regarded as a form of information summarization, called granulation, because it serves to approximate a characterization of ill-defined or complex phenomena (Zadeh, 1975). In these circumstances, fuzzy sets provide a way to map a finite term set to a linguistic scale whose values are fuzzy sets. In general, it is difficult to find incontestable thresholds, such as 15 and 30 °C for instance, which allows us to assign cold = [0,15], comfortable = [15,30], and warm = [30,40]. Cold, comfortable, and warm are fuzzy sets instead of single numbers or sets (intervals). Since fuzzy sets concern the representation of collections with unclear boundaries by means of membership functions taking values in an ordered set of membership values, they provide a means to interface numerical and linguistic quantities, a way to link computing with words and granular computing.

In contrast to the idea of numerical variables as commonly used, the notion of linguistic variable can be regarded as a variable whose values are fuzzy sets. In general, linguistic variables may assume values consisting of words or sentences expressed in a certain language (Zadeh, 1999). Formally, a linguistic variable is characterized by a quintuple $\langle X, T(X), \mathbf{X}, G, M \rangle$ where its components are as follows:

X – the name of the variable;

$T(X)$ – a term set of X whose elements are labels L of linguistic values of X;

G – a grammar that generates the names of X;

M – a semantic rule that assigns to each label $L \in T(\mathbf{X})$ a meaning whose realization is a fuzzy set on the universe $\mathbf{X}$ whose base variable is X.

Example 2.1. Let us consider the linguistic variable of temperature. Here the linguistic variable is formalized by explicitly identifying all the components of the formal definition:

$X =$ temperature, $\mathbf{X} = [0,40]$.

T (temperature) $= \{$*cold*, *comfortable*, *warm*$\}$.

M (cold) $\rightarrow C$, M (comfortable) $\rightarrow F$, and M (warm) $\rightarrow W$ where C, F, and W are fuzzy sets whose membership functions are illustrated in Figure 2.6.

The notion of the linguistic variable plays a major role in applications of fuzzy sets. In fuzzy logic and approximate reasoning, truth values can be viewed as linguistic variables whose truth values form the term set as, for example, *true*, *very true*, *false*, *more or less true*, and the like.

Fuzzy sets can be cast in the more general setting of granular computing (Bargiela and Pedrycz, 2003; Pedrycz, 2005; Zadeh, 1997) in which processing is realized in terms of

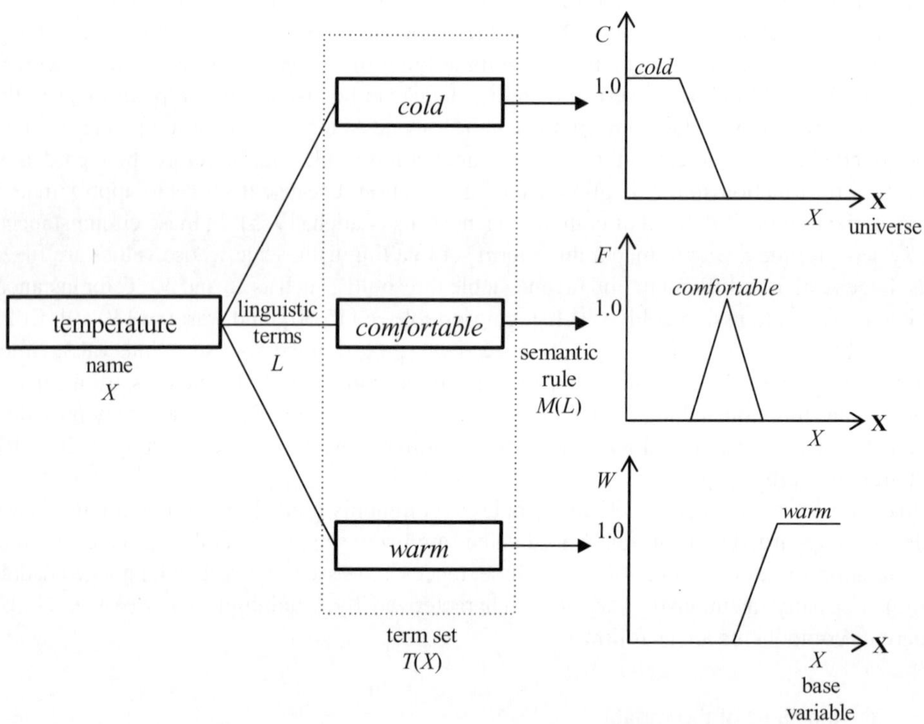

Figure 2.6 An example of the linguistic variable of temperature.

information granules – conceptual entities being a result of a certain abstraction we exercise to perceive real-world phenomena and build their effective models. The granules can be formally represented as sets, fuzzy sets, rough sets (Pawlak, 1982), or shadowed sets where the representation depends upon the nature of the problem itself.

2.6 A Generic Characterization of Fuzzy Sets: Some Fundamental Descriptors

In principle, any function $A : \mathbf{X} \rightarrow [0,1]$ becomes potentially eligible to represent the membership function of fuzzy set A. In practice, however, the type and shape of membership functions should fully reflect the nature of the underlying phenomenon we are interested in describing. Thus we require that fuzzy sets should be semantically sound, which implies that the selection of membership functions needs to be guided by the character of the application and the nature of the problem we intend to solve.

Given the enormous diversity of potentially useful (namely, semantically sound) membership functions, there are certain common characteristics (descriptors) that are conceptually and operationally qualified to capture the essence of the granular constructs represented in terms of fuzzy sets. In what follows, we provide a list of the descriptors commonly encountered in practice.

Normality: We say that the fuzzy set A is *normal* if its membership function attains one, that is,

$$\sup_{x \in \mathbf{X}} A(x) = 1 \tag{2.10}$$

If this property does not hold, we call the fuzzy set *subnormal*. An illustration of the corresponding fuzzy set is shown in Figure 2.7. The supremum (sup) in the above expression is also referred to as the height of the fuzzy set A, $\text{hgt}(A) = \sup_{x \in \mathbf{X}} A(x) = 1$.

The normality of A has a simple interpretation: by determining the height of the fuzzy set, we identify an element with the highest membership degree. The value of the height being equal to one states that there is at least one element in $\mathbf{X}$ whose typicality with respect to A is the highest one and which could be sought as fully compatible with the semantic category presented by A. A subnormal fuzzy set whose height is lower than one, namely $\text{hgt}(A) < 1$,

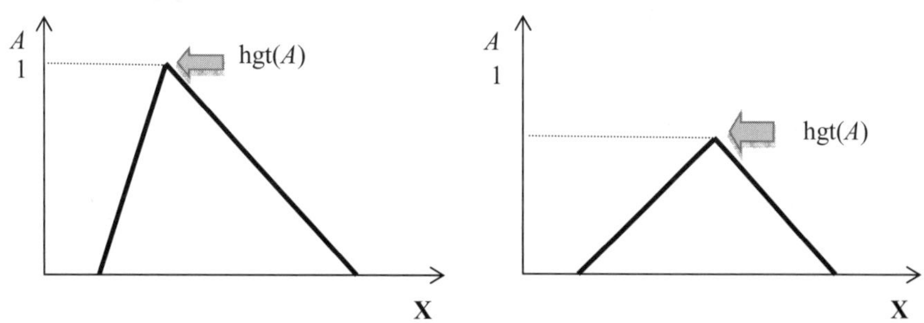

Figure 2.7 Examples of normal and subnormal fuzzy sets.

means that the degree of typicality of elements in this fuzzy set is somewhat lower (weaker) and we cannot identify any element in **X** which is fully compatible with the underlying concept. Generally, while forming a fuzzy set we expect its normality (otherwise, why would such a fuzzy set for which there are no typical elements come into existence in the first place?).

Normalization: The normalization operation, Norm(A), is a transformation mechanism that is used to convert a subnormal nonempty fuzzy set A into its normal counterpart. This is done by dividing the original membership function by the height of this fuzzy set, that is,

$$\text{Norm}(A) = \frac{A(x)}{\text{hgt}(A)} \tag{2.11}$$

While the height describes the global property of the membership grades, the following notions offer an interesting characterization of the elements of **X** vis-à-vis their membership degrees.

Support: The support of a fuzzy set A, denoted by Supp(A), is a set of all elements of **X** with nonzero membership degrees in A

$$\text{Supp}(A) = \{x \in \mathbf{X} | A(x) > 0\} \tag{2.12}$$

In other words, the support identifies all elements of **X** that exhibit some association with the fuzzy set under consideration (by being allocated to A with nonzero membership degrees).

Core: The core of a fuzzy set A, Core(A), is a set of all elements of the universe that are typical for A, that is, they come with membership grades equal to one,

$$\text{Core}(A) = \{x \in \mathbf{X} | A(x) = 1\} \tag{2.13}$$

The support and core are related in the sense that they identify and collect elements belonging to the fuzzy set but at two different levels of membership. Given the character of the core and support, we note that all elements of the core of A are subsumed by the elements of the support of this fuzzy set. Note that both the support and core are sets, not fuzzy sets, Figure 2.8. We refer to them as the set-based characterizations of fuzzy sets.

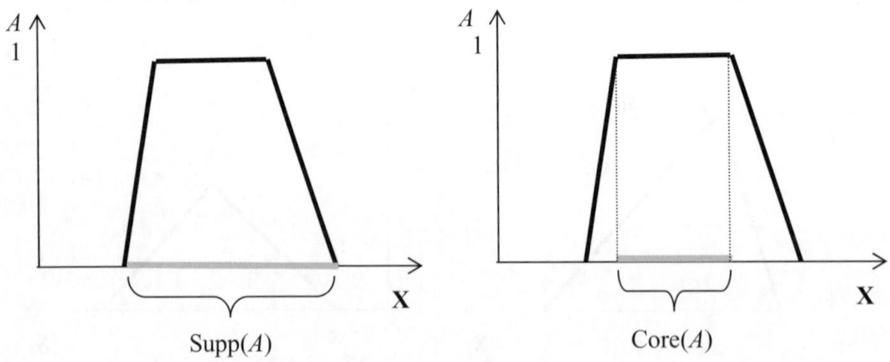

Figure 2.8 Support and core of A.

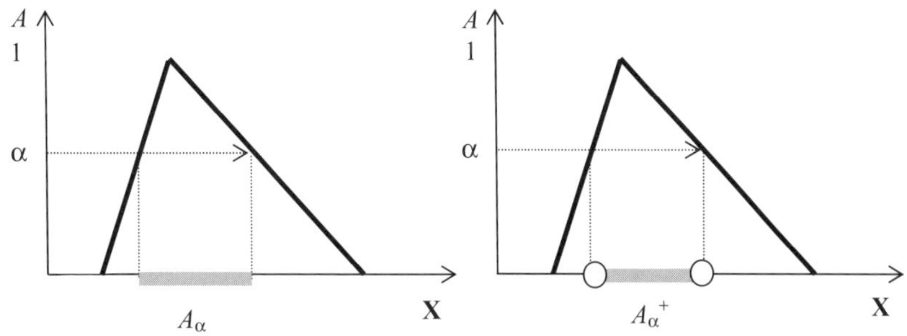

Figure 2.9 Examples of α-cut and strong α-cut.

While the core and support are somewhat extreme (in the sense that they identify the elements of A that exhibit the strongest and the weakest linkages with A), we may also be interested in characterizing sets of elements that come with some intermediate membership degrees. A notion of a so-called α-cut offers here an interesting insight into the nature of fuzzy sets.

α-cut: The α-cut of a fuzzy set A, denoted by A_α, is a set consisting of the elements of the universe whose membership values are equal to or exceed a certain threshold level α where $\alpha \in [0,1]$ (Zadeh, 1975; Nguyen and Walker, 1999). Formally speaking, we have $A_\alpha = \{x \in \mathbf{X} \mid A(x) \geq \alpha\}$. A strong α-cut differs from the α-cut in the sense that it identifies all elements in $\mathbf{X}$ for which we have the "greater than" relationship satisfied, $A_\alpha = \{x \in \mathbf{X} | A(x) > \alpha\}$. An illustration of the concept of the α-cut and strong α-cut is presented in Figure 2.9. Both the support and core are limit cases of α-cuts and strong α-cuts. For $\alpha = 0$ and the strong α-cut, we arrive at the concept of the support of A. The threshold $\alpha = 1$ means that the corresponding α-cut is the core of A.

Convexity: We say that a fuzzy set is convex if its membership function satisfies the following condition:

for all $x_1, x_2 \in \mathbf{X}$ and all $\lambda \in [0,1]$,

$$A(\lambda x_1 + (1 - \lambda)x_2) \geq \min(A(x_1), A(x_2)) \tag{2.14}$$

This relationship states that, whenever we choose a point x on a line segment between x_1 and x_2, the point $(x, A(x))$ is always located above or on the line passing through the two points $(x_1, A(x_1))$ and $(x_2, A(x_2))$, Figure 2.10. Note that the membership function is not a convex function in the traditional sense (Klir and Yuan, 1995).

Let us recall that a set S is convex if, for all $x_1, x_2 \in \mathbf{S}$, then $x = \lambda x_1 + (1 - \lambda)x_2 \in S$ for all $\lambda \in [0,1]$. In other words, convexity means that any line segment identified by any two points in $\mathbf{S}$ is also contained in $\mathbf{S}$. For instance, intervals of real numbers are convex sets. Therefore, if a fuzzy set is convex, then all of its α-cuts are convex, and, conversely, if a fuzzy set has all its α-cuts convex, then it is a convex fuzzy set, Figure 2.11. Thus we may say that a fuzzy set is convex if all its α-cuts are convex (intervals).

Fuzzy sets can be characterized by counting their elements and bringing a single numeric quantity as a meaningful descriptor of this count. While in the case of sets this sounds convincing, here we have to take into account different membership grades. In its simplest form this counting comes under the name of cardinality.

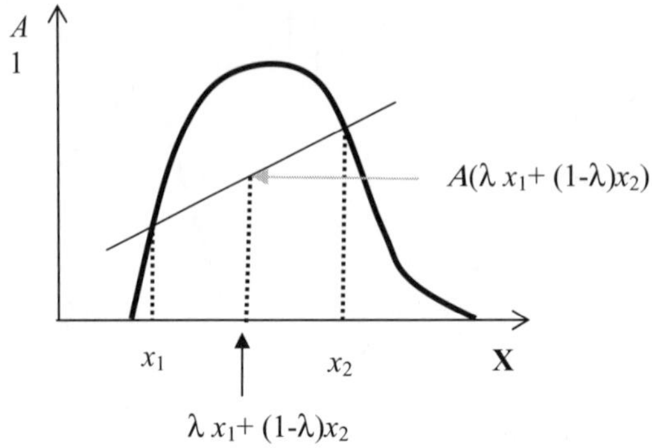

Figure 2.10 An example of a convex fuzzy set A.

Cardinality: Given a fuzzy set A defined in a finite or countable universe **X**, its cardinality, denoted by card(A), is expressed as the following sum:

$$\mathrm{Card}(A) = \sum_{x \in \mathbf{X}} A(x) \tag{2.15}$$

or alternatively as the following integral:

$$\mathrm{Card}(A) = \int_{\mathbf{X}} A(x)\,\mathrm{d}x \tag{2.16}$$

(we assume that the integral shown above does make sense). The cardinality produces a count of the number of elements in the given fuzzy set. As there are different degrees of membership, the use of the sum here makes sense as we keep adding contributions coming from the individual elements of this fuzzy set. Note that, in the case of sets, we count the number of elements belonging to the corresponding sets. We also use the alternative notation of Card(A) = $|A|$, and refer to it as a sigma count (σ-count).

The cardinality of fuzzy sets is explicitly associated with the concept of granularity of information granules realized in this manner. More descriptively, the more elements of A we encounter, the higher the level of abstraction supported by A and the lower the granularity of the

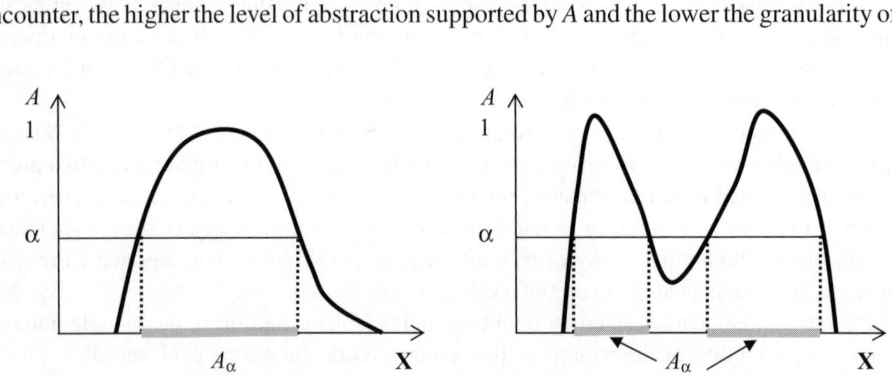

Figure 2.11 Examples of convex and nonconvex fuzzy sets.

construct. Higher values of cardinality come with a higher level of abstraction (generalization) and lower values of granularity (specificity).

Example 2.2. Consider fuzzy sets $A = (1.0, 0.6, 0.8, 0.1)$, $B = (0.1, 0.8, 1.0, 0.1)$, and $C = (0.6, 0.9, 1.0, 1.0)$ defined in the same space. We can order them in a linear fashion by computing their cardinalities. Here we obtain: Card(A) = 2.5, Card(B) = 2.0, and Card(C) = 3.5. In terms of the levels of abstraction, C is the most general, A lies in between, and B is the least general.

So far we have discussed the properties of a single fuzzy set. The operations to be studied look into the characterization of relationships between two fuzzy sets.

Equality: We say that two fuzzy sets A and B defined in the same universe **X** are equal if and only if their membership functions are identical, meaning that

$$A(x) = B(x)\ \forall x \in \mathbf{X} \tag{2.17}$$

Inclusion: Fuzzy set A is a subset of B (A is included in B), denoted by $A \subseteq B$, if and only if every element of A also is an element of B. This property expressed in terms of membership degrees means that the following inequality is satisfied:

$$A(x) \leq B(x)\ \forall x \in \mathbf{X} \tag{2.18}$$

An illustration of these two relationships in the case of sets is shown in Figure 2.12. In order to satisfy the relationship of inclusion, we require that the characteristic functions adhere to (2.18) for all elements of **X**. If the inclusion is not satisfied, even for a single point of **X**, the inclusion property does not hold.

If A and B are fuzzy sets in **X**, we have adopted the same definition of inclusion as that available from set theory.

Interestingly, the definitions of equality and inclusion exhibit an obvious dichotomy as the property of equality (or inclusion) is satisfied or not satisfied. While this quantification could be acceptable in the case of sets, fuzzy sets require more attention in this regard given the membership degrees involved in expressing the corresponding definitions.

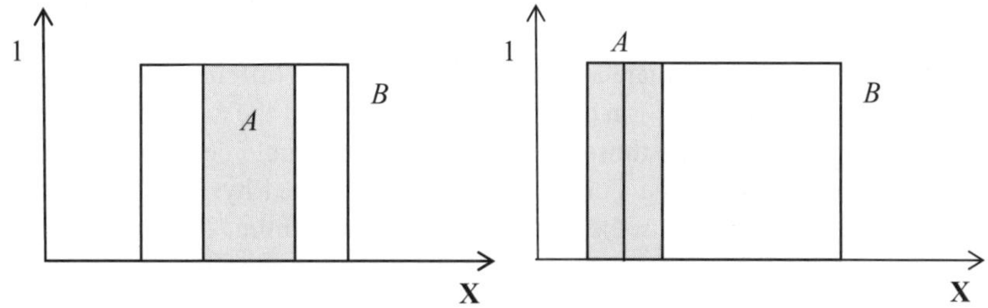

Figure 2.12 Set inclusion: (a) $A \subset B$ and (b) inclusion not satisfied as $A \not\subset B$.

Energy measure of fuzziness: For a fuzzy set A in $\mathbf{X}$, denoted by $E(A)$, this measure is a functional of the membership degrees

$$E(A) = \sum_{i=1}^{n} e[A(x_i)] \tag{2.19}$$

if Card($\mathbf{X}$) $= n$. In the case of infinite space, the energy measure of fuzziness is expressed as the integral

$$E(A) = \int_{\mathbf{X}} e(A(x))\,\mathrm{d}x \tag{2.20}$$

The mapping $e : [0,1] \to [0,1]$ is a functional monotonically increasing over [0,1] with the boundary conditions $e(0) = 0$ and $e(1) = 1$.

As the name of this measure stipulates, its role is to quantify a sort of energy associated with the given fuzzy set. The higher the membership degrees, the more essential their contributions to the overall energy measure. In other words, by computing the energy measure of fuzziness we can compare fuzzy sets in terms of their overall count of membership degrees.

A particular form of the above functional comes with the identity mapping $e(u) = u$ for all u in [0,1]. We can see that, in this case, expressions (2.19) and (2.20) reduce to the cardinality of A:

$$E(A) = \sum_{i=1}^{n} A(x_i) = \mathrm{Card}(A) \tag{2.21}$$

The energy measure of fuzziness forms a convenient way of expressing a total mass of the fuzzy set. Since Card($\oslash$) $= 0$ and Card($\mathbf{X}$) $= n$, the more a fuzzy set differs from the empty set, the larger is its mass. Indeed, rewriting (2.21) we obtain

$$E(A) = \sum_{i=1}^{n} A(x_i) = \sum_{i=1}^{n} |A(x_i) - \oslash(x_i)| = d(A, \oslash) = \mathrm{Card}(A) \tag{2.22}$$

where $d(A, \oslash)$ is the Hamming distance between fuzzy set A and the empty set.

While the identity mapping (e) is the simplest alternative one could think of, in general, we can envision an infinite number of possible options. For instance, one could consider functionals such as $e(u) = u^p,\ p > 0$, and $e(u) = \sin(\pi u/2)$. Note that by choosing a certain form of the functional, we accentuate a varying contribution of different membership grades. For instance, depending upon the form of "e", the contribution of the membership grades close to one could be emphasized while those located close to zero could be very much reduced.

Entropy measure of fuzziness: For A this measure, denoted by $H(A)$, is built upon the entropy functional (h) and comes in the form (De Luca and Termini, 1972)

$$H(A) = \sum_{i=1}^{n} h[A(x_i)] \tag{2.23}$$

or in the continuous case of **X**

$$H(A) = \int_{\mathbf{X}} h(A(x))\,\mathrm{d}x \tag{2.24}$$

where $h : [0,1] \rightarrow [0,1]$ is a functional such that: (1) it is monotonically increasing in $[0,{}^1\!/_2]$ and monotonically decreasing in $[{}^1\!/_2,1]$; and (2) it comes with the boundary conditions $h(0) = h(1) = 0$ and $h({}^1\!/_2) = 1$. This functional emphasizes membership degrees around ${}^1\!/_2$; in particular, the value of ${}^1\!/_2$ is stressed to be the most "unclear" (causing the highest level of hesitation with its quantification by means of the proposed functional).

Specificity of fuzzy sets: Quite often, we face the issue of quantifying how much a single element of a universe could be regarded as representative of a fuzzy set. If this fuzzy set is a singleton

$$A(x) = \begin{cases} 1 & \text{if} \quad x = x_0 \\ 0 & \text{if} \quad x \neq x_0 \end{cases} \tag{2.25}$$

then there is no hesitation in selecting x_0 as the sole representative of A. We say that A is very *specific* and its choice comes with no hesitation. At the other extreme, if A covers the entire universe **X** and embraces all elements with the membership grade being equal to one, the choice of the only one representative of A comes with a great deal of hesitation which is triggered by a lack of specificity faced in this problem. These two extreme situations are portrayed in Figure 2.13. Intuitively, we sense that the specificity is a concept that relates quite visibly to the cardinality of a set (Yager, 1983). The higher the cardinality of the set (that is, the more evident its abstraction), the lower its specificity. Having said that, we are interested in developing a measure which might be able to capture this effect of hesitation.

2.7 Geometric Interpretation of Sets and Fuzzy Sets

In the case of finite universes of discourse **X**, we can arrive at an interesting and geometrically appealing interpretation of sets and fuzzy sets (Kosko, 1992). Such an interpretation is also helpful in contrasting sets and fuzzy sets. It visualizes the interrelationships between them.

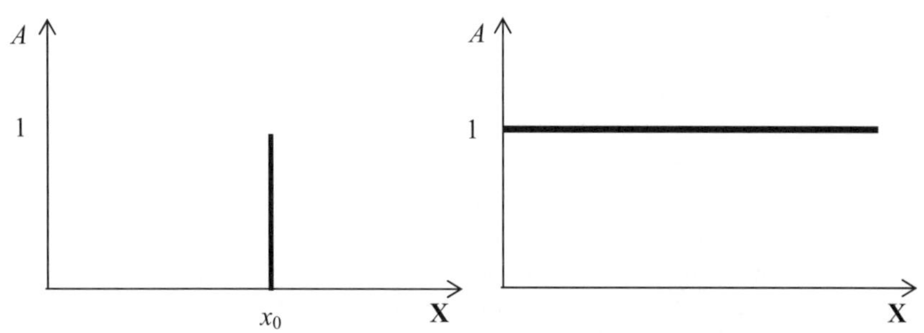

Figure 2.13 Examples of two extreme cases of sets exhibiting distinct levels of specificity.

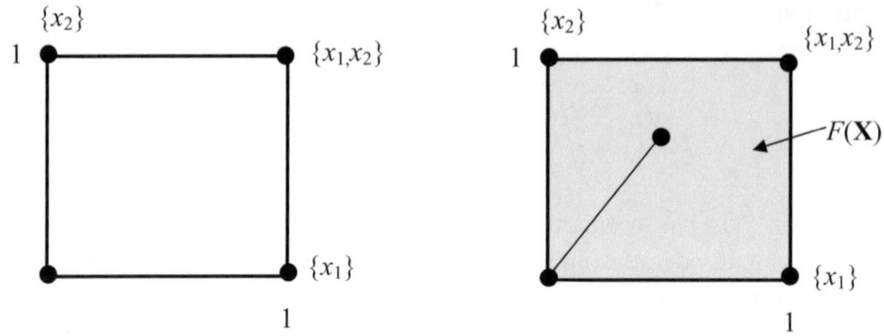

Figure 2.14 Sets and fuzzy sets represented as points in the unit square.

The geometric interpretation is also helpful in casting the decision-making problems and their solutions in some illustrative geometric setting. For the n-element space $\mathbf{X}$, any set there can be represented as an n-dimensional vector $\boldsymbol{x}$ with 0–1 values. The cardinality of the family of all sets defined in $\mathbf{X}$ is 2^n. The ith component of vector $\boldsymbol{x}$ is the value of the corresponding characteristic function of the ith element in the respective set. In the simplest case when $\mathbf{X} = \{x_1, x_2\}$; $n = 2$, the family of sets comprises the following elements: $\varnothing$, $\{x_1\}$, $\{x_2\}$, and $\{x_1, x_2\}$. The cardinality of $\mathbf{X}$ is $2^2 = 4$. Thus each of the four elements of this family can be represented by a two-dimensional vector, say $\varnothing = (0, 0)$, $\{x_1\} = (1, 0)$, $\{x_2\} = (0, 1)$, and $\{x_1, x_2\} = (1, 1)$. Those sets are located at the corners of the unit square, as illustrated in Figure 2.14.

Due to the values of the membership grades assuming any values in [0,1], fuzzy sets, being two-dimensional vectors, are distributed throughout the entire unit square. For instance, referring to Figure 2.14, fuzzy set A is represented as vector $\boldsymbol{x} = (0.25, 0.75)$. A family of fuzzy sets over $\mathbf{X} = \{x_1, x_2\}$ occupies the whole shaded area, including the borders and corners of the unit square. In general, proceeding with higher dimensionality of the space, we end up with a unit cube ($n = 3$) and unit hypercubes (for the dimensionality of the space of dimensionality higher than three).

2.8 Fuzzy Sets and the Family of α-cuts

Fuzzy sets offer an important conceptual and operational feature of information granules by endowing their formal models with gradual degrees of membership. We are interested in exploring relationships between fuzzy sets and sets. While sets come with the binary (yes–no) model of membership, it could be worth investigating whether they are indeed some special cases of fuzzy sets and, if so, in which sense a set could be treated as a suitable approximation of some given fuzzy set. This could shed light on some related processing aspects. To gain a detailed insight into this matter, we recall here a concept of an α-cut and a family of α-cuts and show that they relate to fuzzy sets in an [0,1] $\rightarrow$ [0,1] intuitive and transparent way. Let us revisit the semantics of α-cuts: an α-cut of A embraces all elements of the fuzzy set whose degrees of belongingness (membership) to this fuzzy set are at least equal to α. In this sense, by selecting a sufficiently high value of α, we identify (tag) elements of A that belong to it to a significant extent and thus could be sought as those highly representative of the concept conveyed by A. Those elements of **X** exhibiting lower values of the membership grades are suppressed, so this allows us to selectively focus on the elements with the highest degrees of membership while dropping the others.

For α-cuts A_α the following properties hold:

$$\begin{aligned} &\text{(a)} \quad A_0 = \mathbf{X} \\ &\text{(b)} \quad \text{If } \alpha \le \beta \text{ then } A_\alpha \supseteq A_\beta \end{aligned} \tag{2.26}$$

The first property shows that if we allow for the zero value of α, then all elements of **X** are included in this α-cut (0-cut, to be more specific). The second property underlines the monotonic character of the construct: higher values of the threshold imply that more elements are accepted in the resulting α-cuts. In other words, we can say that the level sets (α -cuts) A_α form a nested family of sets indexed by some parameter (α). If we consider the limit value of α, that is, $\alpha = 1$, the corresponding α -cut is nonempty if and only if A is a normal fuzzy set.

It is also worth remembering that α-cuts, in contrast to fuzzy sets, are sets. We showed how, for some given fuzzy set, its α-cut could be formed. An interesting question arises as to the construction that could be realized when moving in the opposite direction. Could we "reconstruct" a fuzzy set on the basis of an infinite family of sets? The answer to this problem is offered in what is known as the representation theorem for fuzzy sets (Klir and Yuan, 1995).

Theorem: Let $\{A_\alpha\}$, $\alpha \in [0,1]$, be a family of sets defined in **X** such that they satisfy the following properties:

(a) $A_0 = \mathbf{X}$.
(b) If $\alpha \le \beta$ then $A_\alpha \supseteq A_\beta$.
(c) For the sequence of threshold values $\alpha_1 \le \alpha_2 \le \ldots$ such that $\lim(\alpha_n) = \alpha$, we have

$$A_\alpha = \bigcap_{n=1}^{\infty} A_{\alpha_n}$$

then there exists a unique fuzzy set B defined in **X** such that $B_\alpha = A_\alpha$ for each $\alpha \in [0,1]$.

In other words, the representation theorem states that any fuzzy set A can be uniquely represented by an infinite family of its α-cuts. The following reconstruction expression shows how the corresponding α-cuts contribute to the formation of the corresponding fuzzy set:

$$A = \bigcup_{\alpha > 0} \alpha A_\alpha \tag{2.27}$$

that is,

$$A(x) = \sup_{\alpha \in (0,1]} (\alpha \cdot A_\alpha(x)) \tag{2.28}$$

where A_α denotes the corresponding α-cut.

The essence of this construct is that any fuzzy set can be uniquely represented by the corresponding family of nested sets (that is, ordered by the inclusion relation). An illustration

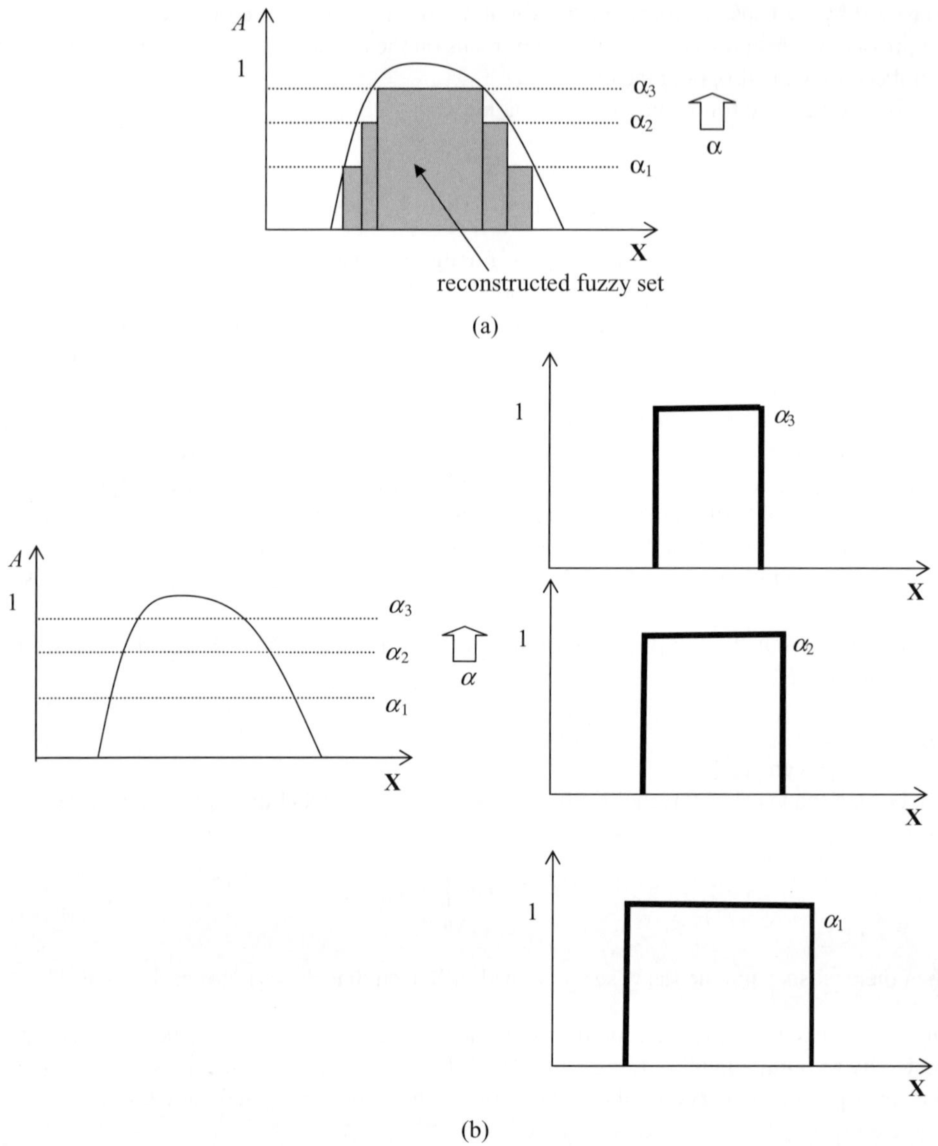

Figure 2.15 Fuzzy set A, examples of some of its α-cuts (a) and a representation of A through the corresponding family of sets (α-cuts) (b).

of the concept of the α-cut and the way in which the representation of the corresponding fuzzy set is realized is shown in Figure 2.15.

More descriptively, we can say that fuzzy sets can be reconstructed by a family of sets. Apparently, we need a family of sets (intervals, in particular) to capture the essence of a single fuzzy set. The reconstruction scheme illustrated in Figure 2.15 is self-explanatory in this regard. In more descriptive terms, we can look at the expression offered by (2.29) as a way of decomposing A into a series of layers (indexed sets) calibrated by the values of the associated levels of α.

For the finite universe of discourse, dim(**X**) $= n$, we encounter a finite number of membership grades and subsequently a finite number of α-cuts. This finite family of α-cuts is then sufficient to fully "represent" or reconstruct the original fuzzy set.

Example 2.3. To illustrate the essence of α-cuts and the ensuing reconstruction, let us consider a fuzzy set with a finite number of membership grades, $A = [0.8\ 1.0\ 0.2\ 0.5\ 0.1\ 0.0\ 0.0\ 0.7]$. The corresponding α-cuts of A are equal to

$$\alpha = 1.0 A_{1.0} = [0\ 1\ 0\ 0\ 0\ 0\ 0\ 0]$$

$$\alpha = 0.8 A_{0.8} = [1\ 1\ 0\ 0\ 0\ 0\ 0\ 0]$$

$$\alpha = 0.7 A_{0.7} = [1\ 1\ 0\ 0\ 0\ 0\ 0\ 1]$$

$$\alpha = 0.5 A_{0.5} = [1\ 1\ 0\ 1\ 0\ 0\ 0\ 1]$$

$$\alpha = 0.2 A_{0.2} = [1\ 1\ 1\ 1\ 0\ 0\ 0\ 1]$$

$$\alpha = 0.1 A_{0.1} = [1\ 1\ 1\ 1\ 1\ 0\ 0\ 1]$$

We can clearly see the layered character of the consecutive α-cuts indexed by the sequence of the increasing values of α. Because of the finite number of membership grades, the reconstruction realized in terms of (2.28) returns the original fuzzy set (which is possible given the finite space over which the original fuzzy set has been defined) $A(x) = \max(1.0A_{1.0}(x), 0.8A_{0.8}(x), 0.7A_{0.7}(x), 0.5A_{0.5}(x), 0.2A_{0.2}(x), 0.1A_{0.1}(x))$.

Considering a finite number of α-cuts, an important question arises as to the choice of values of such thresholds (values of α) so that a fuzzy set could be approximated to the highest extent. This is a problem formulated and solved with the aid of population-based optimization in Pedrycz, Dong, and Hirota (2009). Refer also to Bodjanova (2006).

2.9 Operations on Fuzzy Sets

Similarly as in set theory, we operate on fuzzy sets to obtain new fuzzy sets. The operations must possess properties to match intuition, to comply with the semantics of the intended operation, and to be flexible to fit application requirements. This section covers set operations beginning with early fuzzy set operations and continuing with their generalization, interpretations, formal requirements, and realizations. We emphasize complements, triangular norms, and triangular conorms as unifying, general constructs of the complement, intersection, and union operations. Combinations of fuzzy sets to provide aggregations are also essential when operating with fuzzy sets. Analyses of the fundamental properties and characteristics of operations with fuzzy sets are discussed thoroughly.

It is instructive to start with the familiar operations of intersection, union, and complement encountered in set theory. They also exhibit some similarities with the operations commonly used in set theory. For instance, consider two sets $A = \{x \in \mathbf{R} \mid 1 \leq x \leq 3\}$ and $B = \{x \in \mathbf{R} \mid$

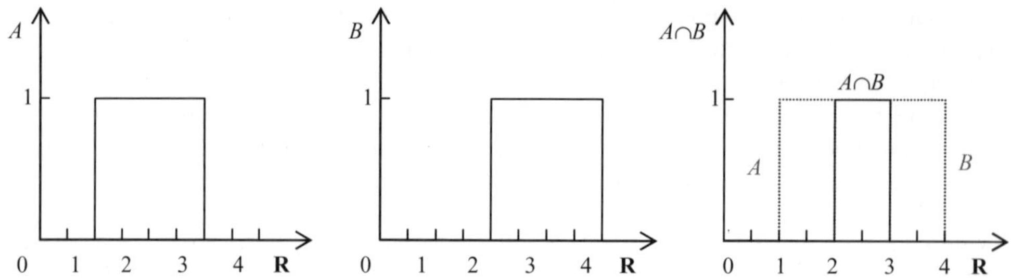

Figure 2.16 Intersection of sets represented in terms of their characteristic functions.

$2 \leq x \leq 4\}$, both being closed intervals on the real line. Their intersection is a set $A \cap B = \{x \in \mathbf{R} \mid 2 \leq x \leq 3\}$. Figure 2.16 illustrates the intersection operation represented in terms of the characteristic functions of A and B. Looking at the values of the characteristic function of $A \cap B$ that results when comparing the individual values of $A(x)$ and $B(x)$ at each $x \in \mathbf{R}$, we note that these are taken as the minimum between the values of $A(x)$ and $B(x)$.

In general, given the characteristic functions of A and B, the characteristic function of their intersection $A \cap B$ is computed in the form

$$(A \cap B)(x) = \min(A(x), B(x)) \quad \forall x \in \mathbf{X} \tag{2.29}$$

where $(A \cap B)(x)$ denotes the characteristic function of the intersection $A \cap B$.

We now consider the union of sets A and B and express its characteristic function in terms of the respective characteristic functions of A and B. For example, if A and B are the same intervals as presented before, then $A \cup B = \{x \in \mathbf{R} \mid 1 \leq x \leq 4\}$. We note that the value of the characteristic function of the union is taken as the maximum of corresponding values of the characteristic functions $A(x)$ and $B(x)$ taken at each point of the universe of discourse, Figure 2.17.

Therefore, given the characteristic functions of A and B, we determine the characteristic function of the union to be computed as

$$(A \cup B)(x) = \max(A(x), B(x)) \quad \forall x \in \mathbf{X} \tag{2.30}$$

where $(A \cup B)(x)$ denotes the characteristic function of the intersection $A \cup B$.

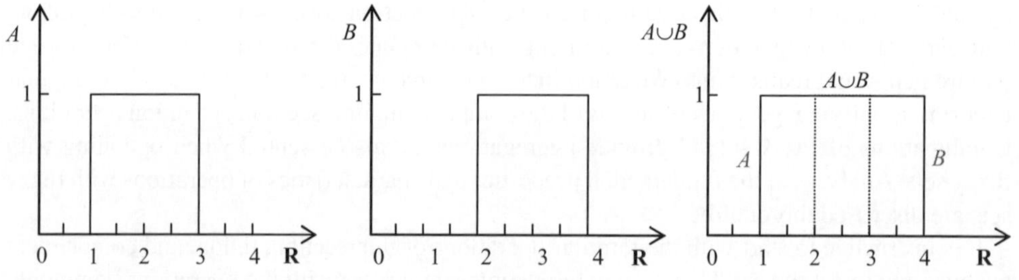

Figure 2.17 Union of two sets expressed in terms of their characteristic functions.

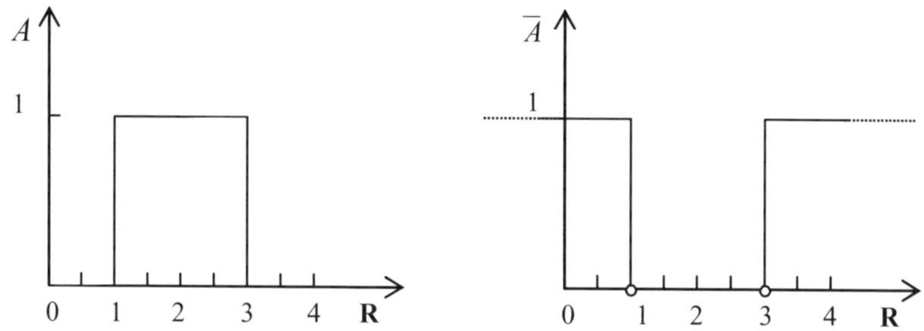

Figure 2.18 Complement of a set in terms of its characteristic function.

Likewise, as Figure 2.18 suggests, the complement $\overline{A}$ of set A, expressed in terms of its characteristic function, is the complement of the characteristic function of A. For instance, if $A = \{x \in \mathbf{R} \mid 1 \leq x \leq 3\}$, which is the same interval as discussed before, then $\overline{A} = \{x \in \mathbf{R}|x < 1\} \cup \{x \in \mathbf{R}|x > 3\}$, see Figure 2.18.

In general, the characteristic function of the complement of a set A is given in the form

$$\overline{A}(x) = 1 - A(x) \quad \forall x \in \mathbf{X} \tag{2.31}$$

One may anticipate that, since sets are particular instances of fuzzy sets, the operations of intersection, union, and complement as previously defined should equally well apply to fuzzy sets. Indeed, when we used membership functions in expressions (2.29)–(2.31), these formulas served as definitions of intersection, union, and complement of fuzzy sets. An illustration of these operations is included in Figure 2.19.

2.9.1 *Triangular Norms and Triangular Conorms as Models of Operations on Fuzzy Sets*

Operations on fuzzy sets concern manipulation of their membership functions. Therefore they are domain dependent and different contexts may require their different realizations. For instance, since operations provide ways to combine information, they can be performed differently in image processing, control, and diagnostic systems, for example. When contemplating

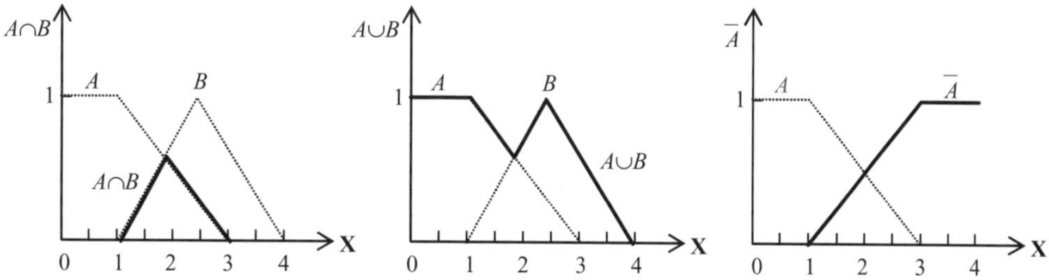

Figure 2.19 Operations on fuzzy sets realized with the use of min, max, and complement functions.

the realization of the operations of intersection and union of fuzzy sets, we should require satisfaction of the following intuitively appealing properties:

(a) commutativity;
(b) associativity;
(c) monotonicity;
(d) identity.

The last requirement of identity takes on a different form depending on the operation. More specifically, in the case of intersection, we anticipate that an intersection of any fuzzy set with the universe of discourse **X** should return this fuzzy set. For the union operations, the identity implies that the union of any fuzzy set and an empty fuzzy set returns the fuzzy set.

Thus any binary operator $[0,1] \times [0,1] \rightarrow [0,1]$, which satisfies the collection of the requirements outlined above, can be regarded as a potential candidate for realizing the intersection or union of fuzzy sets. Note also that identity acts as a boundary condition, meaning that when confined to sets, the above stated operations return the same results as encountered in set theory. In general, idempotency is not required; however, the realizations of union and intersection could be idempotent as this happens for the operations of minimum and maximum where $\min(a, a) = a$ and $\max(a, a) = a$.

In the theory of fuzzy sets, triangular norms offer a general class of operators of intersection and union. For instance, t-norms generalize the intersection of fuzzy sets. Given a t-norm, a dual operator called a t-conorm (or s-norm) can be derived using the relationship $x\ S\ y = 1 - (1 - x)\ T\ (1 - y)$, $\forall\ x, y \in [0,1]$, the De Morgan law, but the t-conorm can also be described by an independent axiomatic system (Valverde and Ovchinnikov, 2008). Triangular conorms provide generic models for the union of fuzzy sets.

A triangular norm, t-norm in brief, is a binary operation T: $[0,1] \times [0,1] \rightarrow [0,1]$ that satisfies the properties

(a) Commutativity: $a\ T\ b = b\ T\ a$
(b) Associativity: $a\ T\ (b\ T\ c) = (a\ T\ b)\ T\ c$
(c) Monotonicity: if $b \leq c$ then $a\ T\ b \leq a\ T\ c$
(d) Boundary conditions: $a\ T\ 1 = a$ and $a\ T\ 0 = 0$

where $a, b, c \in [0,1]$.

Let us elaborate on the meaning of these requirements vis-à-vis the use of t-norms as models of operators of the union and intersection of fuzzy sets. There is a one-to-one correspondence between the general requirements outlined in the previous section and the properties of t-norms. The first three reflect the general character of set operations. Boundary conditions stress the fact that all t-norms attain the same values at boundaries of the unit square $[0,1] \times [0,1]$. Thus, for sets, any t-norm produces the same result that coincides with the one we could have expected in set theory when dealing with the intersection of sets, that is, $A \cap \mathbf{X} = A$ and $A \cap \varnothing = \varnothing$. Some commonly encountered examples of t-norms include the following operations:

(a) Minimum: $a\ T_m\ b = \min(a, b) = a \wedge b$
(b) Product: $a\ T_p\ b = ab$

(c) Lukasiewicz: $a\,T_l\,b = \max(a + b - 1, 0)$
(d) Drastic product:

$$a\,T_d\,b = \begin{cases} a & \text{if} \quad b = 1 \\ b & \text{if} \quad a = 1 \\ 0 & \text{otherwise} \end{cases}$$

In general, t-norms cannot be linearly ordered. One can demonstrate that the min (T_m) t-norm is the largest t-norm, while the drastic product is the smallest one. They form the lower and upper bounds of the t-norms in the following sense:

$$a\,T_d\,b \le a\,T\,b \le a\,T_m\,b = \min(a, b) \tag{2.32}$$

Triangular conorms are functions S: [0,1] × [0,1] → [0,1] that serve as generic realizations of the union operator on fuzzy sets. Similar to triangular norms, conorms provide the highly desirable modeling flexibility needed to construct fuzzy models. Triangular conorms can be viewed as dual operators to the t-norms and, as such, are explicitly defined with the use of the De Morgan laws. We may characterize them in a fully independent fashion by offering the following definition.

A triangular conorm (s-norm) is a binary operation S: [0,1] × [0,1] → [0,1] that satisfies the following requirements:

(a) Commutativity: $a\,S\,b = b\,S\,a$
(b) Associativity: $a\,S\,(b\,S\,c) = (a\,S\,b)\,S\,c$
(c) Monotonicity: if $b \le c$ then $a\,S\,b \le a\,S\,c$
Boundary conditions: $a\,S\,0 = a$ and $a\,S\,1 = 1$

where $a, b, c \in [0,1]$.

One can show that S: [0,1] × [0,1] → [0,1] is a t-conorm if and only if (iff) there exists a t-norm (dual t-norm) such that, $\forall a, b \in [0,1]$, we have

$$a\,S\,b = 1 - (1 - a)\,T\,(1 - b) \tag{2.33}$$

For the corresponding dual t-norm we have

$$a\,T\,b = 1 - (1 - a)\,S\,(1 - b) \tag{2.34}$$

The duality expressed by (2.33) and (2.34) can be viewed as an alternative definition of t-conorms. This duality allows us to deduce the properties of t-conorms on the basis of the analogous properties of t-norms. Note that after rewriting (2.33) and (2.34), we obtain

$$(1 - a)\,T\,(1 - b) = 1 - a\,S\,b \tag{2.35}$$

$$(1 - a)\,S\,(1 - b) = 1 - a\,T\,b \tag{2.36}$$

These two relationships can be expressed symbolically as

$$\overline{A} \cap \overline{B} = \overline{A \cup B} \tag{2.37}$$

$$\overline{A} \cup \overline{B} = \overline{A \cap B} \tag{2.38}$$

that are nothing more than the De Morgan laws.

The boundary conditions mean that all t-conorms behave similarly at the boundary of the unit square [0,1] × [0,1]. Thus, for sets, any t-conorm returns the same result as encountered in set theory.

A list of commonly used t-conorms includes the following examples:

(a) Maximum:

$$a\, S_m\, b = \max(a, b) = a \vee b \tag{2.39}$$

(b) Probabilistic sum:

$$a\, S_p\, b = a + b - ab \tag{2.40}$$

(c) Lukasiewicz:

$$a\, S_l\, b = \min(a + b, 1) \tag{2.41}$$

(d) Drastic sum:

$$a\, S_d\, b = \begin{cases} a & \text{if} \quad b = 0 \\ b & \text{if} \quad a = 0 \\ 1 & \text{otherwise} \end{cases} \tag{2.42}$$

2.9.2 Negations

Negation is a single-argument operation which is a generalization of the complement operation encountered in set theory. More formally, by a negation we mean a function N: [0,1] → [0,1] satisfying the following conditions:

Monotonicity:	N is nonincreasing
Boundary conditions:	$N(0) = 1$ and $N(1) = 0$

If the function N is continuous and decreasing, the negation is called *strict* (Fodor, 1993). If, in addition, a strict negation is involutive, that is,

$$N(N(x)) = x \quad \forall x \in [0, 1] \tag{2.43}$$

it is called *strong*.

Two realizations of the negation operator are

$$N(x) = \sqrt[w]{1 - x^w} \qquad w \in (0, \infty) \tag{2.44}$$

$$N(x) = \frac{1 - x}{1 + \lambda x} \qquad \lambda \in (-1, \infty) \tag{2.45}$$

Interestingly, if in the above expressions we set $w = 1$ or $\lambda = 0$, these realizations of the negation return the standard complement function, that is, $N(x) = 1 - x$.

It is worth noting that negation is a logic operation but not a model of antonyms encountered in natural language (Kim, Kim, and Park, 2000).

2.10 Fuzzy Relations

Relations represent and quantify associations between objects. They provide a fundamental vehicle to describe interactions and dependencies between variables, components, modules, and so on. Fuzzy relations generalize the concept of relations in the same way as fuzzy sets generalize the fundamental idea of sets (Kandel and Yelowitz, 1974; Naessens, Meyer, and De Baets, 2002; Tsabadze, 2008). Fuzzy relations are highly instrumental in problems of information retrieval, pattern classification, control, and decision-making. In particular, in decision-making, the notion of fuzzy relations is significant visibility. In what follows, we introduce the idea of fuzzy relations, present some illustrative examples, discuss the main properties of fuzzy relations, and provide some interpretation. The discussed properties exhibit interesting linkages with the essentials of decision-making where they come with some useful characterizations of the underlying decision processes and decision-makers involved there.

2.10.1 The Concept of Relations

Before proceeding with fuzzy relations, we present a few introductory lines on relations. Relations capture associations between objects. For instance, consider the space of documents $\mathbf{X}$ and a space of keywords $\mathbf{Y}$ that these documents contain. Now form a Cartesian product of $\mathbf{X}$ and $\mathbf{Y}$, that is, $\mathbf{X} \times \mathbf{Y}$. Recall that the Cartesian product of $\mathbf{X}$ and $\mathbf{Y}$, denoted $\mathbf{X} \times \mathbf{Y}$, is the set of all pairs (x, y) such that $x \in \mathbf{X}$ and $y \in \mathbf{Y}$. We define a relation R as the set of pairs of documents and keywords, $R = \{(d_i, w_j) \mid d_i \in \mathbf{X} \text{ and } w_j \in \mathbf{Y}\}$. In terms of the characteristic function we express this as follows: $\boldsymbol{R}(d_i, w_j) = 1$ if keyword w_j is in document d_i, and $\boldsymbol{R}(d_i, w_j) = 0$ otherwise. In decision-making, situations and actions are related: to each situation (state of nature) we assign a collection of pertinent actions which are of interest.

More generally, a relation $\boldsymbol{R}$ defined over the Cartesian product of $\mathbf{X}$ and $\mathbf{Y}$ is a collection of selected pairs (x, y) where $x \in \mathbf{X}$ and $y \in \mathbf{Y}$. Equivalently, it is a mapping

$$\boldsymbol{R} : \mathbf{X} \times \mathbf{Y} \rightarrow \{0, 1\} \tag{2.46}$$

The characteristic function of $\boldsymbol{R}$ is such that if $\boldsymbol{R}(x, y) = 1$, then we say that the two elements x and y are related. If $\boldsymbol{R}(x, y) = 0$, we say that these two elements (x and y) are unrelated. For example, suppose that $\mathbf{X} = \mathbf{Y} = \{2, 4, 6, 8\}$. The relation "equal to" formed over $\mathbf{X} \times \mathbf{X}$ is

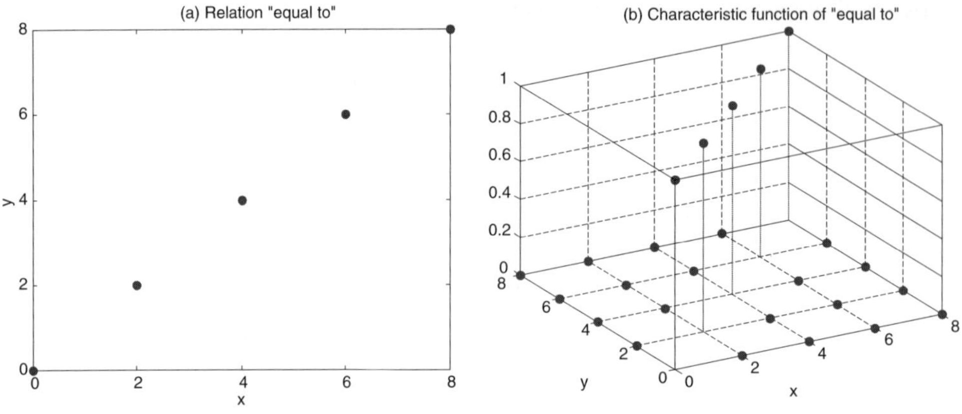

Figure 2.20 Relation "equal to" and its characteristic function.

the set of pairs $\boldsymbol{R} = \{(x, y) \in \mathbf{X} \times \mathbf{X} \mid x = y\} = \{(2,2), (4,4), (6,6), (8,8)\}$, Figure 2.20(a). Its characteristic function is equal to

$$\boldsymbol{R}(x, y) = \begin{cases} 1 & \text{if} \quad x = y \\ 0 & \text{otherwise} \end{cases} \tag{2.47}$$

A plot of this characteristic function is included in Figure 2.20(b).

Depending on the nature of the universe, which could be either finite or infinite, relations are represented in a tabular, matrix form, or described analytically. For instance, the set $\mathbf{X} = \{2, 4, 6, 8\}$ is finite and the relation "equal to" in $\mathbf{X} \times \mathbf{X}$ has a representation as a 4×4 matrix

$$\boldsymbol{R} = \begin{bmatrix} 1 & 0 & 0 & 0 \\ 0 & 1 & 0 & 0 \\ 0 & 0 & 1 & 0 \\ 0 & 0 & 0 & 1 \end{bmatrix}$$

In general, if $\mathbf{X}$ and $\mathbf{Y}$ are finite, say Card($\mathbf{X}$) $= n$ and Card($\mathbf{Y}$) $= m$, then $\boldsymbol{R}$ is a $n \times m$ matrix $\boldsymbol{R} = [r_{ij}]$ with the entries r_{ij} being equal to one if and only if $(x_i, y_j) \in \boldsymbol{R}$. Elementary geometry provides examples of relations on infinite universes such as $\mathbf{R} \times \mathbf{R} = \mathbf{R}^2$. In these cases characteristic functions can, in general, be expressed analytically:

$$\boldsymbol{R}(x, y) = \begin{cases} 1 & \text{if} \quad |x| \leq 17 \quad \text{and} \quad |y| \leq 1 \\ 0 & \text{otherwise} \end{cases} \qquad \text{square}$$

$$\boldsymbol{R}(x, y) = \begin{cases} 1 & \text{if} \quad x^2 + y^2 = r^2 \\ 0 & \text{otherwise} \end{cases} \qquad \text{circle}$$

Relations subsume functions but not vice versa: all functions are relations, but not all relations are functions. For instance, the relation "equal to" shown above is a function but the relations "square" and "circle" are not. A relation is a function if and only if for every x in $\mathbf{X}$ there is only a single element $y \in \mathbf{Y}$ such that $\boldsymbol{R}(x, y) = 1$. Therefore, functions are directional constructs, clearly implying a certain direction, for example, from $\mathbf{X}$ to $\mathbf{Y}$, say $f : \mathbf{X} \rightarrow \mathbf{Y}$.

If the mapping "f" is a function, there is no guarantee that the mapping f^{-1}: $\mathbf{Y} \rightarrow \mathbf{X}$ is also a function, except in some case when f^{-1} exists. In contrast, relations are direction free as there is no specific direction identified. Being more descriptive, they can be accessed from any direction. This makes a significant conceptual and computational difference.

When a space under discussion involves "n" universes as its coordinate, an n-ary relation is any subset of the Cartesian product of these universes,

$$\boldsymbol{R} : \mathbf{X}_1 \times \mathbf{X}_2 \times \cdots \times \mathbf{X}_n \rightarrow \{0, 1\} \tag{2.48}$$

If $\mathbf{X}_1, \mathbf{X}_2, \ldots, \mathbf{X}_n$ are finite and $\text{Card}(\mathbf{X}_1) = n_1 \ldots \text{Card}(\mathbf{X}_n) = n_p$, then $\boldsymbol{R}$ can be written as an $(n_1 \times \cdots \times n_p)$ matrix $\boldsymbol{R} = [r_{ij\ldots k}]$ with $r_{ij\ldots k} = 1$ if and only if $(x_i, x_j, \ldots, x_k) \in \boldsymbol{R}$.

2.10.2 *Fuzzy Relations*

Fuzzy relations generalize the concept of relations by admitting the notion of partial association between elements of universes. Given two universes $\mathbf{X}$ and $\mathbf{Y}$, a fuzzy relation $\boldsymbol{R}$ is any fuzzy subset of the Cartesian product of $\mathbf{X}$ and $\mathbf{Y}$ (Zadeh, 1971). Equivalently, a fuzzy relation on $\mathbf{X} \times \mathbf{Y}$ is a mapping

$$\boldsymbol{R} : \mathbf{X} \times \mathbf{Y} \rightarrow [0, 1] \tag{2.49}$$

The membership function of $\boldsymbol{R}$ for some pair (x, y), $\boldsymbol{R}(x, y) = 1$, denotes that the two elements x and y are fully related. On the other hand, $\boldsymbol{R}(x, y) = 0$ means that these elements are unrelated, while the values in between, $0 < \boldsymbol{R}(x, y) < 1$, underline a partial association. For instance, if d_{fs}, d_{nf}, d_{ns}, and d_{gf} are documents whose subjects concern mainly fuzzy systems, neural fuzzy systems, neural systems, and genetic fuzzy systems, with keywords w_f, w_n, and w_g, respectively, then a relation $\boldsymbol{R}$ on $\mathbf{D} \times \mathbf{W}$, $\mathbf{D} = \{d_{fs}, d_{nf}, d_{ns}, d_{gf}\}$ and $\mathbf{W} = \{w_f, w_n, w_g\}$, can assume a matrix form with the following entries:

$$\boldsymbol{R} = \begin{bmatrix} 1 & 0 & 0.6 \\ 0.8 & 1 & 0 \\ 0 & 1 & 0 \\ 0.8 & 0 & 1 \end{bmatrix}$$

Since the universes are discrete, $\boldsymbol{R}$ can be represented as a 4×3 matrix (four documents and three keywords), and entries, for example, $\boldsymbol{R}(d_{fs}, w_f) = 1$, mean that the document content d_{fs} is fully compatible with the keyword w_f whereas $\boldsymbol{R}(d_{fs}, w_n) = 0$ and $\boldsymbol{R}(d_{fs}, w_g) = 0.6$ indicate that d_{fs} does not mention neural systems, but does have genetic systems as part of its content, Figure 2.21(a). As with relations, when $\mathbf{X}$ and $\mathbf{Y}$ are finite with $\text{Card}(\mathbf{X}) = n$ and $\text{Card}(\mathbf{Y}) = m$, then $\boldsymbol{R}$ can be can be arranged into a certain $n \times m$ matrix $\boldsymbol{R} = [r_{ij}]$, with $r_{ij} \in [0,1]$ being the corresponding degrees of association between x_i and y_j.

Fuzzy relations defined on some continuous spaces such as $\mathbf{R}^2$, say "much smaller than", "approximately equal", and "similar", could, for instance, be characterized by the following

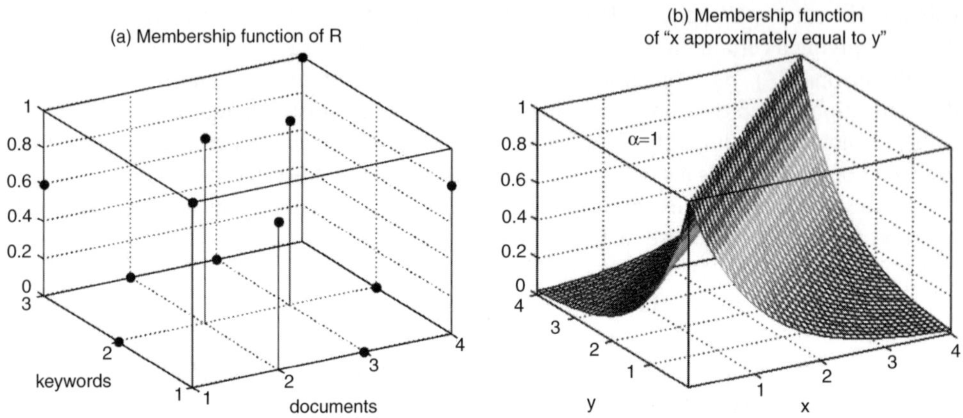

Figure 2.21 Membership functions of the relation $\boldsymbol{R}$ (a) and "x approximately equal to y" (b).

membership functions:

$$\boldsymbol{R}_m(x, y) = \begin{cases} 1 - \exp(-|y - x|) & \text{if } \; x \leq y \\ 0 & \text{otherwise} \end{cases} \qquad x \text{ much smaller than } y$$

$$\boldsymbol{R}_e(x, y) = \exp\left(\frac{-|x - y|}{\alpha}\right) \qquad \alpha > 0 \qquad x \text{ and } y \text{ approximately equal}$$

$$\boldsymbol{R}_s(x, y) = \begin{cases} \exp(-(x - y)/\beta) & \text{if } \; |x - y| \leq 5 \\ 0 & \text{if } \; |x - y| \geq 5 \end{cases} \qquad \beta > 0 \qquad x \text{ and } y \text{ similar}$$

Figure 2.21(b) displays the membership function of the relation "x approximately equal to y" on $\mathbf{X} = \mathbf{Y} = [0,4]$ assuming that $\alpha = 1$.

2.10.3 Properties of the Fuzzy Relations

Fuzzy relations come with a number of properties which capture the nature of the relationships conveyed by relations.

2.10.4 Domain and Codomain of Fuzzy Relations

The domain, dom $\boldsymbol{R}$, of a fuzzy relation $\boldsymbol{R}$ defined in $\mathbf{X} \times \mathbf{Y}$ is a fuzzy set whose membership function is equal to

$$\text{dom}\, R(x) = \sup_{y \in \mathbf{Y}} \boldsymbol{R}(x, y) \tag{2.50}$$

while its codomain, cod $\boldsymbol{R}$, is a fuzzy set whose membership function is given as

$$\text{cod}\, R(y) = \sup_{x \in \mathbf{X}} \boldsymbol{R}(x, y) \tag{2.51}$$

Considering finite universes of discourse, domain and codomain can be viewed as the height of the rows and columns of the fuzzy relation matrix (Zadeh, 1971).

2.10.5 Representation of Fuzzy Relations

Similar to the case of fuzzy sets, fuzzy relations can be represented by their α-cuts, that is,

$$\boldsymbol{R} = \bigcup_{\alpha \in [0,1]} \alpha \boldsymbol{R}_\alpha \tag{2.52}$$

or, in terms of the membership function $\boldsymbol{R}(x, y)$ of $\boldsymbol{R}$,

$$\boldsymbol{R}(x, y) = \sup_{\alpha \in [0,1]} (\min \alpha, \boldsymbol{R}(x, y)) \tag{2.53}$$

2.10.6 Equality of Fuzzy Relations

We say that two fuzzy relations $\boldsymbol{P}$ and $\boldsymbol{Q}$ defined in the same Cartesian product of spaces $\mathbf{X} \times \mathbf{Y}$ are equal if and only if their membership functions are identical, that is,

$$\boldsymbol{P}(x, y) = \boldsymbol{Q}(x, y) \quad \forall (x, y) \in \mathbf{X} \times \mathbf{Y} \tag{2.54}$$

2.10.7 Inclusion of Fuzzy Relations

A fuzzy relation $\boldsymbol{P}$ is included in $\boldsymbol{Q}$, denoted by $\boldsymbol{P} \subseteq \boldsymbol{Q}$, if and only if

$$\boldsymbol{P}(x, y) \leq \boldsymbol{Q}(x, y) \quad \forall (x, y) \in \mathbf{X} \times \mathbf{Y} \tag{2.55}$$

Similar to the presented case of relations, given an n-fold Cartesian product of these universes we define the fuzzy relation in the form

$$\boldsymbol{R} : \mathbf{X}_1 \times \mathbf{X}_2 \times \cdots \times \mathbf{X}_n \to [0, 1] \tag{2.56}$$

If the spaces $\mathbf{X}_1, \mathbf{X}_2, \ldots, \mathbf{X}_n$ are finite with Card($\mathbf{X}_1$) $= n_1 \ldots$ Card($\mathbf{X}_n$) $= n_n$, then $\boldsymbol{R}$ can be expressed as an n-fold $(n_1 \times \cdots \times n_p)$ matrix $\boldsymbol{R} = [r_{ij\ldots k}]$ with $r_{ij\ldots k} \in [0,1]$ being the degree of association assigned to the n-tuple $(x_i, x_j, \ldots, x_k) \in \mathbf{X}_1 \times \mathbf{X}_2 \times \cdots \times \mathbf{X}_n$. If $\mathbf{X}_1, \mathbf{X}_2, \ldots, \mathbf{X}_n$ are infinite, then the membership function of $\boldsymbol{R}$ is a certain function of many variables. The concepts of equality and inclusion of fuzzy relations could be easily extended for relations defined in multidimensional spaces.

2.10.8 Operations on Fuzzy Relations

The basic operations on fuzzy relations, say union, intersection, and complement, conceptually follow the corresponding operations on fuzzy sets once fuzzy relations are fuzzy sets formed on multidimensional spaces. For illustrative purposes the definitions of union, intersection, and complement below involve two-argument fuzzy relations. Without any loss of generality, we can focus on binary fuzzy relations $\boldsymbol{P}, \boldsymbol{Q}, \boldsymbol{R}$ defined in $\mathbf{X} \times \mathbf{Y}$. As in the case of fuzzy sets, all definitions are defined pointwise.

2.10.9 Union of Fuzzy Relations

The union $\boldsymbol{R}$ of two fuzzy relations $\boldsymbol{P}$ and $\boldsymbol{Q}$ defined in $\mathbf{X} \times \mathbf{Y}$, $\boldsymbol{R} = \boldsymbol{P} \cup \boldsymbol{Q}$, is defined with the use of the following membership function:

$$\boldsymbol{R}(x, y) = \boldsymbol{P}(x, y)\ S\ \boldsymbol{Q}(x, y) \quad \forall (x, y) \in \mathbf{X} \times \mathbf{Y} \tag{2.57}$$

(Recall that "S" stands for some t-conorm.)

2.10.10 Intersection of Fuzzy Relations

The intersection $\boldsymbol{R}$ of fuzzy relations $\boldsymbol{P}$ and $\boldsymbol{Q}$ defined in $\mathbf{X} \times \mathbf{Y}$, $\boldsymbol{R} = \boldsymbol{P} \cap \boldsymbol{Q}$, is defined in the following form:

$$\boldsymbol{R}(x, y) = \boldsymbol{P}(x, y)\ T\ \boldsymbol{Q}(x, y) \quad \forall (x, y) \in \mathbf{X} \times \mathbf{Y} \tag{2.58}$$

2.10.11 Complement of Fuzzy Relations

The complement $\overline{\boldsymbol{R}}$ of the fuzzy relation $\boldsymbol{R}$ is defined by the membership function

$$\overline{\boldsymbol{R}}(x, y) = 1 - \boldsymbol{R}(x, y) \quad \forall (x, y) \in \mathbf{X} \times \mathbf{Y} \tag{2.59}$$

2.10.12 Transposition of Fuzzy Relations

Given a fuzzy relation $\boldsymbol{R}$, its transpose or inverse relation, denoted by $\boldsymbol{R}^{-1}$, is a fuzzy relation on $\mathbf{Y} \times \mathbf{X}$ such that the following relationship holds:

$$\boldsymbol{R}^{\mathrm{T}}(y, x) = \boldsymbol{R}(x, y) \quad \forall (x, y) \in \mathbf{X} \times \mathbf{Y} \tag{2.60}$$

If $\boldsymbol{R}$ is a relation defined in some finite space, then $\boldsymbol{R}^{-1}$ is the transpose of the corresponding $n \times m$ matrix representation of $\boldsymbol{R}$. Therefore the form of $\boldsymbol{R}^{-1}$ is an $m \times n$ matrix whose columns are now the rows of $\boldsymbol{R}$.

The following properties are direct consequences of the definitions provided above:

$$(\boldsymbol{R}^{-1})^{-1} = \boldsymbol{R} \tag{2.61}$$

$$(\overline{\boldsymbol{R}})^{-1} = \overline{(\boldsymbol{R}^{-1})} \tag{2.62}$$

2.10.13 Cartesian Product of Fuzzy Relations

Given fuzzy sets $A_1, A_2, \ldots, A_n$ defined in universes $\mathbf{X}_1, \mathbf{X}_2, \ldots, \mathbf{X}_n$, respectively, their Cartesian product $A_1 \times A_2 \times \cdots \times A_n$ is a fuzzy relation $\boldsymbol{R}$ on $\mathbf{X}_1 \times \mathbf{X}_2 \times \cdots \times \mathbf{X}_n$ with the following membership function:

$$\boldsymbol{R}(x_1, x_2, \ldots, x_n) = \min(A_1(x_1), A_2(x_2), \ldots, A_n(x_n)) \quad \forall x_1 \in \mathbf{X}_1, \forall x_2 \in \mathbf{X}_2, \ldots, \forall x_n \in \mathbf{X}_n \tag{2.63}$$

In general, we can generalize the concept of this Cartesian product by using some t-norms

$$\boldsymbol{R}(x_1, x_2, \ldots, x_n) = A_1(x_1)\ T\ A_2(x_2)\ T \ldots T\ A_n(x_n) \quad \forall x_1 \in \mathbf{X}_1, \forall x_2 \in \mathbf{X}_2, \ldots, \forall x_n \in \mathbf{X}_n \tag{2.64}$$

2.10.14 Projection of Fuzzy Relations

In contrast to the concept of the Cartesian product, the idea of projection is to construct fuzzy relations on some subspaces of the original relation. Projection reduces the dimensionality of the original space over which the original fuzzy relation as defined.

Given $\boldsymbol{R}$ as a fuzzy relation defined in $\mathbf{X}_1 \times \mathbf{X}_2 \times \cdots \times \mathbf{X}_n$, its projection on $\mathbf{X} = \mathbf{X}_i \times \mathbf{X}_j \times \cdots \times \mathbf{X}_k$, where $\mathbf{I} = \{i, j, \ldots, k\}$ is a subsequence of the set of indexes $\mathbf{Q} = \{1, 2, \ldots, n\}$, is a fuzzy relation $\boldsymbol{R}_\mathbf{X}$ with the membership function (Zadeh, 1971)

$$\boldsymbol{R}_X(x_i, x_j, \ldots, x_k) = \text{Proj}_X \boldsymbol{R}(x_1, x_2, \ldots, x_n) = \sup_{x_t, x_u, \ldots, x_v} \boldsymbol{R}(x_1, x_2, \ldots, x_n) \tag{2.65}$$

where $\mathbf{J} = \{t, u, \ldots, v\}$ is a subsequence of $\mathbf{Q}$ such that $\mathbf{I} \cup \mathbf{J} = \mathbf{Q}$ and $\mathbf{I} \cap \mathbf{J} = \varnothing$. In other words, J is the complement of I with respect to N. Notice that the above expression is computed for all values of $(x_1, x_2, \ldots, x_n) \in \mathbf{X}_i \times \mathbf{X}_j \times \cdots \times \mathbf{X}_k$.

For instance, Figure 2.22 shows the projections $\boldsymbol{R}_\mathbf{X}$ and $\boldsymbol{R}_\mathbf{Y}$ of a certain Gaussian binary fuzzy relation $\boldsymbol{R}$ defined in $\mathbf{X} \times \mathbf{Y}$ with $\mathbf{X} = [0,8]$ and $\mathbf{Y} = [0,10]$, whose membership function is equal to $\boldsymbol{R}(x, y) = \exp\{-\alpha[(x-4)^2 + (y-5)^2]\}$. In this case the projections are formed as

$$\boldsymbol{R}_\mathbf{X}(x) = \text{Proj}_\mathbf{X} \boldsymbol{R}(x, y) = \sup_y \boldsymbol{R}(x, y) \tag{2.66}$$

$$\boldsymbol{R}_\mathbf{Y}(y) = \text{Proj}_\mathbf{Y} \boldsymbol{R}(x, y) = \sup_x \boldsymbol{R}(x, y) \tag{2.67}$$

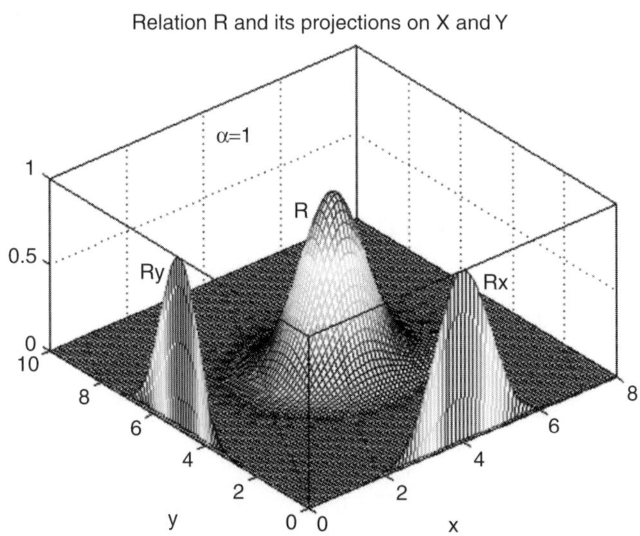

Figure 2.22 Fuzzy relation $\boldsymbol{R}$ along with its projections on $\mathbf{X}$ and $\mathbf{Y}$.

To find projections of the fuzzy relations defined in some finite spaces, the maximum operation replaces the sup operation occurring in the definition provided above. For example, for the fuzzy relation $\boldsymbol{R}$: $\mathbf{X} \times \mathbf{Y} \to [0,1]$ with $\mathbf{X} = \{1, 2, 3\}$ and $\mathbf{Y} = \{1,2, 3, 4, 5\}$,

$$\boldsymbol{R}(x, y) = \begin{bmatrix} 1.0 & 0.6 & 0.8 & 0.5 & 0.2 \\ 0.6 & 0.8 & 1.0 & 0.2 & 0.9 \\ 0.8 & 0.6 & 0.8 & 0.3 & 0.9 \end{bmatrix}$$

The three elements of the projection $\boldsymbol{R}_{\mathbf{X}}$ are taken as the maximum computed for each of the three rows of $\boldsymbol{R}$

$$\begin{aligned} \boldsymbol{R}_{\mathbf{X}} &= [\max(1,0, 0.6, 0.8, 0.5, 0.2) \quad \max(0.6, 0.8, 1.0, 0.2, 0.9) \quad \max(0.8, 0.6, 0.8, 0.3, 0.9)] \\ &= [1.0 \quad 1.0 \quad 0.9] \end{aligned}$$

Similarly, the five elements of $\boldsymbol{R}_{\mathbf{Y}}$ are taken as the maximum among the entries of the five columns of $\boldsymbol{R}$. Figure 2.23 shows $\boldsymbol{R}$ and its projections $\boldsymbol{R}_{\mathrm{x}}$ and $\boldsymbol{R}_{\mathbf{Y}}$:

$$\boldsymbol{R}_{\mathbf{Y}} = [1.0 \quad 0.8 \quad 1.0 \quad 0.5 \quad 0.9]$$

Note that the domain and codomain of the fuzzy relation are examples of its projections.

2.10.15 Cylindrical Extension

The cylindrical extension increases the number of coordinates of the Cartesian product over which the fuzzy relation is formed. In this sense, the cylindrical extension is an operation that is complementary to the already discussed projection operation (Zadeh, 1971).

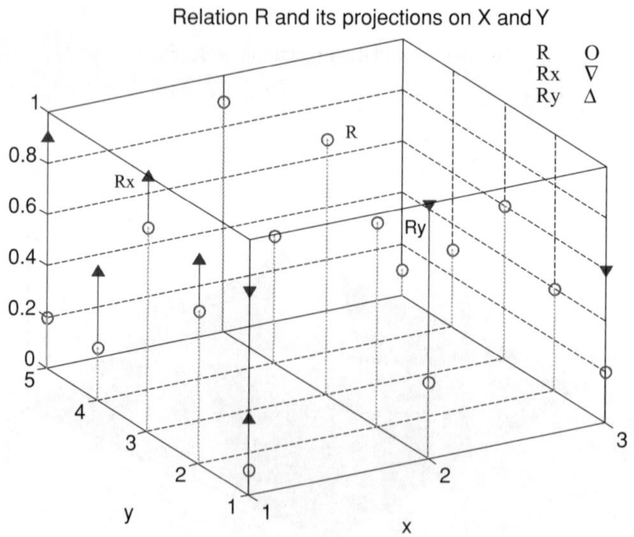

Figure 2.23 Fuzzy relation $\boldsymbol{R}$ and its projections on $\mathbf{X}$ and $\mathbf{Y}$.

The cylindrical extension on $\mathbf{X} \times \mathbf{Y}$ of a fuzzy set of $\mathbf{X}$ is a fuzzy relation cyl$A$ whose membership function is equal to

$$\text{cyl}\, \boldsymbol{A}(x, y) = A(x) \qquad \forall x \in \mathbf{X}, \quad \forall y \in \mathbf{Y} \tag{2.68}$$

If the fuzzy relation is viewed as a two-dimensional matrix, the operation of cylindrical extension forms identical columns indexed by the successive values of $y \in \mathbf{Y}$. The main intent of cylindrical extensions is to achieve compatibility of spaces over which fuzzy sets and fuzzy relations are formed. For instance, let A be a fuzzy set of $\mathbf{X}$ and $\boldsymbol{R}$ a fuzzy relation on $\mathbf{X} \times \mathbf{Y}$. Suppose we attempt to compute the union and intersection of A and $\boldsymbol{R}$. Because the universes over which A and $\boldsymbol{R}$ are defined are different, we cannot carry out any set-based operations on A and $\boldsymbol{R}$. The cylindrical extension of A, denoted by cyl$A$, provides the compatibility required. Then the operations such as $(\text{cyl}\,A) \cup \boldsymbol{R}$ and $(\text{cyl}\,A) \cap \boldsymbol{R}$ make sense. The concept of cylindrical extension can be easily generalized to multidimensional cases.

2.10.16 Reconstruction of Fuzzy Relations

Projections do not retain complete information conveyed by the original fuzzy relation. This means that, in general, one cannot faithfully reconstruct a relation from its projections. In other words, projections $\text{Proj}_X \boldsymbol{R}$ and $\text{Proj}_Y \boldsymbol{R}$ of some fuzzy relation $\boldsymbol{R}$ do not necessarily lead to the original fuzzy relation $\boldsymbol{R}$. In general, the reconstruction of a relation via the Cartesian product of its projections is a relation that includes the original relation, that is,

$$\text{Proj}_{\mathbf{X}}\, \boldsymbol{R} \times \text{Proj}_{\mathbf{Y}}\, \boldsymbol{R} \supseteq \boldsymbol{R} \tag{2.69}$$

If, however, the equality holds in the relationship above, then we call the relation $\boldsymbol{R}$ noninteractive.

2.10.17 Binary Fuzzy Relations

A binary fuzzy relation $\boldsymbol{R}$ on $\mathbf{X} \times \mathbf{X}$ is defined as follows:

$$\boldsymbol{R} : \mathbf{X} \times \mathbf{X} \to [0, 1] \tag{2.70}$$

There are several important features of binary fuzzy relations:

(a) **Reflexivity:** $\boldsymbol{R}(x, x) = 1\ \forall x \in \mathbf{X}$, Figure 2.24(a). When $\mathbf{X}$ is finite $\boldsymbol{R} \supseteq \boldsymbol{I}$, where $\boldsymbol{I}$ is the identity matrix, $\boldsymbol{I}(x, y) = 1$ if $x = y$ and $\boldsymbol{I}(x, y) = 0$ otherwise. Reflexivity can be relaxed by admitting a concept of so-called ε-reflexivity, $\varepsilon \in [0,1]$. This means that $\boldsymbol{R}(x, x) \geq \varepsilon$. When $\boldsymbol{R}(x, x) = 0$ the fuzzy relation is irreflexive. A fuzzy relation is locally reflexive if, for any $x, y \in \mathbf{X}$, $\max(\boldsymbol{R}(x, y), \boldsymbol{R}(y, x)) \leq \boldsymbol{R}(x, x)$.

(b) **Symmetry:** $\boldsymbol{R}(x, y) = \boldsymbol{R}(y, x)\ \forall (x, y) \in \mathbf{X} \times \mathbf{X}$, Figure 2.24(b). For finite $\mathbf{X}$, the matrix representing $\boldsymbol{R}$ has entries distributed symmetrically along the main diagonal. Clearly, if $\boldsymbol{R}$ is symmetric, then $\boldsymbol{R}^T = \boldsymbol{R}$.

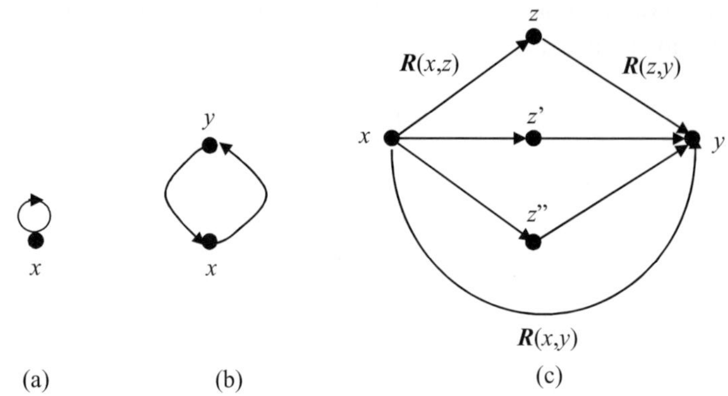

Figure 2.24 Main characteristics of binary fuzzy relations; see the details in the text.

(c) **Transitivity:** Here, $\sup_{z\in\mathbf{X}}(\boldsymbol{R}(x, z)\ T\ \boldsymbol{R}(z, y)) \leq \boldsymbol{R}(x, y)\ \forall x, y, z \in \mathbf{X}$. In particular, if this relationship holds for t = min, then the relation is called sup-min transitive. Looking at the levels of associations $\boldsymbol{R}(x, z)$ and $\boldsymbol{R}(z, y)$ occurring between x, and z, and z and y, the property of transitivity reflects the maximal strength among all possible links arranged in series (such as ($\boldsymbol{R}(x, z)$ and $\boldsymbol{R}(z, y)$)) that does not exceed the strength of the direct link $\boldsymbol{R}(x, z)$, Figure 2.24(c).

2.10.18 Transitive Closure

Given a binary fuzzy relation in a finite universe **X**, there exists a unique fuzzy relation $\overleftrightarrow{\boldsymbol{R}}$ on **X**, called the transitive closure of $\boldsymbol{R}$, that contains $\boldsymbol{R}$ and is itself included in any transitive fuzzy relation on **X** that contains $\boldsymbol{R}$. Therefore, if $\boldsymbol{R}$ is defined on a finite universe of cardinality n, the transitive closure is given by

$$\text{trans}(\boldsymbol{R}) = \overleftrightarrow{\boldsymbol{R}} = \boldsymbol{R} \cup \boldsymbol{R}^2 \cup \cdots \cup \boldsymbol{R}^n \tag{2.71}$$

where, by definition,

$$\boldsymbol{R}^2 = \boldsymbol{R} \circ \boldsymbol{R} \circ \cdots \circ \boldsymbol{R}^p = \boldsymbol{R} \circ \boldsymbol{R}^{p-1} \tag{2.72}$$

$$\boldsymbol{R} \circ \boldsymbol{R}(x, y) = \max_z\{\boldsymbol{R}(x, z)\ \boldsymbol{T}\, \boldsymbol{R}(z, y)\} \tag{2.73}$$

Note that the composition $\boldsymbol{R} \circ \boldsymbol{R}$ can be computed similarly as encountered in matrix algebra by replacing the ordinary multiplication by some t-norm and the sum by the max operations. In other words, if $r_{ij}^2 = [\boldsymbol{R}^2]_{ij} = [\boldsymbol{R} \circ \boldsymbol{R}]_{ij}$, then

$$r_{ij}^2 = \max_k(r_{ik}\ T\ r_{kj}) \tag{2.74}$$

If $\boldsymbol{R}$ is reflexive, then

$$\boldsymbol{I} \subseteq \boldsymbol{R} \subseteq \boldsymbol{R}^2 \subseteq \cdots \subseteq \boldsymbol{R}^{n-1} = \boldsymbol{R}^n \tag{2.75}$$

The transitive closure of the fuzzy relation $\boldsymbol{R}$ can be found by computing the successive k max T products of $\boldsymbol{R}$ until $\boldsymbol{R}^k = \boldsymbol{R}^{k-1}$, a procedure whose complexity is $O(n^3 \log_2 n)$ in time and $O(n^2)$ in space (Naessens, Meyer, and De Baets, 2002; De Baets and Meyer, 2003). See also Wallace, Acrithis, and Kollias (2006).

2.10.19 Equivalence and Similarity Relations

Equivalence relations are relations that are reflexive, symmetric, and transitive (Foulloy and Benoit, 2006). Suppose that one of the arguments of $\boldsymbol{R}(x, y)$, "x" for example, has been fixed. Thus, all elements related to x constitute a set called an equivalence class of $\boldsymbol{R}$ with respect to "x" and denoted by

$$A_x = \{y \in \mathbf{Y} | \boldsymbol{R}(x, y) = 1\} \tag{2.76}$$

The family of all equivalence classes of $\boldsymbol{R}$, denoted $\mathbf{X}/\boldsymbol{R}$, is a partition of $\mathbf{X}$. In other words, $\mathbf{X}/\boldsymbol{R}$ is a family of pairwise disjoint nonempty subsets of $\mathbf{X}$ whose union is $\mathbf{X}$. Equivalence relations can be viewed as a generalization of the equality relations in the sense that members of an equivalence class can be considered equivalent to each other under the relation $\boldsymbol{R}$.

Similarity relations are fuzzy relations that are reflexive, symmetric, and transitive. Like any fuzzy relation, a similarity relation can be represented by a nested family of its α-cuts, $\boldsymbol{R}_\alpha$. Each α-cut constitutes an equivalence relation and forms a partition of $\mathbf{X}$. Therefore, each similarity relation is associated with a set $P(\boldsymbol{R})$ of partitions of $\mathbf{X}$,

$$P(\boldsymbol{R}) = \{\mathbf{X}/\boldsymbol{R}_\alpha | \alpha \in [0, 1]\} \tag{2.77}$$

Partitions are nested in the sense that, if $\alpha > \beta$, then the partition $\mathbf{X}/\boldsymbol{R}_\alpha$ is finer than the partition $\mathbf{X}/\boldsymbol{R}_\beta$. For example, consider the relation defined on $\mathbf{X} = \{a, b, c, d, e\}$ in the following way:

$$\boldsymbol{R} = \begin{bmatrix} 1.0 & 0.8 & 0 & 0 & 0 \\ 0.8 & 1.0 & 0 & 0 & 0 \\ 0 & 0 & 1.0 & 0.9 & 0.5 \\ 0 & 0 & 0.9 & 1.0 & 0.5 \\ 0 & 0 & 0.5 & 0.5 & 1.0 \end{bmatrix}$$

One can easily verify that $\boldsymbol{R}$ is a symmetric matrix, has values of one at its main diagonal, and is max–min transitive. Therefore $\boldsymbol{R}$ is a similarity relation. The levels of refinement of the similarity relation $\boldsymbol{R}$ can be represented in the form of a partition tree in which each node corresponds to a fuzzy relation on $\mathbf{X}$ whose degree of association between the elements is

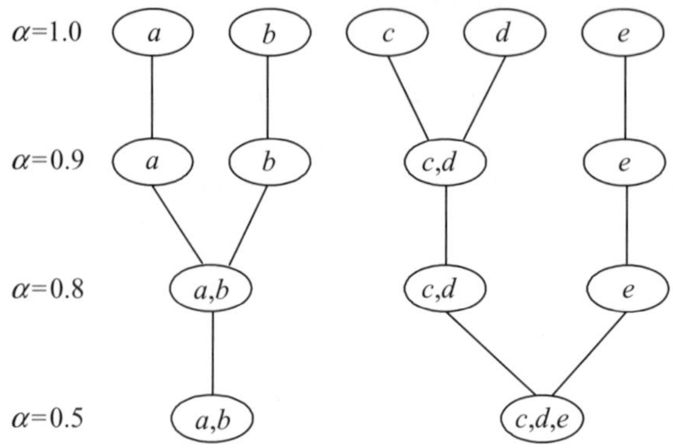

Figure 2.25 Partition tree induced by binary fuzzy relation $\boldsymbol{R}$.

greater than or equal to the threshold value α. For instance, we have the following fuzzy relations for $\alpha = 0.5, 0.8$, and 0.9, respectively:

$$\boldsymbol{R}_{0.5} = \begin{bmatrix} 1 & 1 & 0 & 0 & 0 \\ 1 & 1 & 0 & 0 & 0 \\ 0 & 0 & 1 & 1 & 1 \\ 0 & 0 & 1 & 1 & 1 \\ 0 & 0 & 1 & 1 & 1 \end{bmatrix}, \quad \boldsymbol{R}_{0.8} = \begin{bmatrix} 1 & 1 & 0 & 0 & 0 \\ 1 & 1 & 0 & 0 & 0 \\ 0 & 0 & 1 & 1 & 0 \\ 0 & 0 & 1 & 1 & 0 \\ 0 & 0 & 0 & 0 & 1 \end{bmatrix}, \quad \boldsymbol{R}_{0.9} = \begin{bmatrix} 1 & 0 & 0 & 0 & 0 \\ 0 & 1 & 0 & 0 & 0 \\ 0 & 0 & 1 & 1 & 0 \\ 0 & 0 & 1 & 1 & 0 \\ 0 & 0 & 0 & 0 & 1 \end{bmatrix}$$

Note that $\boldsymbol{R} = \cup_{\alpha \in \Lambda} \alpha \boldsymbol{R}_\alpha$ where $\cup = \max$ and $\Lambda = \{0.5, 0.8, 0.9, 1.0\}$ is the level set of $\boldsymbol{R}$. Also note that the greater the value of α, the finer the classes, as shown in Figure 2.25.

2.10.20 *Compatibility and Proximity Relations*

Compatibility relations are reflexive and symmetric relations. Associated with any compatibility relation are sets called compatibility classes. A compatibility class is a subset A of a universe $\mathbf{X}$ such that $\boldsymbol{R}(x, y) = 1$ for all $x, y \in A$.

Proximity relations are reflexive and symmetric fuzzy relations. Let A be a subset of a universe $\mathbf{X}$. Thus, A is an ε-proximity class of $\boldsymbol{R}$ if $\boldsymbol{R}(x, y) \geq \varepsilon$ for all $x, y \in A$. For instance, the relation $\boldsymbol{R}$ on $\mathbf{X} = \{1, 2, 3, 4, 5\}$

$$\boldsymbol{R} = \begin{bmatrix} 1.0 & 0.7 & 0 & 0 & 0.6 \\ 0.7 & 1.0 & 0.6 & 0 & 0 \\ 0 & 0.6 & 1.0 & 0.7 & 0.4 \\ 0 & 0 & 0.7 & 1.0 & 0.5 \\ 0.6 & 0 & 0.4 & 0.5 & 1.0 \end{bmatrix}$$

has unity on its main diagonal and is symmetric. Therefore, $\boldsymbol{R}$ is a proximity relation. Compatibility classes and α-compatibility classes do not necessarily induce partitions of $\mathbf{X}$ (Klir and Yuan, 1995).

Proximity is an important concept in pattern recognition, being used in contexts such as visual images because, under these circumstances, human subjectivity leads to some useful information that could be represented in the form of proximity relations.

2.11 Conclusions

Fuzzy sets provide a conceptual and operational framework to deal with granular information. Particular cases of degenerate membership functions (singletons) represent numeric information. Fuzzy sets come with a well-defined semantics whose formal description is conveyed in the form of membership functions. The variety of available membership functions offers a great deal of flexibility in capturing the meaning of the information granule. We presented the relationships between fuzzy sets and sets by stressing that sets are subsumed as particular cases of sets. The reconstruction of fuzzy sets through a finite or infinite family of α-cuts emphasizes the computational linkages between fuzzy sets and sets and indicates that the set-theoretic methods can be effectively utilized when processing fuzzy sets. Fuzzy relations and their important classes present an interesting view on the characterization of dependencies captured by membership grades.

Exercises

Problem 2.1. There is an interesting problem posed by Borel (1950) that could now be conveniently handled in the setting of fuzzy sets:

> One seed does not constitute a pile nor two or three. From the other side, everybody will agree that 100 million seeds constitute a pile. What therefore is the appropriate limit?

Given this description, suggest a membership function for the concept discussed here. What type of membership function would you consider in this problem? Why?

Problem 2.2. Consider two situations: (a) the number of people expected to ride on a bus on a certain day; (b) the number of people that could ride on a bus at any one time. Both situations describe an uncertain scenario. Which of these two situations involves randomness? Which one involves fuzziness? What is the nature of fuzziness: similarity, possibility, or preference?

Problem 2.3. We are interested in describing the state of an environment by quantifying temperature as *very cold*, *cold*, *comfortable*, *warm*, and *hot*. Choose an appropriate universe of discourse. Represent state values using (a) sets and (b) fuzzy sets.

Problem 2.4. Suppose that allowed speeds on city streets range between 0 and 60 km/h. Describe the speed values such as low, medium, and high using sets and fuzzy sets. Would this description be adequate also for highways? Justify the answer.

Problem 2.5. Given the fuzzy set A with the membership function

$$A(x) = \begin{cases} x - 4 & \text{if } 4 \leq x \leq 5 \\ -x + 6 & \text{if } 5 < x \leq 6 \\ 0 & \text{otherwise} \end{cases}$$

(a) Plot the membership function and identify its type.
(b) What type of linguistic label (semantics) could be associated with the concept conveyed by A?

Problem 2.6. Consider the fuzzy set A with the following membership function:

$$A(x) = \begin{cases} x - 4/2 & \text{if} \quad 4 \leq x \leq 5 \\ -x + 6/2 & \text{if} \quad 5 < x \leq 6 \\ 0 & \text{otherwise} \end{cases}$$

(a) Plot this membership function.
(b) Is A normal? Does A have a core? What is the height of this fuzzy set?
(c) Find the support of A. Is A a convex fuzzy set?

Problem 2.7. Assume a fuzzy set A whose membership function is defined in the following form:

$$A(x) = \begin{cases} x - 4 & \text{if} \quad 4 \leq x \leq 5 \\ 1 & \text{if} \quad 5 < x \leq 6 \\ -x + 7 & \text{if} \quad 6 < x \leq 7 \\ 0 & \text{otherwise} \end{cases}$$

(a) Sketch the graph of the membership function.
(b) Find an analytic expression for its α-cuts.
(c) Is A a convex fuzzy set?

Problem 2.8. Demonstrate that if a fuzzy set is convex, then all its α-cuts are convex.

Problem 2.9. Consider the following fuzzy sets defined in the finite universe of discourse $\mathbf{X} = \{1, 2, 3, \ldots, 10\}$:

$$A = (0, 0, 0, 0, 0.4, 0.6, 0.8, 1, 0.8, 0.6)$$
$$B = (0, 0, 0, 0, 0.4, 0.5, 0.6, 1, 0.6, 0.4)$$
$$C = (0, 1, 0.2, 0.3, 0.4, 0.5, 0.6, 1, 0.5, 0)$$

(a) Is $A \subseteq B$? Is $B \subseteq A$?
(b) Is $C \subseteq A$? Is $C \subseteq B$?
(c) Quantify the findings obtained in (a) and (b).

Problem 2.10. Suppose that fuzzy sets A and B defined in $\mathbf{X} = \{x_1, x_2, x_3\}$ are represented as vectors whose components are the membership degrees of x_1, x_2, and x_3 in A and B. Plot A and B in the unit cube for each of the following cases:

(a) $A = (1, 0, 0)$ and $B = (0, 1, 1)$,
(b) $A = (0, 1, 0)$ and $B = (1, 0, 1)$,
(c) $A = (0, 0, 1)$ and $B = (1, 1, 0)$,
(d) $A = (0.5, 0.5, 0.5)$ and $B = (0.5, 0.5, 0.5)$.

Problem 2.11. Let $\boldsymbol{R}_\alpha = \{(x, y) \in \mathbf{X} \times \mathbf{Y} \mid \boldsymbol{R}(x, y) \geq \alpha\}$ be the α-cut of the fuzzy relation $\boldsymbol{R}$. Show that any fuzzy relation $\boldsymbol{R} : \mathbf{X} \times \mathbf{Y} \rightarrow [0,1]$ can be represented in the following canonical form:

$$\boldsymbol{R} = \bigcup_{\alpha \in (0,1]} \alpha \boldsymbol{R}_\alpha$$

where $\cup$ denotes the standard union operation and $\alpha \boldsymbol{R}_\alpha$ is a subnormal fuzzy set whose membership function is α if $(x, y) \in \boldsymbol{R}_\alpha$ and zero otherwise.

Problem 2.12. How can the algorithm to compute the transitive closure of a fuzzy relation be used to verify if a fuzzy relation is transitive or not?

Problem 2.13. Show that if $\boldsymbol{R}$ is a similarity relation, then each of its α-cuts $\boldsymbol{R}_\alpha$ is an equivalence relation.

Problem 2.14. Verify that the transitive closure of a fuzzy proximity relation is a similarity relation.

Problem 2.15. A tolerance relation $\boldsymbol{R}$ in $\mathbf{X} \times \mathbf{Y}$ is a reflexive and symmetric ordinary relation. Show that if $\boldsymbol{R}$ is a proximity relation, then, for any $0 < \alpha \leq 1$, $\boldsymbol{R}_\alpha$ is a tolerance relation.

References

Bargiela, A. and Pedrycz, W. (2003) *Granular Computing: An Introduction*, Kluwer, Dordrecht.

Bodjanova, S. (2006) Median alpha-levels of a fuzzy number. *Fuzzy Sets and Systems*, **157** (7), 879–891.

Borel, E. (1950) *Probabilité e Certitude*, Press Université de France, Paris.

De Baets, B. and Meyer, H. (2003) On the existence and construction of T-transitive closures. *Information Sciences*, **152** (1), 167–179.

De Luca, A. and Termini, S. (1972) A definition of nonprobabilistic entropy in the setting of fuzzy sets. *Information and Control*, **20** (3), 301–312.

Dubois, D. and Prade, H. (1979) Outline of fuzzy set theory: an introduction, in *Advances in Fuzzy Set Theory and Applications* (eds M.M. Gupta, R.K. Ragade, and R.R. Yager), North-Holland, Amsterdam, pp. 27–39.

Dubois, D. and Prade, H. (1997) The three semantics of fuzzy sets. *Fuzzy Sets and Systems*, **90** (2), 141–150.

Dubois, D. and Prade, H. (1998) An introduction to fuzzy sets. *Clinica Chimica Acta*, **70** (1), 3–29.

Fodor, J. (1993) A new look at fuzzy connectives. *Fuzzy Sets and Systems*, **57** (2), 141–148.

Foulloy, L. and Benoit, E. (2006) Building a class of fuzzy equivalence relations. *Fuzzy Sets and Systems*, **157** (11), 1417–1437.

Kandel, A. and Yelowitz, L. (1974) Fuzzy chains. *IEEE Transactions on Systems, Man, and Cybernetics*, **SMC-4** (5), 472–475.

Kim, C.S., Kim, D.S., and Park, J.S. (2000) A new fuzzy resolution principle based on the antonym. *Fuzzy Sets and Systems*, **13** (2), 299–307.

Klir, G. and Yuan, B. (1995) *Fuzzy Sets and Fuzzy Logic: Theory and Applications*, Prentice Hall, Upper Saddle River, NJ.

Kosko, B. (1992) *Neural Networks and Fuzzy Systems*, Prentice Hall International, Englewood Cliffs, NJ.

Naessens, H., Meyer, H., and De Baets, B. (2002) Algorithms for the computation of T-transitive closures. *IEEE Transactions on Fuzzy Systems*, **10** (4), 541–551.

Nguyen, H. and Walker, E. (1999) *A First Course in Fuzzy Logic*, Chapman and Hall/CRC Press, Boca Raton, FL.

Pawlak, Z. (1982) Rough sets. *International Journal of Information and Computer Science*, **11** (15), 341–356.

Pedrycz, A., Dong, F., and Hirota, K. (2009) Finite α cut-based approximation of fuzzy sets and its evolutionary optimization. *Fuzzy Sets and Systems*, **160** (24), 3550–3564.

Pedrycz, W. (2005) From granular computing to computational intelligence and human-centric systems. *IEEE Connections*, **3** (2), 6–11.

Tsabadze, T. (2008) The reduction of binary fuzzy relations and its applications. *Information Sciences*, **178**, (2), 562–572.

Valverde, L. and Ovchinnikov, S. (2008) Representations of T-similarity relations. *Fuzzy Sets and Systems*, **159** (17), 2211–2220.

Wallace, M., Acrithis, Y., and Kollias, S. (2006) Computationally efficient sup-t transitive closure for sparse fuzzy binary relations. *Fuzzy Sets and Systems*, **157** (3), 341–372.

Yager, R. (1983) Entropy and specificity in a mathematical theory of evidence. *International Journal of General Systems*, **9** (1), 249–260.

Zadeh, L.A. (1965) Fuzzy sets. *Information and Control*, **8** (3), 33–353.

Zadeh, L.A. (1971) Similarity relations and fuzzy orderings. *Information Sciences*, **3** (2), 177–200.

Zadeh, L.A. (1975) The concept of linguistic variables and its application to approximate reasoning I, II, III. *Information Sciences*, **8** (3), **8** (4), **9** (1), 199–249, 301–357, 43–80.

Zadeh, L.A. (1978) Fuzzy sets as a basis for a theory of possibility. *Fuzzy Sets and Systems*, **1** (1), 3–28.

Zadeh, L.A. (1997) Toward a theory of fuzzy information granulation and its centrality in human reasoning and fuzzy logic. *Fuzzy Sets and Systems*, **90** (2), 111–127.

Zadeh, L.A. (1999) From computing with numbers to computing with words: from manipulation of measurements to manipulation of perceptions. *IEEE Transactions on Circuits and Systems*, **45** (1), 105–119.

3

Selected Design and Processing Aspects of Fuzzy Sets

In this chapter, we continue the discussion on the fundamentals of fuzzy sets by concentrating on three main issues: (1) the design of fuzzy sets (membership functions); (2) logic operations and aggregation of fuzzy sets; and (3) transformations (mappings) of fuzzy sets including the fundamentals of fuzzy arithmetic. These are the essentials which make the framework of fuzzy sets fully operational when supporting a wide range of applications.

3.1 The Development of Fuzzy Sets: Elicitation of Membership Functions

The issue of elicitation and interpretation of fuzzy sets (their membership functions) is significant from the conceptual, algorithmic, and application-oriented standpoints (Klir and Yuan, 1995; Nguyen and Walker, 1999). In the literature, we can find a large number of methods that support the construction of membership functions. In general, we distinguish here between *user-driven* and *data-driven* approaches, with a number of techniques that share some features specific to both data- and user-driven methods and hence located somewhere in between. The determination of membership functions has been a debatable issue for a long time, almost since the very inception of fuzzy sets. In contrast to interval analysis and set theory where the estimation of bounds of the interval constructs has not attracted a great deal of attention and seemed to be somewhat taken for granted, an estimation of membership degrees (and membership functions, in general) became essential and over time has led us to sound, well-justified, and algorithmically appealing estimation techniques (Civanlar and Trussell, 1986; Dombi, 1990; Turksen, 1991; Chen and Wang, 1999; Medaglia *et al.*, 2002).

3.1.1 Semantics of Fuzzy Sets: Some General Observations

Fuzzy sets are constructs that come with a well-defined meaning. They capture the semantics of the framework they intend to operate within. Fuzzy sets are the conceptual building blocks

Fuzzy Multicriteria Decision-Making: Models, Methods and Applications Witold Pedrycz, Petr Ekel and Roberta Parreiras

(generic constructs) that are used in problem description, modeling, decision-making, control, and pattern classification tasks. Before discussing specific techniques of membership function estimation, it is worth casting the overall presentation in a certain framework by emphasizing the aspect of the use of a finite number of fuzzy sets leading to some essential vocabulary reflective of the underlying domain knowledge. In particular, we are concerned with the related semantics and calibration capabilities of membership functions and the locality properties of fuzzy sets.

The limited capacity of a short-term memory, as identified by Miller (1956), suggests that we could easily and comfortably handle and process five to nine items. This implies that the number of fuzzy sets to be considered as meaningful conceptual entities should be kept more or less at the same level. The observation sounds reasonable – quite commonly in practice we witness situations in which this assumption holds. For instance, when describing linguistically quantified variables, say error or change of error, or quantifying temperature (*warm*, *hot*, *cold*, etc.), we can use seven generic concepts (descriptors) labeling them as positive *large*, positive *medium*, positive *small*, *around* zero, negative *small*, negative *medium*, negative *large*. When characterizing speed on a highway, we can talk about its quite intuitively appealing descriptors such as *low*, *medium*, and *high* speed. In the description of an approximation error, we can typically use the concept of a *small* error around a point of linearization (in all these examples, the terms are set in italics to emphasize the granular character of the constructs and the role being played there by fuzzy sets). While embracing very different tasks, these descriptors exhibit a striking similarity. All of them are information granules, not numbers. We can stress that the descriptive power of numbers is very much limited and numbers themselves are not used to describe abstract concepts. In general, the use of an excessive number of terms does not offer any advantage. On the contrary, it clutters our description of the phenomenon and hampers further effective usage of such concepts that we intend to establish to capture the essence of the domain knowledge. With the increase in the number of fuzzy sets, their semantics and interpretation capabilities also become negatively impacted. Fuzzy sets may be built into a hierarchy of terms (descriptors) but at each level of the hierarchy (when moving down toward higher specificity that is an increasing level of detail), the number of fuzzy sets is kept relatively small.

Although fuzzy sets capture the semantics of the concepts, they may require some calibration depending upon the specification of the problem at hand. This flexibility of fuzzy sets should not be treated as a shortcoming but rather viewed as a certain and fully exploited advantage. For instance, the term *low* temperature comes with a clear meaning, yet it requires a certain calibration depending upon the environment and the context it was placed into. The concept of *low* temperature is used in different climate zones and is of relevance in any communication between people, yet for each member of the community the meaning of the term is different, thereby requiring some calibration. This could be realized, for example, by shifting the membership function along the universe of discourse of temperature, affecting the universe of discourse by some translation, dilation, and so on. As a communication vehicle, linguistic terms are fully legitimate and as such they appear in different settings. They require some refinement so that their meaning is fully understood and shared by the community of the users.

When discussing the methods aimed at the determination of membership functions or membership grades, it is worthwhile underlining the existence of the two main categories of approaches that are reflective of the origin of the numeric values of membership. The first one is reflective of the domain knowledge and opinions of experts. In the second one, we consider experimental data whose global characteristics become reflected in the form and parameters of the membership functions. In the first group, we can refer to the pairwise comparison (for instance, Saaty's approach, as discussed later in this chapter) as one of the quite visible and representative examples, while fuzzy clustering is usually presented as a typical example of the data-driven methods of membership function estimation. In what follows, we elaborate on several representative methods, which will help us to appreciate the relevance and flexibility of fuzzy sets.

3.1.2 Fuzzy Set as a Descriptor of Feasible Solutions

The aim of the method is to relate the membership function to the level of feasibility of individual elements of a family of solutions associated with the problem at hand. Let us consider a certain function $F(x)$ defined in $\mathbf{L}$, that is, $F : \mathbf{L} \rightarrow \mathbf{R}^+$, where $\mathbf{L} \subset \mathbf{R}$. Our intent is to determine its maximum, namely $x^0 = \arg\max_{x\in\mathbf{L}} F(x)$. On the basis of the values of $F(x)$, we can form a fuzzy set A describing a collection of feasible solutions that could be labeled as optimal. More specifically, we use the fuzzy set to represent an extent (degree) to which some specific values of "x" could be sought as potential (optimal) solutions to the problem. Taking this into consideration, we relate the membership function of A to the corresponding value of $F(x)$ cast in the context of the boundary values assumed by "F". For instance, the membership function of A could be expressed in the following form:

$$A(x) = \frac{F(x) - \min\limits_{x\in\mathbf{L}} F(x)}{\max\limits_{x\in\mathbf{L}} F(x) - \min\limits_{x\in\mathbf{L}} F(x)} \tag{3.1}$$

The boundary conditions are intuitively associated with values of $\min_{x\in\mathbf{L}} F(x)$ and $\max_{x\in\mathbf{L}} F(x)$. For other values of "$x$" where F attains is maximal value, $A(x)$ is equal to one and, around this point, the membership values become lower when "x" is likely to be a solution to the problem $F(x) < \max_{x\in\mathbf{L}} F(x)$. The form of the membership function depends upon the character of the function under consideration.

If the fuzzy set is used to quantify the quality (performance) of the solution to the minimization problem, then the resulting membership function reads as follows:

$$A(x) = 1 - \frac{F(x) - \min\limits_{x\in\mathbf{L}} F(x)}{\max\limits_{x\in\mathbf{L}} F(x) - \min\limits_{x\in\mathbf{L}} F(x)} \tag{3.2}$$

If the function of interest assumes values in $\mathbf{R}$, then the two formulas are modified by including the absolute values of the differences, that is,

$$A(x) = \frac{\left|F(x) - \min\limits_{x\in\mathbf{L}} F(x)\right|}{\left|\max\limits_{x\in\mathbf{L}} F(x) - \min\limits_{x\in\mathbf{L}} F(x)\right|} \text{ and } A(x) = 1 - \frac{\left|F(x) - \min\limits_{x\in\mathbf{L}} F(x)\right|}{\left|\max\limits_{x\in\mathbf{L}} F(x) - \min\limits_{x\in\mathbf{L}} F(x)\right|}$$

Linearization, its quality, and the description of such quality fall under the same banner as the optimization problem. We show how the membership function could be formed in this case. When linearizing a function around some predetermined point, a quality of the resulting linearization scheme can be quantified in the form of some fuzzy set. Its membership function attains one for all these points where the linearization error is equal to zero (in particular, this holds at the point around which the linearization is carried out). The following example illustrates this idea.

Example 3.1. We are interested in the linearization of the function $y = g(x) = x^2$ around $x_0 = 1$ and assessing the quality of this linearization in the range [0,4]. The linearization formula reads $y - y_0 = g'(x_0)(x - x_0)$ where $y_0 = g(x_0)$ and $g'(x_0)$ is the derivative of $g(x)$ at x_0. Given the form of the function under consideration, its linearized version comes in the form $(2x_0)$ $(x - x_0) = 2(x - 1)$. We define the quality of this linearization by taking the absolute value of the difference between the original function and its linearized version, $f(x) = |g(x) - 2(x - 1)| = |x^2 - 2(x - 1)|$. As the fuzzy set A describes the quality of linearization, its membership function has to take into consideration the expression

$$A(x) = 1 - \frac{\left|F(x) - \min\limits_{x \in \mathbf{L}} F(x)\right|}{\left|\max\limits_{x \in \mathbf{L}} F(x) - \min\limits_{x \in \mathbf{L}} F(x)\right|} \tag{3.3}$$

where $\max_{x \in \mathbf{L}} F(x) = F(4) = 10$ and $\min_{x \in \mathbf{L}} F(x) = 0.0$. When, at some z, $F(z) = \min_{x \in \mathbf{L}} F(z)$, this means that $A(z) = 1$, which in the sequel indicates that the linearization at this point is perfect; no linearization error has been generated. We note that the higher quality of approximation is achieved for the arguments positioned closer to x_0; however, the form of the membership function depends on the form of the function to be linearized and the position of the point around which this linearization takes place.

3.1.3 Fuzzy Set as a Descriptor of the Notion of Typicality

Fuzzy sets address an issue of gradual *typicality* of elements to a given concept whose essence is being captured by the fuzzy set. They stress the fact that there are elements that fully satisfy the concept (are typical of it) and there are various elements that are allowed only with partial membership degrees. The form of the membership function is reflective of the semantics of the concept. Its details could be conveniently captured by adjusting the parameters of the membership function or choosing its form depending upon the available experimental data. For instance, consider a fuzzy set of circles. Formally, an ellipsoid includes a circular shape as its very special example, which satisfies the condition of equal axes, that is, $a = b$, see Figure 3.1. What if we have $a = b + \varepsilon$ where ε is a very small positive number? Could this figure be perceived as a circle? Very likely so, but perhaps not a circle in the straight mathematical sense, which we may note by assigning a membership grade that is very close to one, say 0.97. Our perception, which comes with some level of tolerance to imprecision, does not allow us to tell apart this figure from the ideal circle.

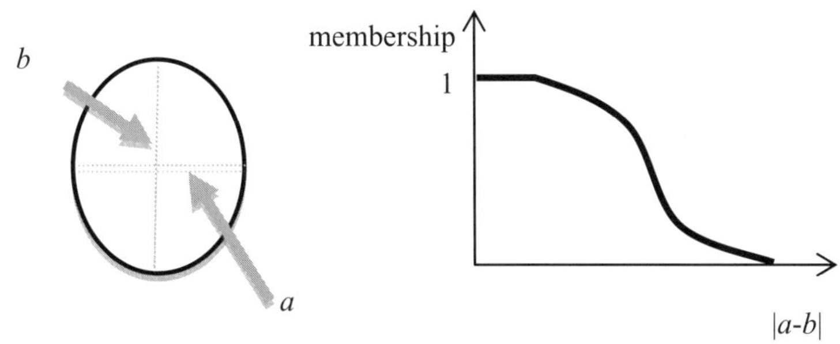

Figure 3.1 Perception of geometry of ellipsoids and quantification of their membership grades to the concept of "fuzzy circles".

It is intuitively appealing to see that higher differences between the values of the axes "a" and "b" result in lower values of the membership function. The definition of the fuzzy set of a circle could be formed in a number of ways. Prior to the definition or even before a visualization of the shape of the membership function, it is important to formulate a universe of discourse over which it is to be defined. There are several sound alternatives worth considering:

(a) For each pair of values of the axes (a and b), collect an experimental assessment of membership of the ellipsoids to the category of circles. Here the membership function is defined over a Cartesian space of the spaces of lengths of axes of the ellipsoids. While selecting a form of the membership we require that it assumes values at $a = b$ and becomes gradually reduced when the arguments start getting more distant.

(b) We can define an absolute distance between "a" and "b", $|a - b|$, and form a fuzzy set over this space $\mathbf{X}$; that is, $\mathbf{X} = \{x \mid x = |a - b|\}$, $\mathbf{X} \subset \mathbf{R}_+$. These semantic constraints translate into the condition of $A(0) = 1$. For higher values of x we may consider monotonically decreasing values of the membership function A.

(c) We can envision ratios of a and b, $x = a/b$, and construct a fuzzy set over the space of $\mathbf{R}_+$ such that $\mathbf{X} = \{x \mid x = a/b\}$. Here we require that $A(1) = 1$. We also anticipate lower values of membership grades when moving to the left and to the right from $x = 1$. Note that the membership function could be asymmetric, so we allow for different membership values for the same length of the sides, say $a = 6$, $b = 5$ and $a = 6$ and $b = 5$ (the effect could be quite apparent due to the occurrence of visual effects when perceiving geometric phenomena). The previous model of $\mathbf{X}$ as outlined in (a) cannot capture this effect.

Once the form of the membership function has been defined, it could be further adjusted by modifying the values of its parameters on the basis of some experimental findings. They come in the form of ordered triples or pairs, say (a, b, μ), $(a/b, \mu)$, or $(|a - b|, \mu)$ depending on the previously accepted definition of the universe of discourse. The membership values μ are those available from the expert offering an assessment of the likeness of the corresponding geometric figure. Note that the resulting membership functions are formulated in different universes of discourse.

3.1.4 Vertical and Horizontal Schemes of Membership Function Estimation

The vertical and horizontal modes of membership estimation are two standard approaches used in the determination of fuzzy sets. They reflect distinct ways of looking at fuzzy sets whose membership functions at some finite number of points are quantified by experts. In the horizontal approach we identify a collection of elements in the universe of discourse **X** and request that an expert answers the following question:

$$\text{Does } x \text{ belong to concept } A? \tag{3.4}$$

The answers are expected to come in a binary (yes–no) format. The concept A defined in **X** could be any linguistic notion, say *high* speed, *low* temperature, and so on. Given "n" experts whose answers for a given point of **X** form a mix of yes–no replies, we count the number of "yes" answers and compute the ratio of the positive answers (p) versus the total number of replies(n), that is, p/n. This ratio (likelihood) is treated as a membership degree of the concept at the given point of the universe of discourse. When all experts accept that the element belongs to the concept, then its membership degree is equal to one. Higher disagreement between the experts (quite divided opinions) results in lower membership degrees. The concept A defined in **X** requires the collection of results for some other elements of **X** and determining the corresponding ratios as outlined in Figure 3.2 (observe a series of estimates that are determined for selected elements of **X**; note also that the elements of **X** need not be evenly distributed).

If replies follow some binomial distribution, for example, then we can determine a confidence interval of the individual membership grade. The standard deviation of the estimate of the ratio of the positive answers associated with the point x, denoted here by σ, is given in the form

$$\sigma = \sqrt{\frac{p/n(1 - p/n)}{n}} \tag{3.5}$$

The associated confidence interval, which describes a range of membership values, is then determined as

$$[p - \sigma, p + \sigma] \tag{3.6}$$

In essence, when the confidence intervals are taken into consideration, the membership estimates become intervals of possible membership values and this leads to the concept of

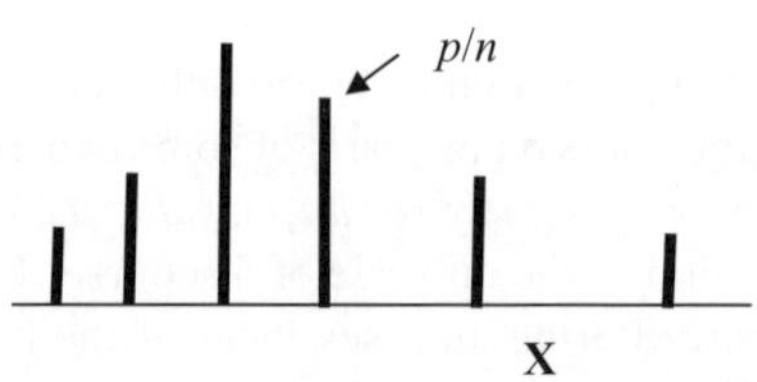

Figure 3.2 A horizontal method of the estimation of the membership function.

so-called interval-valued fuzzy sets. By assessing the width of the estimates, we can control the execution of the experiment: when the ranges are too long, one could redesign the experiment and monitor closely the consistency of the responses collected during its realization.

Example 3.2. Let us consider the responses of 10 experts who came up with the following assessment of the concept *high* interest rate (%) with the number of "yes" responses collected:

x (%)	2	3	5	8	10
No. of "yes" replies	0	2	4	7	10

Following these responses, the membership function and its confidence values σ producing confidence intervals are given as

x (%)	2	3	5	8	10
$A(x)$ (*high* interest rate)	0.0	0.2	0.4	0.7	1.0
σ	0.0	0.126	0.155	0.144	0.0

The advantage of the method lies in its simplicity as the technique relies explicitly upon the direct counting of responses. The concept is also intuitively appealing. The probabilistic nature of the replies helps us to construct confidence intervals that are essential to the assessment of the specificity of the membership quantification. A certain drawback is related to the local character of the construct: as the estimates of the membership function are completed separately for each element of the universe of discourse, they could exhibit a lack of continuity when moving from a certain point to its neighbor. This concern is particularly valid in the case when **X** is a subset of real numbers.

The vertical mode of membership estimation is concerned with the estimation of the membership function by focusing on the determination of the successive α-cuts. The experiment focuses on the unit interval of membership grades. The experts involved in the experiment are asked questions of the following form:

$$\text{What are the elements of } \mathbf{X} \text{ which belong to fuzzy set } A \text{ at degree no lower than } \alpha? \quad (3.7)$$

where α is a certain level (threshold) of membership grades in [0,1]. The essence of the method is illustrated in Figure 3.3. Note that satisfaction of the inclusion constraint is obvious: we envision that for higher values of α, the expert is going to provide more limited (smaller) subsets of **X**; the vertical approach leads to the fuzzy set by combining the estimates of the corresponding α-cuts. Given the nature of this method, we refer to the collection of random sets as these estimates appear in the successive stages of the estimation process.

These elements are identified by the expert as those forming the corresponding α-cuts of A. By repeating the process for several selected values of α we end up with the α-cuts and, using them, we reconstruct the fuzzy set. The simplicity of the method is its genuine advantage. As in the horizontal method of membership estimation, a possible lack of continuity is a

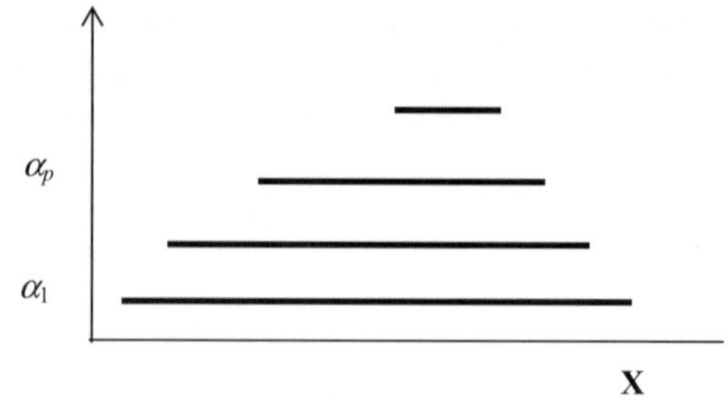

Figure 3.3 A vertical approach of membership estimation through the reconstruction of a fuzzy set through its estimated α-cuts.

certain disadvantage one has to be aware of. Here the selection of suitable levels of α needs to be carefully investigated. Similarly, an order at which different levels of α are used in the experiment could impact the estimate of the membership function. The discussion on the optimization of a series of α-cuts (which might be of relevance in the context of the estimation of membership functions) is given in Pedrycz, Dong, and Hirota (2009).

3.1.5 Saaty's Priority Approach of Pairwise Membership Function Estimation

The priority approach introduced by Saaty (Saaty, 1980; Saaty, 1986b) forms another interesting alternative used to estimate the membership function, which helps to alleviate the limitations associated with the horizontal and vertical schemes of membership function estimation. To explain the essence of the method, let us consider a collection of elements $X_1, X_2, \ldots, X_n$ (those could be, for instance, some alternatives whose allocation to a certain fuzzy set is sought) for which membership grades $A(X_1), A(X_2), \ldots, A(X_n)$ are given. Let us organize them into a so-called reciprocal matrix (or multiplicative preference relation) of the following form:

$$\boldsymbol{M} = [\boldsymbol{M}(X_k, X_i)] = \begin{bmatrix} \frac{A(X_1)}{A(X_1)} & \frac{A(X_1)}{A(X_2)} & \cdots & \frac{A(X_1)}{A(X_n)} \\ \frac{A(X_2)}{A(X_1)} & \frac{A(X_2)}{A(X_2)} & \cdots & \frac{A(X_2)}{A(X_n)} \\ \vdots & \vdots & \ddots & \vdots \\ \frac{A(X_n)}{A(X_1)} & \frac{A(X_n)}{A(X_2)} & \cdots & \frac{A(X_n)}{A(X_n)} \end{bmatrix} = \begin{bmatrix} 1 & \frac{A(X_1)}{A(X_2)} & \cdots & \frac{A(X_1)}{A(X_n)} \\ \frac{A(X_2)}{A(X_1)} & 1 & \cdots & \frac{A(X_2)}{A(X_n)} \\ \vdots & \vdots & \ddots & \vdots \\ \frac{A(X_n)}{A(X_1)} & \frac{A(X_n)}{A(X_2)} & \cdots & 1 \end{bmatrix} \tag{3.8}$$

Noticeably, the diagonal values of $\boldsymbol{M}$ are equal to one. The entries that are symmetrically positioned with respect to the diagonal satisfy the condition of reciprocality, that is, $\boldsymbol{M}(X_i, X_j) = 1/\boldsymbol{M}(X_j, X_i)$. We will be referring to this form of reciprocality as *multiplicative reciprocality* as opposed to so-called *additive reciprocality* for which $\boldsymbol{M}(X_i, X_j) + \boldsymbol{M}(X_j, X_i) = 1$. Furthermore, an important transitivity property (the multiplicative transitivity) holds, namely $\boldsymbol{M}(X_i, X_k)$

$M(X_k, X_j) = M(X_i, X_j)$, for all indexes i, j, and k. This property holds because of the way in which the matrix has been constructed. By plugging in the corresponding ratios, one obtains

$$M(X_i, X_k)\,M(X_k, X_j) = \frac{A(X_i)}{A(X_k)}\frac{A(X_k)}{A(X_j)} = \frac{A(X_i)}{A(X_j)} = M(X_i, X_j)$$

Let us now multiply the matrix by the vector of the membership grades $A = [A(X_1)\,A(X_2)\ldots A(X_n)]^T$. For the ith row of M (that is, the ith entry of the resulting vector of results) we obtain

$$[M\,A]_i = \left[\frac{A(X_i)}{A(X_1)}\frac{A(X_i)}{A(X_2)}\cdots\frac{A(X_i)}{A(X_n)}\right]\begin{bmatrix}A(X_1)\\A(X_2)\\\vdots\\A(X_n)\end{bmatrix}, \quad i = 1, 2, \ldots, n \tag{3.9}$$

Thus the ith element of the vector is equal to $nA(X_i)$. Overall, completing the calculations once for all "i" leads to the expression $MA = nA$. In other words, we conclude that A is the eigenvector of M associated with the largest eigenvalue of M, which is equal to "n". In the above scenario, we have assumed that the membership values $A(x_i)$ are given and then showed what form of results could they lead to. In practice the membership grades are not given and have to be estimated.

The starting point of the estimation process is the entries of the reciprocal matrix which are obtained through collecting results of pairwise evaluations offered by an expert, designer, or user (depending on the character of the task at hand). Prior to making any assessment, the expert is provided with a finite scale with values spread between one and nine. Some other alternatives of the scales, such as those involving five or nine levels, could be sought as well. If X_i is strongly preferred over X_j when considered in the context of the fuzzy set whose membership function we would like to estimate, then this judgment is expressed by assigning high values of the available scale, say six or seven. If we still sense that X_i is preferred over X_j, yet the strength of this preference is lower in comparison to the previous case, then this is quantified using some intermediate values of the scale, say three or four. If no difference is sensed, the values close to one are the preferred choice, say two or one. The value of one indicates that X_i and X_j are equally preferred. The general quantification of preferences positioned on the scale of 1–9 can be described as in Table 3.1 (Saaty, 1986a) .

Table 3.1 Scale of intensities of relative importance

Intensity of relative importance	Description
1	Equal importance (equal significance)
3	Moderate importance of one element over another (weak superiority)
5	Essential or strong importance (strong superiority)
7	Demonstrated importance (evident superiority)
9	Extreme importance (absolute superiority)
2, 4, 6, 8	Intermediate values between the two adjacent judgments

On the other hand, if X_j is preferred over X_i, the corresponding entry assumes values below one. Given the reciprocal nature of the assessment, once the preference of X_i over X_j has been quantified, the inverse of this number is plugged into the entry of the matrix that is located at the (j, i)th coordinate. As indicated earlier, the elements on the main diagonal are equal to one. Next, the maximal eigenvalue is computed along with its corresponding eigenvector. The normalized version of the eigenvector is then the membership function of the fuzzy set we considered when realizing all pairwise assessments of the elements of its universe of discourse. The effort to complete pairwise evaluations is far more manageable in comparison to any experimental overhead we encounter when assigning membership grades to all elements (alternatives) of the universe in a single step. Practically, the pairwise comparison helps the expert focus on only two elements once at a time, thus reducing uncertainty and hesitation while leading to a higher level of consistency. The assessments are not free of bias and could exhibit some inconsistent evaluations. In particular, we cannot expect the transitivity requirement to be fully satisfied. Fortunately, the lack of consistency could be quantified and monitored. The largest eigenvalue computed for $\boldsymbol{M}$ is always greater than the dimensionality of the reciprocal matrix (recall that in reciprocal matrices the elements positioned symmetrically along the main diagonal are the inverse of each other), that is, $\lambda_{\max} > n$ where the equality $\lambda_{\max} = n$ occurs only if the results are fully consistent. The ratio

$$\nu = (\lambda_{\max} - n)/(n - 1) \tag{3.10}$$

can be regarded as an index of inconsistency of the data: the higher its value, the less consistent the collected experimental results. This expression can be sought as an indicator of the quality of the pairwise assessments provided by the expert. If the value of ν is too high, exceeding a certain superimposed threshold, the experiment may need to be repeated. Typically, if ν is less than 0.1 the assessment is sought to be consistent, while higher value of ν call for the re-examination of the experimental data and a rerun of the experiment. To quantify how much the experimental data deviate from the transitivity requirement, we calculate the absolute differences between the corresponding experimentally obtained entries of the reciprocal matrix, namely $\boldsymbol{M}(X_i, X_k)$ and $\boldsymbol{M}(X_i, X_j)\,\boldsymbol{M}(X_j, X_k)$. The sum expressed in the form

$$V(i,k) = \sum_{j=1}^{n} |\boldsymbol{M}(X_i, X_j)\boldsymbol{M}(X_j, X_k) - \boldsymbol{M}(X_i, X_k)| \tag{3.11}$$

serves as a useful indicator of the lack of transitivity of the experimental data for the given pair of elements (i, k). If required, we may repeat the experiment when the above sum takes high values. The overall sum $\sum_{i,k}^{n} V(i,k)$ then becomes a global evaluation of the lack of transitivity of the experimental assessment.

Example 3.3. Let us estimate the membership function of the concept *hot* temperature for the space of temperatures consisting of 10, 20, 30, 45 degrees Celsius. The scale in which the pairs of these elements are evaluated consists of five levels (say, 1, 2, . . . , 5). The experimental results of the pairwise comparison are collected in the reciprocal matrix $\boldsymbol{M}$,

$$\boldsymbol{M} = \begin{bmatrix} 1 & 1/2 & 1/4 & 1/5 \\ 2 & 1 & 1/3 & 1/4 \\ 4 & 3 & 1 & 1/3 \\ 5 & 4 & 3 & 1 \end{bmatrix} \tag{3.12}$$

Calculating the maximal eigenvalue, we obtain $\lambda_{\max} = 4.114$ which is slightly higher than the dimension ($n = 4$) of the reciprocal matrix. The corresponding eigenvector is equal to [0.122 0.195 0.438 0.869] which after normalization gives rise to the membership function of *hot* temperature equal to [0.14 0.22 0.50 1.00]. The value of the inconsistency index n is equal to $(4.114 - 4)/3 = 0.038$ and is far lower than the threshold of 0.1.

Example 3.4. Let us consider some modified version of the previously discussed reciprocal matrix with the following entries:

$$\boldsymbol{M} = \begin{bmatrix} 1 & 1/2 & 1/4 & 1/5 \\ 2 & 1 & 1/3 & 4 \\ 4 & 3 & 1 & 1/3 \\ 5 & 1/4 & 3 & 1 \end{bmatrix} \tag{3.13}$$

Now the maximal eigenvalue is far higher than the dimensionality of the problem, $\lambda_{\max} = 5.426$. In this case, given the high value of the inconsistency index, $\nu = (5.426 - 4)/3 = 0.475$, there is no point in computing the corresponding eigenvector. To fix the problem we could compute the lack of transitivity for the triples of indexes (i, j, k) and in this way highlight those assessments that tend to be highly inconsistent. These are the candidates whose evaluation has to be revised.

The method of pairwise comparison has been generalized in many different ways by allowing estimates to be expressed as fuzzy sets (van Laarhoven and Pedrycz, 1983). One can refer to a number of applications in which the technique of pairwise comparison has been directly applied (Kulak and Kahraman, 2005).

3.1.6 Fuzzy Sets as Granular Representatives of Numeric Data – A Principle of Justifiable Granularity

In general, a fuzzy set is reflective of the nature of numeric data that are put together and interpreted in some context. Using its membership function we attempt to embrace them in a concise manner. The development of the fuzzy set is supported by the following experiment-driven and intuitively appealing rationale:

(a) first, we expect that A reflects (or matches) the available experimental data to the highest extent;
(b) second, the fuzzy set is kept specific enough so that it comes with a well-defined semantics.

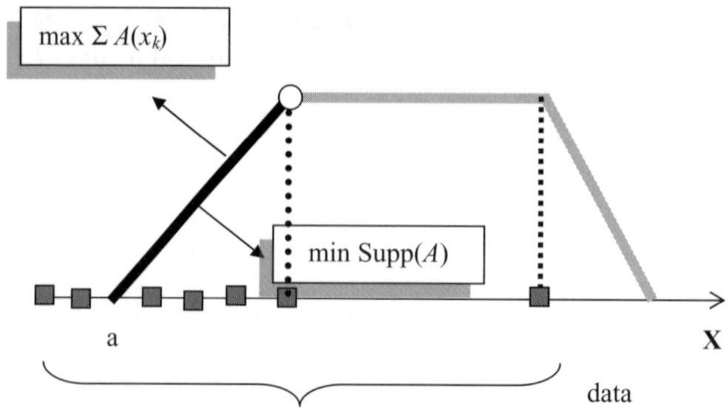

Figure 3.4 Optimization of the slope of the linearly increasing section of the membership function of *A*.

These two requirements point at the multiobjective nature of the construct: we want to maximize the coverage of experimental data (as articulated by (a)) and minimize the spread of the fuzzy set (as captured by (b)). These two requirements give rise to a certain optimization problem. Furthermore, quite legitimately, we assume that the fuzzy set to be constructed has a unimodal membership function or its maximal membership grades occupy a contiguous region in the universe of discourse in which this fuzzy set has been defined. This helps us to build a membership function separately for its rising and declining sections. The core of the fuzzy set is determined first. Next, assuming the simplest scenario when using the linear type of membership functions, the essence of the optimization problem boils down to a rotation of the linear segment of the membership function around the upper point of the core of A; see Figure 3.4. The point of rotation of the linear segment of this membership function is marked by an empty circle. By rotating this segment around this point, we intend to maximize (a) and minimize (b). Note that these two criteria are conflicting; that is, the increase of the experimental evidence comes with reduced specificity of the fuzzy set. The optimization of the decreasing section of the membership function is realized in the same way.

Before moving on to the determination of the membership function, we concentrate on the location of its numeric representative. Typically, one could view an average of the experimental data $x_1, x_2, \ldots, x_n$ to be a sound representative. While its usage is quite common in practice, a better representative of the numeric data is a median value. There is a sound reason for this choice. The median is a *robust* statistic, meaning that it allows for a high level of tolerance to potential noise existing in the data. Its important ability is to ignore outliers. Given that the fuzzy set is sought to be a granular and "stable" representation of the numeric data, our interest is in the robust development not being affected by noise. Undoubtedly, the use of the median is a good starting point. Let us recall that the median is an order statistic and is formed on the basis of an ordered set of numeric values. In the case of an odd number of data points in the data set, the point located in the middle of this ordered sequence is the median. When we encounter an even number of data points in the granulation window, instead of picking an average of the two points located in the middle, we consider these two points to form the core of a fuzzy set. Thus, depending upon the number of data points, we end up with either a triangular or trapezoidal membership function.

Having fixed the modal value of A (that could be a single numeric value, "m", or a certain interval $[m, n]$), the optimization of the spreads of the linear portions of the membership functions is carried out separately for their increasing and decreasing portions. We consider the increasing part of the membership function (the decreasing part is handled analogously). Referring to Figure 3.4, the two requirements guiding the design of the fuzzy set are transformed into the corresponding multiobjective optimization problem outlined as follows:

(a) Maximize the experimental evidence of the fuzzy set; this implies that we tend to "cover" as many numeric data as possible, that is, the coverage has to be made as high as possible. Graphically, in the optimization of this requirement, we rotate the linear segment up (clockwise) as illustrated in Figure 3.4. Formally, the sum of the membership grades $A(x_k)$, $\sum_k A(x_k)$, where A is the linear membership function to be optimized and x_k is located to the left to the modal value, has to be maximized.
(b) Simultaneously, we would like to make the fuzzy set as specific as possible so that is comes with some well-defined semantics. This requirement is met by making the support of A as small as possible, that is, $\min_a |m - a|$.

These two requirements form the crux of the *principle of justifiable granularity* (Bortolan and Pedrycz, 2002; Pedrycz and Vukovich, 2002). To accommodate the two conflicting requirements, we combine (a) and (b) in the form of a ratio that is maximized with respect to the unknown parameter of the linear section of the membership function

$$\max_a \frac{\sum\limits_k A(x_k)}{|m - a|} \tag{3.14}$$

The linearly decreasing portion of the membership function is optimized in the same way. The overall optimization returns the parameters of the fuzzy number in the form of the lower and upper bound (a and b, respectively) and its support (m or $[m, n]$). We can write down such fuzzy numbers as $A(a, m, n, b)$. We exclude the trivial solution of $a = m$, in which case the fuzzy set collapses to a single numeric entity.

Example 3.5. Let us consider a geometric figure that resembles a fuzzy circle, Figure 3.5. The coordinates of the central point are given as (x_0, y_0). Let us represent the figure as a fuzzy circle, that is, a circle whose radius is a fuzzy set (fuzzy number).

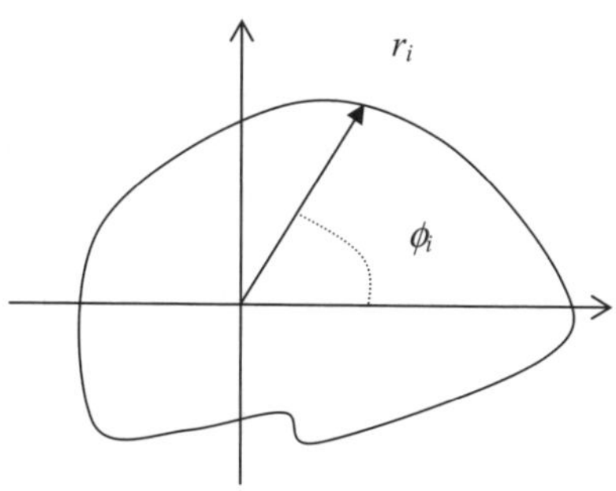

Figure 3.5 Example of figures to be represented as fuzzy circles (a) and a way of generating the numeric values of radius at successive angles of the figure (b).

The membership of the fuzzy radius is determined on the basis of numeric values of the radii obtained for several successive discrete values of the angle ϕ_i, thus giving rise to the values of the corresponding distance $r_1, r_2, \ldots, r_n$. Next, the determination of the fuzzy set of the radius (fuzzy circle) is realized following the optimization scheme governed by (3.12).

3.1.7 *Design of Fuzzy Sets through Fuzzy Clustering: From Data to their Granular Abstraction*

Fuzzy sets can be formed on the basis of numeric data through their clustering (groupings). The groups of data give rise to membership functions that convey a global, more abstract view of the available data. In this regard the fuzzy *c*-means (FCM, for short) algorithm is one of the commonly used mechanisms of fuzzy clustering (Bezdek, 1981).

Let us review its formulation, develop the algorithm, and highlight the main properties of the fuzzy clusters. Given a collection of *n*-dimensional data sets $\{\mathbf{x}_k\}$, $k = 1, 2, \ldots, N$, the task of determining its structure – a collection of "*c*" clusters – is expressed as a minimization of the following objective function (performance index) Q regarded as a sum of the squared distances:

$$Q = \sum_{i=1}^{c} \sum_{k=1}^{N} u_{ik}^{m} \|\mathbf{x}_k - \mathbf{v}_i\|^2 \tag{3.15}$$

where $\mathbf{v}_1, \mathbf{v}_2, \ldots, \mathbf{v}_c$ are *n*-dimensional prototypes of the clusters and $\boldsymbol{U} = [u_{ik}]$ stands for a partition matrix expressing a way of allocating the data to the corresponding clusters; u_{ik} is the membership degree of data $\mathbf{x}_k$ in the *i*th cluster. The distance between the data $\mathbf{z}_k$ and prototype $\mathbf{v}_i$ is denoted by $\|.\|$. The fuzzification coefficient m (> 1.0) expresses the impact of the membership grades on the individual clusters.

A partition matrix satisfies two important properties:

$$\text{(a) } 0 < \sum_{k=1}^{N} u_{ik} < N, \quad i = 1, 2, \ldots, c \tag{3.16}$$

$$\text{(b) } \sum_{i=1}^{c} u_{ik} = 1, \quad k = 1, 2, \ldots, N \tag{3.17}$$

Let us denote by $\mathbf{U}$ a family of matrices satisfying these two requirements (a) and (b). The first requirement states that each cluster has to be nonempty and different from the entire set. The second requirement states that the sum of the membership grades should be confined to one.

The minimization of Q is completed with respect to $U \in \mathbf{U}$ and the prototypes $\mathbf{v}_i$ of $V = \{\mathbf{v}_1, \mathbf{v}_2, \ldots, \mathbf{v}_c\}$ of the clusters. More explicitly, we write this as follows:

$$\min Q \text{ with respect to } U \in \mathbf{U},\ \mathbf{v}_1,\ \mathbf{v}_2, \ldots, \mathbf{v}_c \in \mathbf{R}^n \tag{3.18}$$

From the optimization standpoint, there are two individual optimization tasks to be carried out separately for the partition matrix and the prototypes. The first one concerns the minimization with respect to the constraints given the requirement of the form (3.17), which holds for each data point $\mathbf{x}_k$. The use of Lagrange multipliers transforms the problem into its constraint-free version. The augmented objective function formulated for each data point, $k = 1, 2, \ldots, N$, reads

$$V = \sum_{i=1}^{c} u_{ik}^m d_{ik}^2 + \lambda \left(\sum_{i=1}^{c} u_{ik} - 1 \right) \tag{3.19}$$

where $d_{ik}^2 = \|\mathbf{x}_k - \mathbf{v}_i\|^2$. Proceeding with the necessary conditions for the minimum of V for $k = 1, 2, \ldots, N$ gives

$$\frac{\partial V}{\partial u_{st}} = 0 \qquad \frac{\partial V}{\partial \lambda} = 0 \tag{3.20}$$

where $s = 1, 2, \ldots, c$, $t = 1, 2, \ldots, N$. Now we calculate the derivative of V with respect to the elements of the partition matrix in the following way:

$$\frac{\partial V}{\partial u_{st}} = m u_{st}^{m-1} d_{st}^2 + \lambda \tag{3.21}$$

From (3.20) and using (3.21) we calculate the membership grade u_{st} to be equal to

$$u_{st} = -\left(\frac{\lambda}{m} \right)^{\frac{1}{m-1}} d_{st}^{\frac{2}{m-1}} \tag{3.22}$$

Given the normalization condition $\sum_{j=1}^{c} u_{jt} = 1$ and substituting it into (3.22) one has

$$-\left(\frac{\lambda}{m} \right)^{\frac{1}{m-1}} \sum_{j=1}^{c} d_{jt}^{\frac{2}{m-1}} = 1 \tag{3.23}$$

After some rearrangements of the above expression by isolating the term including the Lagrange multiplier, one obtains

$$-\left(\frac{\lambda}{m} \right)^{\frac{1}{m-1}} = \frac{1}{\sum_{j=1}^{c} d_{jt}^{\frac{2}{m-1}}} \tag{3.24}$$

Inserting this expression into (3.22), we obtain the successive entries of the partition matrix

$$u_{st} = \frac{1}{\sum_{j=1}^{c} \left(\frac{d_{st}^2}{d_{jt}^2} \right)^{\frac{1}{m-1}}} \tag{3.25}$$

The optimization of the prototypes $\mathbf{v}_i$ is carried out assuming the Euclidean distance between the data and the prototypes, that is,

$$\|\mathbf{x}_k - \mathbf{v}_i\|^2 = \sum_{j=1}^{n} (x_{kj} - v_{ij})^2$$

The objective function now reads

$$Q = \sum_{i=1}^{c} \sum_{k=1}^{N} u_{ik}^m \sum_{j=1}^{n} (x_{kj} - v_{ij})^2$$

and its gradient with respect to $\mathbf{v}_i$, $\nabla_{v_i} Q$, made equal to zero yields the system of linear equations

$$\sum_{k=1}^{N} u_{ik}^m (x_{kt} - v_{st}) = 0, \quad s = 1, 2, \ldots, c, t = 1, 2, \ldots, n \tag{3.26}$$

Thus

$$v_{st} = \frac{\sum_{k=1}^{N} u_{ik}^m x_{kt}}{\sum_{k=1}^{N} u_{ik}^m} \tag{3.27}$$

We should emphasize that the use of some other distance functions different from the Euclidean one carries some computational complexity and the formula for the prototype cannot be presented in the concise manner as given above.

Overall, the FCM clustering is completed through a sequence of iterations where we start from some random allocation of data (a certain randomly initialized partition matrix) and carry out the following updates by successively adjusting the values of the partition matrix and the prototypes. The iterative process is repeated until a certain termination criterion has been satisfied. Typically, the termination condition is quantified by looking at the changes in the membership values of the successive partition matrices. Let us denote by $U(t)$ and $U(t+1)$ the two partition matrices produced in two consecutive iterations of the algorithm. If the distance $\|U(t+1) - U(t)\|$ is less than a small predefined threshold ε, then we terminate the algorithm. Typically, one considers the Chebyshev distance between the partition matrices, meaning that the termination criterion is as follows:

$$\max_{i,k} |u_{ik}(t+1) - u_{ik}(t)| \leq \varepsilon \tag{3.28}$$

The key components of the FCM and a quantification of their impact on the form of the produced results are summarized in Table 3.2.

The fuzzification coefficient has a direct impact on the geometry of fuzzy sets generated by the algorithm. Typically, the value of "m" is assumed to be equal to 2.0. Lower values of m (that

Table 3.2 The main features of the FCM clustering algorithm

Feature of the FCM algorithm	Representation and optimization aspects
Number of clusters (c)	Structure of the data set and the number of fuzzy sets estimated by the method; the increase in the number of clusters produces lower values of the objective function; however, given the semantics of fuzzy sets, one should keep this number quite low (5–9 information granules)
Objective function Q	Develops the structure aimed at the minimization of Q; iterative process supports the determination of the local minimum of Q
Distance function $\|.\|$	Reflects (or imposes) a geometry of the clusters one is looking for; essential design parameter affecting the shape of membership functions
Fuzzification coefficient (m)	Implies a certain shape of membership functions present in the partition matrix; essential design parameter. Low values of "m" (being close to 1.0) induce characteristic function. Values higher than 2.0 yield spiky membership functions
Termination criterion	Distance between partition matrices in two successive iterations; the algorithm terminates once the distance is below some assumed positive threshold (ε), that is, $\|U(\text{iter}+1) - U(\text{iter})\| < \varepsilon$

are closer to one) yield membership functions that start resembling characteristic functions of sets; most of the membership values become localized around one or zero. The increase of the fuzzification coefficient ($m = 3$, 4, etc.) produces "spiky" membership functions with the membership grades equal to one at the prototypes and a rapid decline of the values when moving away from the prototypes. Some illustrative examples of the membership functions are included in Figure 3.6. Here the prototypes are equal to 1, 3.5, and 5 while the fuzzification coefficient assumes values of 1.2 (a), 2.0 (b) and 3.5 (c). In addition to the varying shape of the membership functions, observe that the requirement put on the sum of membership grades imposed on the fuzzy sets yields some rippling effect: the membership functions are not unimodal but may exhibit some ripples whose intensity depends upon the distribution of the prototypes and the values of the fuzzification coefficient. The intensity of the rippling effect is also affected by the values of m and increases with its higher values.

The membership functions offer the interesting feature of evaluating the extent to which a certain data point is shared between different clusters and in this sense becomes difficult to allocate to a single cluster (fuzzy set). Let us introduce the following index, which serves as a certain separation measure:

$$\phi(u_1, u_2, \ldots, u_c) = 1 - c^c \prod_{i=1}^{c} u_i \tag{3.29}$$

where $u_1, u_2, \ldots, u_c$ are the membership degrees for some data point. If only one of the membership degrees, say $u_i = 1$, and the remainder are equal to zero, then the separation index attains its maximum equal to one. At the other extreme, when the data point is shared by all clusters to the same degree, being equal to $1/c$, then the value of this index is reduced

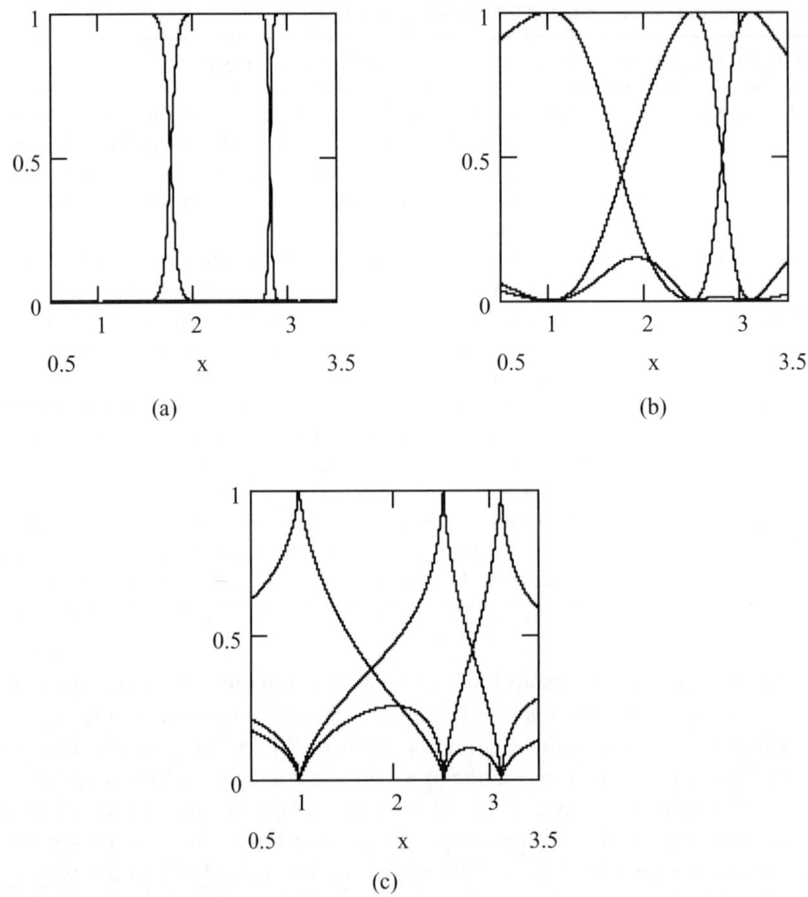

Figure 3.6 Examples of membership functions of fuzzy sets; see the detailed description in the text.

to zero. This means that there is no separation between the clusters as reported for this specific point.

It is worth emphasizing that the FCM algorithm is a highly representative method of membership estimation that profoundly dwells on the use of experimental data. In contrast to some other techniques presented so far that are also data driven, the FCM approach can easily cope with multivariable experimental data.

3.1.8 Fuzzy Equalization as a Way of Building Fuzzy Sets Supported by Experimental Evidence

The underlying principle of this approach is based on the observation that while fuzzy sets are reflective of the perception of systems or phenomena, quite often there is some experimental evidence in the form of data whose nature could be captured in a more synthetic manner through the underlying probability function or probability density function $p(x)$. The essence

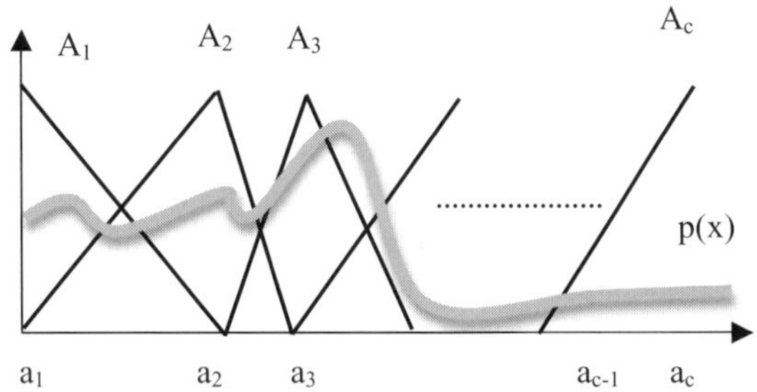

Figure 3.7 A collection of fuzzy sets complying with the equalization rule; note the increased specificity of fuzzy sets in the regions of high density ($p(x)$) of experimental data.

of fuzzy equalization can be articulated as follows. A collection of fuzzy sets $\{A_1, A_2, \ldots, A_c\}$ used to granulate some variable (say, inflation, profit, length, pressure, etc.) is formed in such a way that each fuzzy set in this family comes with the same level of experimental evidence. Put more formally, we consider that the integral (or sum) of the form

$$\int_{\mathbf{X}} A_i(x)p(x)\,\mathrm{d}x \tag{3.30}$$

assumes the same value for all fuzzy sets A_i, $i = 1, 2, \ldots, c$.

In other words, we require that the expected value expressed by (3.30) and computed for each fuzzy set is approximately the same, $\int_{\mathbf{X}} A_i(x)p(x)\,\mathrm{d}x = g$, where g is some constant. The essence of this construct is illustrated in Figure 3.7. This way of developing fuzzy sets is in agreement with our intuition: the less experimental evidence we have, the broader (less specific) the corresponding fuzzy set should be.

The underlying optimization task is concerned with determination of the parameters of the fuzzy sets (assuming that their form has been already specified) so that (3.26) becomes satisfied. For triangular fuzzy sets with a half-overlap between neighboring fuzzy sets, this optimization requires an adjustment of the vector of the modal values of the fuzzy sets.

Note that while fuzzy sets and probability are two orthogonal concepts, there are a number of methods of membership function estimation that invoke some probabilistic information (Dishkant, 1981; Civanlar and Trussell, 1986; Hong and Lee, 1996; Masson and Denoeux, 2006).

3.1.9 Several Design Guidelines for the Formation of Fuzzy Sets

The considerations presented above give rise to a number of general guidelines supporting the development of fuzzy sets:

(a) Highly visible and well-defined semantics of information granules. No matter what the determination technique is, one has to become familiar with the semantics of the resulting fuzzy sets. Fuzzy sets are interpretable information granules with a well-defined meaning and this aspect needs to be fully captured. Given this, the number of information granules has to be kept quite small, being restricted to 5–9 fuzzy sets.

(b) There are several fundamental views of fuzzy sets and, depending upon them, we could consider the use of various estimation techniques (for example, by accepting the horizontal or vertical view of fuzzy sets and adopting a pertinent technique).

(c) Fuzzy sets are context-sensitive constructs and as such require careful calibration. This feature of fuzzy sets should be treated as a genuine advantage. The semantics of fuzzy sets can be adjusted through shifting fuzzy sets or/and adjusting their membership functions. The nonlinear transformation we introduced here helps complete an effective adjustment of the membership functions, making use of some "standard" membership functions. The calibration mechanisms being used in the design of the membership function are reflective of the human centricity of fuzzy sets.

(d) We have delineated two major categories of approaches supporting the design of membership functions, that is, data-driven and expert (user)-based approaches. These are very different in the sense of the origin of the supporting evidence. Fuzzy clustering is a fundamental mechanism of the development of fuzzy sets. It is important in the sense that the method is equally suitable for one-dimensional and multivariable cases. The expert or simply user-based methods of membership estimation are important in that they offer some systematic and coherent mechanisms of elicitation of membership grades. With regard to consistency of the elicited membership grades, the pairwise estimation technique is of particular interest in providing well-quantifiable mechanisms for the assessment of the consistency of the produced membership grades. The estimation procedures underline some need for further development of higher types of constructs, such as fuzzy sets of type 2 or higher, and fuzzy sets of higher order that may be ultimately associated with constructs such as type 2 fuzzy sets or interval-valued fuzzy sets (this particular construct is visible when dealing with the horizontal method of membership estimation that comes with the associated confidence intervals).

(e) User-driven membership estimation uses the statistics of data but in an *implicit* manner. The granular terms – fuzzy sets come into existence once there is some experimental evidence behind them (otherwise there is no point forming such fuzzy sets).

(f) The development of fuzzy sets can be carried out in an stepwise manner. For instance, a certain fuzzy set can be further refined, if required for the problem at hand. This could lead to several more specific fuzzy sets that are associated with the fuzzy set formed at the higher level. Being aware of the complexity of the granular descriptors, we should resist the temptation to form an excessive number of fuzzy sets at a single level as such fuzzy sets could easily lack any sound interpretation.

3.2 Aggregation Operations

Several fuzzy sets can be combined together (aggregated) thus leading to a single fuzzy set forming the result of such an aggregation operation. For instance, when we compute the intersection and union of fuzzy sets, the result is a fuzzy set whose membership function captures information conveyed by the original fuzzy sets. This fact suggests a general view of aggregation of fuzzy sets as a certain transformation performed on their membership functions. In general, we encounter a wealth of aggregation operations (Dubois and Prade, 1980, Dubois and Prade, 1985).

Formally, an aggregation operation is an n-ary function $g : [0,1]^n \rightarrow [0,1]$ satisfying the following requirements:

$$\text{Monotonicity}: g(x_1, x_2, \ldots, x_n) > g(y_1, y_2, \ldots, y_n) \text{ if } x_i > y_j \quad (3.31)$$

$$\text{Boundary conditions}: g(0, 0, \ldots, 0) = 0 \text{ and } g(1, 1, \ldots, 1) = 1 \quad (3.32)$$

An element $e \in [0,1]$ is called a neutral element of the aggregation operation "g" and an element $l \in [0,1]$ is called an annihilator (absorbing element) of the aggregation operation "g" if, for each $i = 1, 2, \ldots, n$, $n \geq 2$ and for all $x_1, x_2, \ldots, x_{i-1}, x_{i+1}, \ldots, x_n \in [0,1]$ we have

$$g(x_1, x_2, \ldots, x_{i-1}, e, x_{i+1}, \ldots, x_n) = g(x_1, x_2, \ldots, x_{i-1}, x_{i+1}, \ldots, x_n) \quad (3.33)$$

$$g(x_1, x_2, \ldots, x_{i-1}, l, x_{i+1}, \ldots, x_n) = l \quad (3.34)$$

Since triangular norms and conorms are monotonic, associative, and satisfy the boundary conditions, they provide a wide class of associative aggregation operations whose neutral elements are equal to one and zero, respectively. We are, however, not restricted to those as the only available alternatives.

3.2.1 Averaging Operations

In addition to monotonicity and the satisfaction of the boundary conditions, averaging operations are idempotent and commutative They can be described in terms of the generalized mean (Dyckhoff and Pedrycz, 1984)

$$g(x_1, x_2, \ldots, x_n) = \sqrt[p]{\frac{1}{n}\sum_{i=1}^{n}(x_i)^p}, \quad p \in \mathbf{R},\ p \neq 0 \quad (3.35)$$

Interestingly, generalized mean subsumes some well-known cases of well-known averages including

$$p = 1 \quad g(x_1, x_2, \ldots, x_n) = \frac{1}{n}\sum_{i=1}^{n} x_i \quad \text{arithmetic mean}$$

$$p \rightarrow 0 \quad g(x_1, x_2, \ldots, x_n) = \sqrt[n]{\prod_{i=1}^{n} x_i} \quad \text{geometric mean}$$

$$p = -1 \quad g(x_1, x_2, \ldots, x_n) = \frac{n}{\sum_{i=1}^{n} 1/x_i} \quad \text{harmonic mean}$$

$$p \rightarrow -\infty \quad g(x_1, x_2, \ldots, x_n) = \min(x_1, x_2, \ldots, x_n) \quad \text{minimum}$$

$$p \rightarrow \infty \quad g(x_1, x_2, \ldots, x_n) = \max(x_1, x_2, \ldots, x_n) \quad \text{maximum}$$

The following inequalities hold:

$$\min(x_1, x_2, \ldots, x_n) \le g(x_1, x_2, \ldots, x_n) \le \max(x_1, x_2, \ldots, x_n) \quad (3.36)$$

Therefore generalized means range over the values not being covered by triangular norms and conorms.

3.3 Transformations of Fuzzy Sets

Transformations of elements (points) through functions are omnipresent. An immediate generalization of such point transformations involves set transformations between spaces. Mappings of fuzzy sets between universes constitute another generalization of mapping sets between spaces. Thus, point transformations can be expanded to cover transformations involving fuzzy sets. Transformations of this nature can be realized using either functions or relations. In both cases these transformations constitute an essential component of various pursuits including system modeling and control applications, pattern recognition, and information retrieval, just to name a few representative areas. This section introduces two important mechanisms to transform fuzzy sets, namely the extension principle and the calculus of fuzzy relations. We elaborate on their essential properties, present the algorithmic aspects, and discuss various interpretations of the resulting constructs.

3.3.1 The Extension Principle

The extension principle is a fundamental construct that enables extensions of point operations to operations involving sets and fuzzy sets. Intuitively, the idea is as follows: given a function (mapping) from some domain **X** to codomain (range) **Y**, the extension principle offers a mechanism to transform a fuzzy set defined in **X** to some fuzzy set defined in **Y**.

Let $f : \mathbf{X} \to \mathbf{Y}$ be a function. Given any $x \in \mathbf{X}$, $y = f(x)$ denotes the image of "x" under "f", that is, the point transformation of "x" under "f", Figure 3.8. This is the straightforward idea that the customary notion of any function conveys. Pointwise transformations can be naturally extended to handle transformations of sets.

Let $P(\mathbf{X})$ and $P(\mathbf{Y})$ be the power sets of **X** and **Y** and $A \in P(\mathbf{X})$ a set. The image of A under f can be determined by realizing point transformations $y = f(x)$ for all $x \in A$. In this sense, the image of A under f is some set B that arises in the following form:

$$B = f(A) = \{y \in \mathbf{Y} | y = f(x), \quad \forall x \in A\} \quad (3.37)$$

Since A and B are sets, they can be expressed in terms of their characteristic functions as follows:

$$B(y) = \sup_{x | y = f(x)} A(x) \quad (3.38)$$

as displayed in Figure 3.9. Note that this mechanism provides a way to extend the notion of functions regarded as point transformations to the notion of set functions. Once viewed in terms of characteristic functions, it is natural to extend this notion to fuzzy sets as follows.

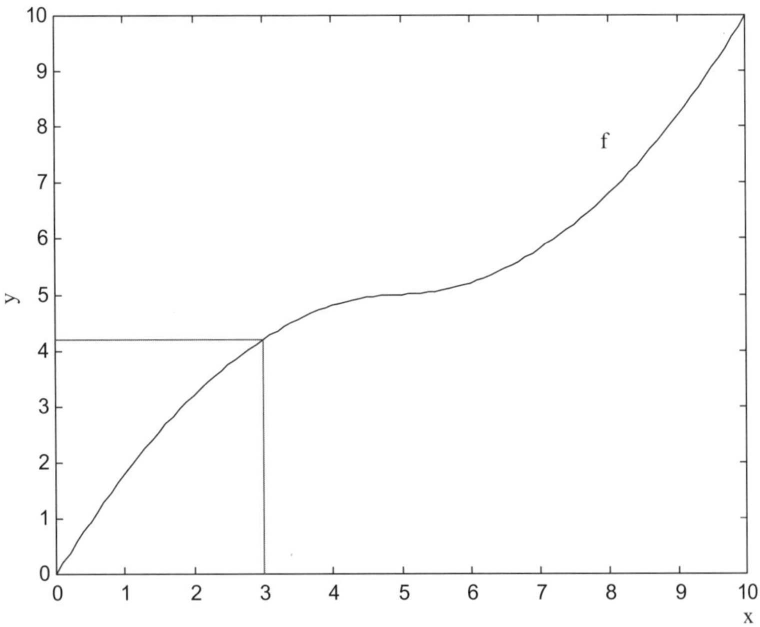

Figure 3.8 An example of function "*f*" along with its point transformation.

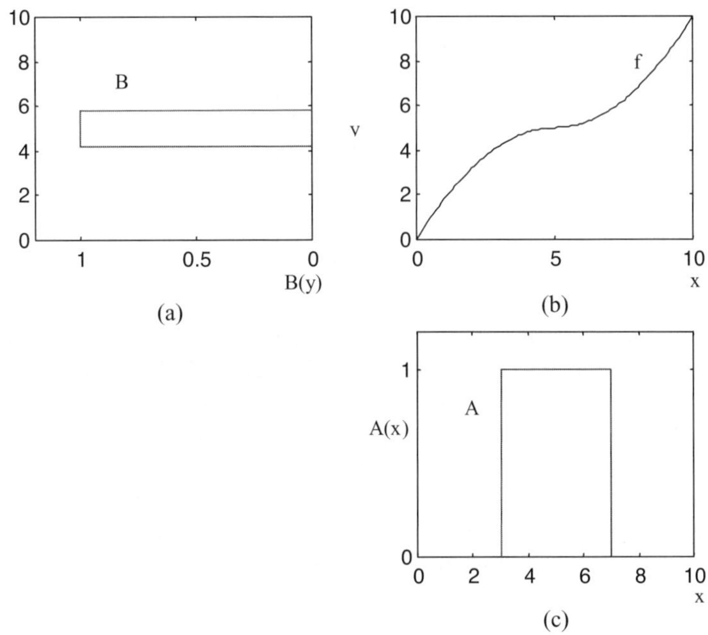

Figure 3.9 Set transformation.

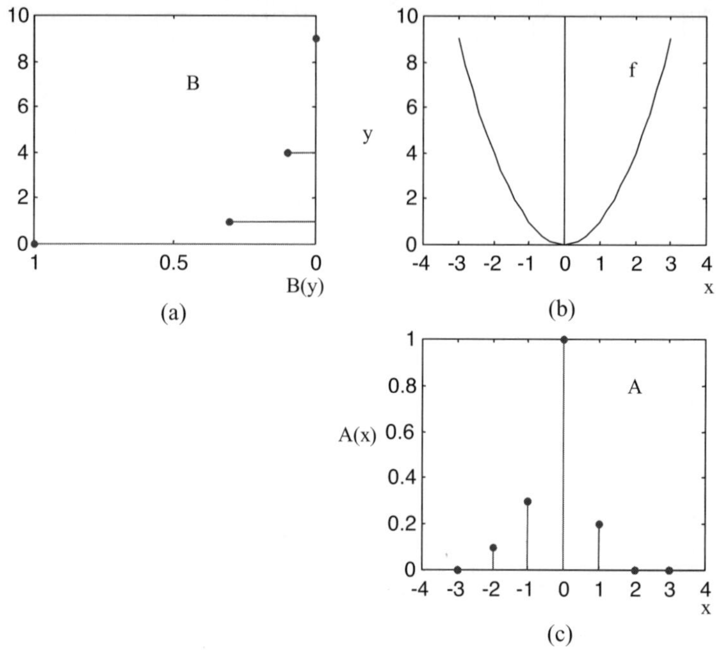

Figure 3.10 Extension principle applied in the case of a certain many-to-one mapping and finite universes.

Let $F(\mathbf{X})$ and $F(\mathbf{Y})$ denote the families of all fuzzy sets defined in $\mathbf{X}$ and $\mathbf{Y}$, respectively, and f: $\mathbf{X} \rightarrow \mathbf{Y}$ be a function. Function "f" induces a mapping $f : F(\mathbf{X}) \rightarrow F(\mathbf{Y})$ such that if A is a fuzzy set in $\mathbf{X}$, then its image under F is a fuzzy set $B = f(A)$ whose membership function is expressed as (Klir and Yuan, 1995; Pedrycz and Gomide, 1998)

$$B(y) = \sup_{x:y=f(x)} A(x) \tag{3.39}$$

For finite universes, consider $\mathbf{X} = \{-3, -2, -1, 0, 1, 2, 3\}$ and $y = f(x) = x^2$. Given the fuzzy set $A = \{0/-3, 0.1/-2, 0.3/-1, 1/0, 0.2/1, 0/2, 0/3\}$ defined in $\mathbf{X}$, the image $B = f(A)$ is a fuzzy set in $\mathbf{Y} = \{y \mid y = x^2\} = \{0, 1, 4, 9\}$ whose membership function is $B = \{1/0, \max(0.2, 0.3)/1, \max(0, 0.1)/4, 0/9\} = \{1/0, 0.3/1, 0.1/4, 0/9\}$, see Figure 3.10.

The extension principle generalizes to functions of many variables as follows. Let $\mathbf{X}_i$, $i = 1, \ldots, n$, and $\mathbf{Y}$ be universes and $\mathbf{X} = \mathbf{X}_1 \times \mathbf{X}_2 \times \cdots \times \mathbf{X}_n$. Consider fuzzy sets A_i on $\mathbf{X}_i$, $i = 1, \ldots, n$, and a function $y = f(\boldsymbol{x})$ with $\boldsymbol{x} = (x_1, x_2, \ldots, x_n)^{\mathrm{T}}$ a point of $\mathbf{X}$. Fuzzy sets A_1, $A_2, \ldots, A_n$ can be transformed through "f" giving rise to a fuzzy set $B = f(A_1, A_2, \ldots, A_n)$ in $\mathbf{Y}$ with the membership function

$$B(y) = \sup_{\mathbf{X} \mid y=f(\mathbf{X})} \min(A_1(x_1), A_2(x_2), \mathrm{K}, A_n(x_n)) \tag{3.40}$$

In (3.40), the min operation is a certain choice coming from the family of triangular norms. Any t-norm can be adopted because each component x_i occurs concurrently in $\boldsymbol{x}$.

3.3.2 Fuzzy Numbers and Fuzzy Arithmetic

We say a membership function $A : \mathbf{X} \rightarrow [0,1]$ is upper semi-continuous if the set $\{x \in \mathbf{X} \mid A(x) > \alpha\}$ is closed, that is, the α-cuts are closed intervals and, therefore, convex sets. If the universe $\mathbf{X}$ is the set $\mathbf{R}$ of real numbers and the membership function is normal, $A(x) = 1\ \forall x \in [b, c]$, then $A(x)$ is a model of a fuzzy interval, with monotone increasing function $f_A : [a, b) \rightarrow [0,1]$, monotone decreasing function g_A: $(c, d] \rightarrow [0,1]$, and null otherwise. Fuzzy intervals $A(x)$ have the following canonical form:

$$A(x) = \begin{cases} f_A(x) & \text{if} \quad x \in [a, b) \\ 1 & \text{if} \quad x \in [b, c] \\ g_A(x) & \text{if} \quad x \in (c, d] \\ 0 & \quad \text{otherwise} \end{cases} \tag{3.41}$$

where $a \leq b \leq c \leq d$, see Figure 3.11(a).

When $b = c$, $A(x) = 1$ for exactly one element of $\mathbf{X}$, and the fuzzy quantity is called a fuzzy number, Figure 3.11(b).

In general, the functions f_A and g_A are semi-continuous from the right and left, respectively. From a practical point of view, fuzzy intervals and numbers are mappings from the real line $\mathbf{R}$ to the unit interval that satisfy a series of properties such as normality, unimodality, continuity, and boundedness of support. As Figure 3.12 suggests, fuzzy intervals and numbers model our intuitive notion of approximate intervals and approximate numbers.

Before we move on to a discussion of operations on fuzzy numbers, let us introduce a few examples that motivate their use.

Consider that you have traveled for 2 hours at a speed of *about* 110 km/h. What was the distance you traveled? The speed is described in the form of some fuzzy set S whose membership function is given.

The next example is a more general version of the above problem.

You have traveled at a speed of *about* 110 km/h for *about* 3 hours. What was the distance traveled? We assume that both the speed and time of travel are described by fuzzy sets.

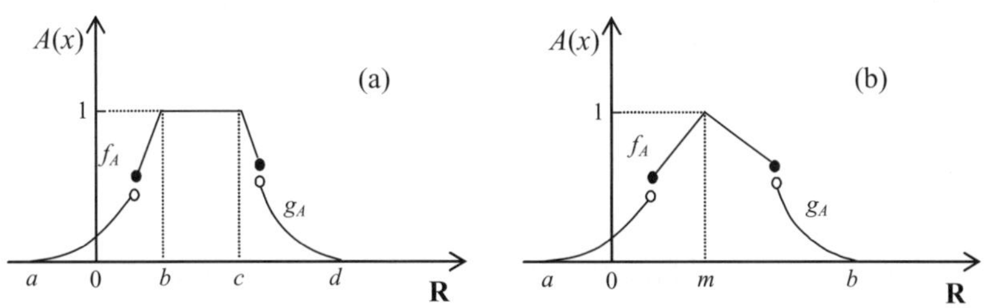

Figure 3.11 Canonical form of a fuzzy interval (a) and fuzzy number (b).

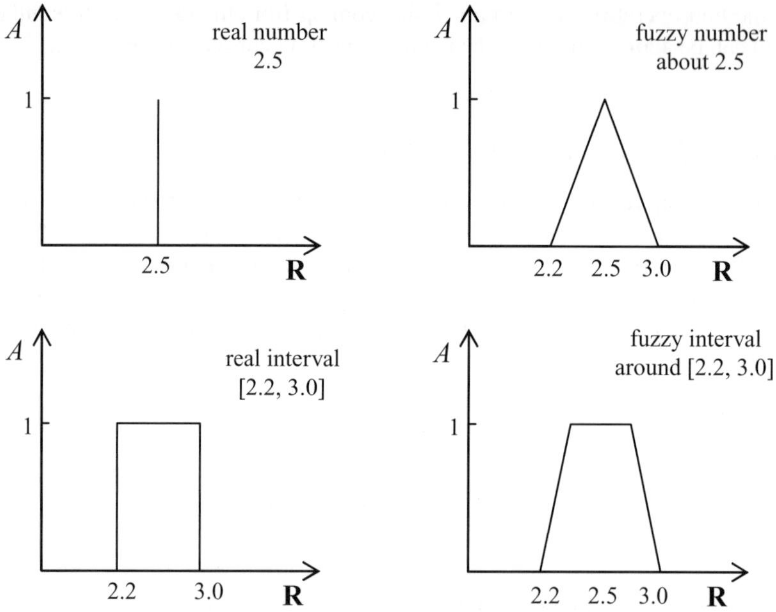

Figure 3.12 Examples of real numbers, fuzzy numbers, and intervals.

In a certain manufacturing process, there are five operations completed in series. Given the nature of the manufacturing activities, the duration of each of them can be characterized by fuzzy sets $T_1, T_2, \ldots, T_5$. What is the time of realization of this process?

Basically, there exist two fundamental methods for carrying out algebraic operations on fuzzy numbers. The first method is based on interval arithmetic and α-cuts while the second employs the extension principle. The fundamentals of these two methods are discussed next.

3.3.3 *Interval Arithmetic and α-cuts*

The first approach to computing with fuzzy numbers has its roots in the framework of interval analysis, a branch of mathematics developed to deal with the calculus of tolerances. In this framework, our interest lies in intervals of real numbers, $[a, b]$, $a, b \in \mathbf{R}$, such as [4,6], [−1.5,3.2], and so forth. The formulas developed to perform the basic arithmetic operations, namely addition, subtraction, multiplication, and division, are as follows (assuming that c, $d \neq 0$ for the division operation):

$$\text{Addition:}\quad [a,b]+[c,d]=[a+c,b+d] \tag{3.42}$$

$$\text{Subtraction:}\quad [a,b]-[c,d]=[a-d,b-c] \tag{3.43}$$

$$\text{Multiplication:}\quad [a,b].[c,d]=[\min(ac,ad,bc,bd),\max(ac,ad,bc,bd)] \tag{3.44}$$

$$\text{Division:}\quad [a,b]/[c,d]=\left[\min\left(\frac{a}{c},\frac{a}{d},\frac{b}{c},\frac{b}{d}\right),\max\left(\frac{a}{c},\frac{a}{d},\frac{b}{c},\frac{b}{d}\right)\right] \tag{3.45}$$

Now, let A and B be two fuzzy numbers and let $*$ be any of the four basic arithmetic operations. Thus, for any $\alpha \in (0,1]$, the fuzzy set $A * B$ is computed via the α-cuts A_α and B_α of A and B, respectively

$$(A * B)_\alpha = A_\alpha * B_\alpha \tag{3.46}$$

Recall that, by definition, the α-cuts A_α and B_α are closed intervals and therefore the formulas of interval operations can be applied for each value of α. When $*$ is / (the division operation), we require that $0 \notin B_\alpha \; \forall\alpha \in (0,1]$.

After the interval operation is performed for α-cuts, the use of the representation theorem leads us to the well-known relationship

$$A * B = \bigcup_{\alpha\in[0,1]} (A * B)_\alpha \tag{3.47}$$

In terms of the membership functions, we obtain

$$(A * B)(x) = \sup_{\alpha\in[0,1]} (\alpha(A * B)_\alpha(x)) = \sup_{\alpha\in[0,1]} ((A * B)_\alpha^f(x)) \tag{3.48}$$

where $(A * B)_\alpha^f(x) = \alpha(A * B)_\alpha(x)$.

Therefore, the interval arithmetic–α-cut method to perform fuzzy arithmetic is a generalization of computing known in interval arithmetic.

Example 3.6. If A and B are two triangular fuzzy numbers, denoted as $A(x, a, m, b)$ and $B(x, c, n, d)$, then their α-cuts are determined as

$$\begin{aligned} A_\alpha &= [(m-a)\alpha + a, (m-b)\alpha + b] \\ B_\alpha &= [(n-c)\alpha + c, (n-d)\alpha + d] \end{aligned}$$

Now let $A = A(x, 1, 2, 3)$ and $B = B(x, 2, 3, 4)$. Then, the corresponding α-cuts are equal to

$$\begin{aligned} A_\alpha &= [\alpha + 1, -\alpha + 3] \\ B_\alpha &= [\alpha + 2, -2\alpha + 5] \end{aligned}$$

Therefore

$$\begin{aligned} (A + B)_\alpha &= [2\alpha + 3, -3\alpha + 8] \\ (A - B)_\alpha &= [3\alpha - 4, -2\alpha + 1] \\ (AB)_\alpha &= [(\alpha + 1)(\alpha + 2), (-\alpha + 3)(-2\alpha + 5)] \\ (A/B)_\alpha &= [(\alpha + 1)/(-2\alpha + 5), (-\alpha + 3)(\alpha + 2)] \\ (A/B)_\alpha &= [(-\alpha + 3)(\alpha + 2), (\alpha + 1)/(-2\alpha + 5)] \end{aligned}$$

Figure 3.13 shows the resulting fuzzy numbers $A + B$, $A - B$, AB and B/A, respectively.

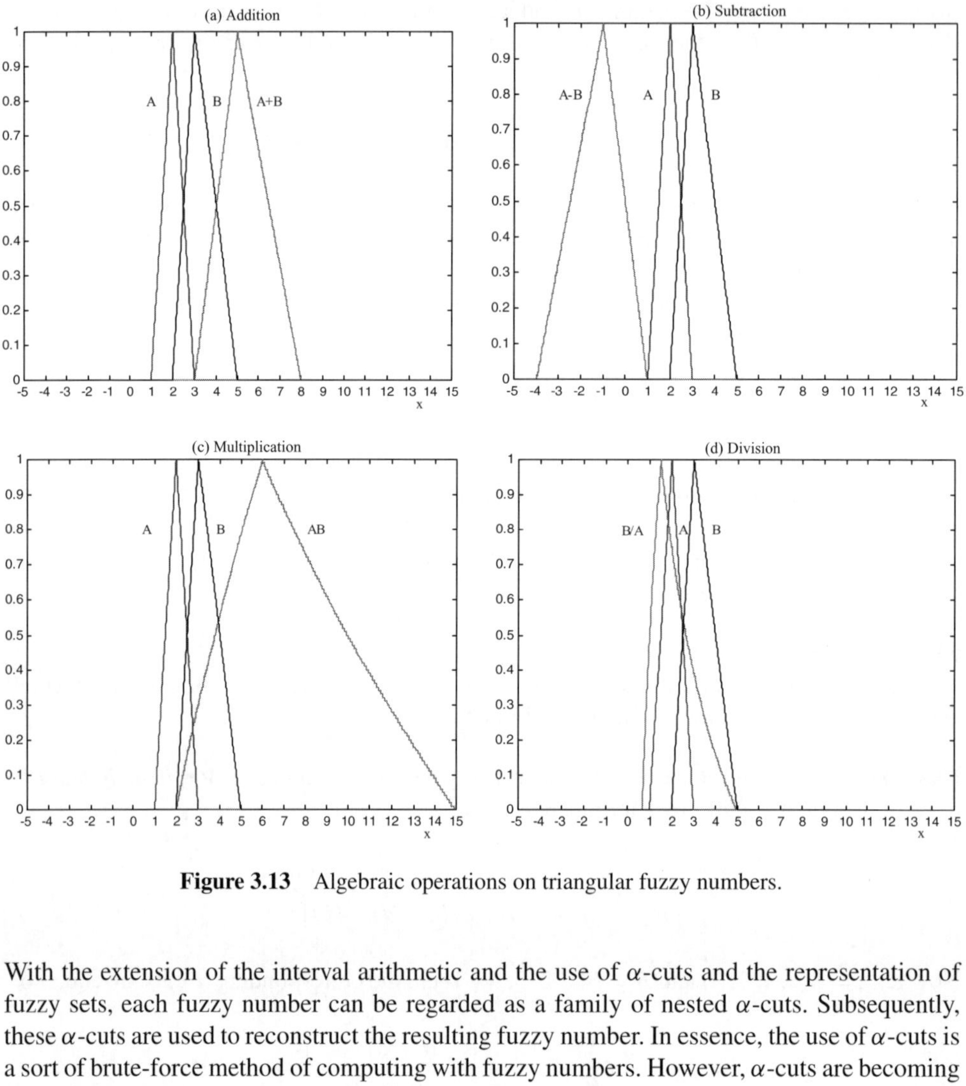

Figure 3.13 Algebraic operations on triangular fuzzy numbers.

With the extension of the interval arithmetic and the use of α-cuts and the representation of fuzzy sets, each fuzzy number can be regarded as a family of nested α-cuts. Subsequently, these α-cuts are used to reconstruct the resulting fuzzy number. In essence, the use of α-cuts is a sort of brute-force method of computing with fuzzy numbers. However, α-cuts are becoming important in developing parametric representations of fuzzy numbers to control their shapes and associated approximation error (Stefanini, Sorini, and Guerra, 2006).

3.3.4 *Fuzzy Arithmetic and the Extension Principle*

The second method of computing with fuzzy numbers uses the extension principle to extend standard operations on real numbers to fuzzy numbers. Here the fuzzy set $A * B$ expressed on $\mathbf{R}$ is defined using the extension principle

$$(A * B)(z) = \sup_{z=x*y} \min(A(x), B(y)), \quad \forall z \in \mathbf{R} \tag{3.49}$$

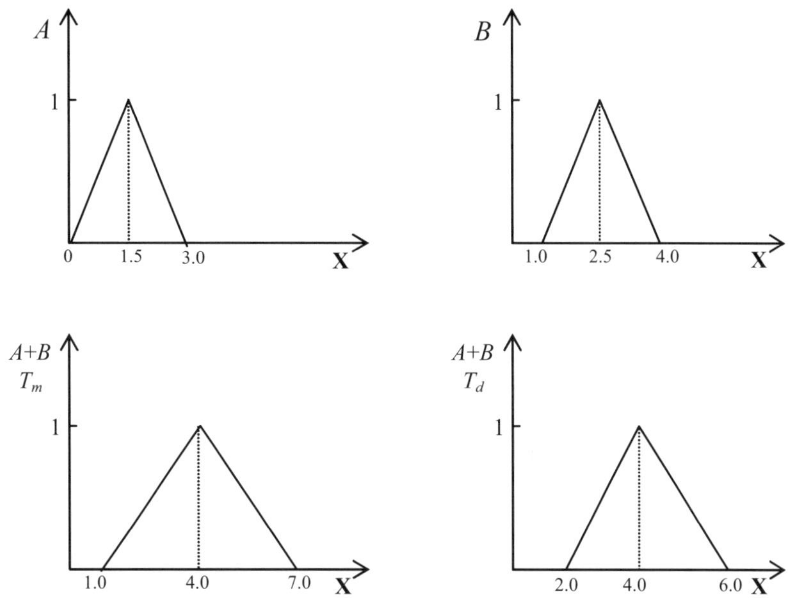

Figure 3.14 Algebraic operations: the use of the extension principle with different triangular norms.

In general, considering the use of some t-norm and treating $* : \mathbf{R}^2 \rightarrow \mathbf{R}$ as an operation on the real line, the operations on fuzzy numbers become

$$(A * B) = \sup_{z=x*y} (A(x)\,TB(y)), \quad \forall z \in \mathbf{R} \tag{3.50}$$

Figure 3.14 illustrates the addition $(A + B)$ of triangular fuzzy numbers A and B when using the minimum t-norm T_m and the drastic product T_d t-norm, respectively. Clearly, different choices of t-norms produce different results. In general, if $T_1 \leq T_2$ in the sense that $a\ T_1\ b \leq a\ T_2\ b, \forall a, b \in [0,1]$, then

$$\sup_{z=x*y} (A(x)\,T_d\,B(y)) \leq \sup_{z=x*y} (A(x)\,TB(y)) \leq \sup_{z=x*y} (A(x)\,T_m\,B(y)) \quad \forall z \in \mathbf{R} \tag{3.51}$$

Therefore

$$^{T_d}(A * B)(z) \leq {}^{T}(A * B)(z) \leq {}^{T_m}(A * B)(z) \forall z \in \mathbf{R} \tag{3.52}$$

In the special case of the largest t-norm, which is minimum, T_m, the one we will concentrate on in the remainder of this section, a fundamental result forming the basis of computing with fuzzy numbers under the framework of the extension principle comes in the following form.

Proposition: For any fuzzy numbers A and B and a continuous monotone binary operation $*$ on **R**, the following equality holds for all α-cuts with $\alpha \in [0,1]$:

$$(A * B)_\alpha = A_\alpha * B_\alpha \tag{3.53}$$

Proof of this proposition is given in Nguyen and Walker (1999). There are important consequences of the proposition:

1. Since A_α and B_α are closed and bounded for all α, $(A * B)_\alpha$ also is closed and bounded.
2. Because A and B are fuzzy numbers, they are normal and therefore $A * B$ is also normal.

These two observations clearly demonstrate that the extension principle produces a transformation that is a fuzzy number and therefore is a sound mechanism for performing algebraic operations with fuzzy numbers. Furthermore we have:

3. Computation of $A * B$ can be done by combining the increasing and decreasing parts of the membership functions of A and B.

Figure 3.15 offers a graphical visualization of the above statement.

The results above can be generalized to broader classes and choices of t-norms and operations with fuzzy quantities (Mares, 1997; Klement, Mesiar, and Pap, 2000; Stefanini, 2010). Moreover, approximation schemes have been developed in the framework of interpolation of a fuzzy function. In what follows we detail the basic operations with triangular fuzzy numbers

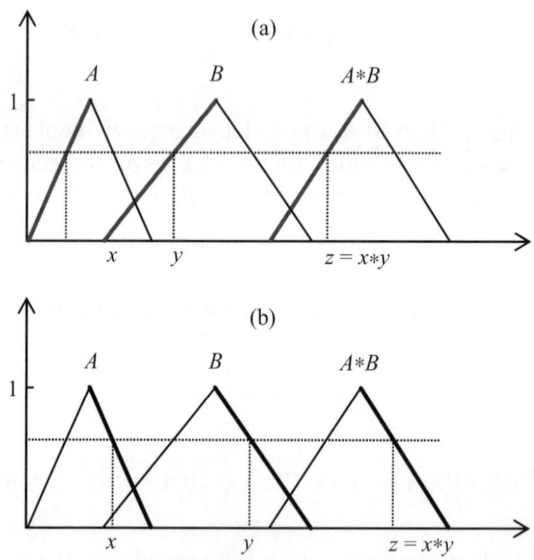

Figure 3.15 Combining increasing and decreasing parts of the membership functions of the fuzzy numbers A and B.

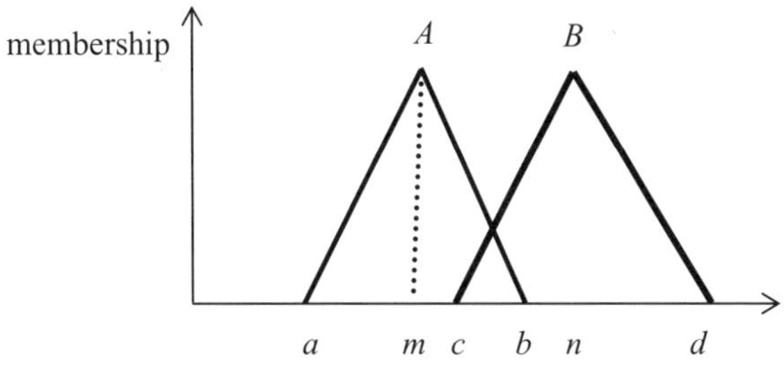

Figure 3.16 Examples of triangular fuzzy numbers A and B.

because they are by far the most commonly used in practice. Moreover, the analysis focusing this class of fuzzy numbers reveals the most visible properties of fuzzy arithmetic.

3.3.5 Computing with Triangular Fuzzy Numbers

Consider two triangular fuzzy numbers $A(x, a, m, b)$ and $B(x, c, n, d)$. More specifically, A and B are described by the following piecewise membership functions:

$$A(x) = \begin{cases} \dfrac{x-a}{m-a} & \text{if} \quad x \in [a, m) \\ \dfrac{b-x}{b-m} & \text{if} \quad x \in [m, b] \\ 0 & \text{otherwise} \end{cases} \qquad B(x) = \begin{cases} \dfrac{x-c}{n-c} & \text{if} \quad x \in [c, n) \\ \dfrac{d-x}{d-n} & \text{if} \quad x \in [n, d] \\ 0 & \text{otherwise} \end{cases} \tag{3.54}$$

Let us recall that the modal values m and n identify a dominant, typical value, while the lower and upper bounds, a or c and b or d, reflect the spread of the numbers. To simplify computing, for the time being we consider fuzzy numbers with positive lower bounds $a, c > 0$.

The plots of examples of triangular fuzzy numbers are given in Figure 3.16. These will be helpful in the clarification of detailed formulas.

3.3.6 Addition

The extension principle (3.49) applied to A and B to compute $C = A + B$ yields

$$C(z) = \sup_{z=x+y} \min(A(x), B(y)), \quad \forall z \in \mathbf{R} \tag{3.55}$$

The resulting fuzzy number is normal, that is, $C(z) = 1$ for $z = m + n$.

Computations of the spreads of C can be done, according to statement 3 above, by treating the increasing and decreasing parts of the membership functions of A and B separately.

Consider first that $z < m + n$. In this situation, the calculation involves the increasing parts of the membership function of A and B. Note that there exist values x and y such that $x < m$

and $y < n$ for which we have

$$A(x) = B(y) = \alpha \quad \alpha \in [0, 1] \tag{3.56}$$

Based on this relationship we derive

$$\frac{x - a}{m - a} = \alpha \tag{3.57}$$

along with

$$\frac{y - c}{n - c} = \alpha \tag{3.58}$$

for $x \in [a, m]$ and $y \in [c, n]$. Expressing x and y as functions of α we get

$$x = (m - \alpha)\alpha + a \tag{3.59}$$

$$y = (n - c)\alpha + c \tag{3.60}$$

which are the same as the lower intervals we get using interval analysis, as should happen.

Replacing the values of x and y in $z = x + y$ we have

$$z = x + y = (m - a)\alpha + a + (n - c)\alpha + c \tag{3.61}$$

that is,

$$\alpha = \frac{z - (a + c)}{(m + n) - (a + c)} \tag{3.62}$$

Notice that z has, as expected, the same lower limit value as the corresponding interval associated with the α-cut we use with interval analysis.

Proceeding similarly for the decreasing portions of the membership functions, we obtain

$$\frac{b - x}{b - m} = \alpha \tag{3.63}$$

along with

$$\frac{d - y}{d - n} = \alpha \tag{3.64}$$

for $x \in [m, b]$ and $y \in [n, d]$. Again, expressing x and y as functions of α we get

$$x = (m - b)\alpha + b \tag{3.65}$$

$$y = (n - d)\alpha + d \tag{3.66}$$

Furthermore, replacing the values of x and y in $z = x + y$ we have

$$z = x + y = (m - b)\alpha + b + (n - d)\alpha + d \tag{3.67}$$

that is,

$$\alpha = \frac{(b + d) - z}{(b + d) - (m + n)} \tag{3.68}$$

As expected, z has the same upper limit value as the corresponding interval associated with the α-cut we use with interval analysis.

Finally, from (3.62) and (3.67) we obtain the membership function of $C = A + B$:

$$C(z) = \begin{cases} \dfrac{z - (a + c)}{(m + n) - (a + c)} & \text{if} \quad z < m + n \\ 1 & \text{if} \quad z = m + n \\ \dfrac{(b + d) - z}{(b + d) - (m + n)} & \text{if} \quad z > m + n \end{cases} \tag{3.69}$$

Interestingly, C is also a triangular fuzzy number. To emphasize this fact, we use a concise notation

$$C = C(x, a + c, m + n, b + d) \tag{3.70}$$

Whenever several triangular fuzzy numbers are added, the result is also a triangular fuzzy number. In general, however, shape preserving does not hold for any shape of fuzzy numbers and t-norms being used in the extension principle.

3.3.7 Multiplication

As with addition, we look first at the increasing parts of the membership functions from which we obtain

$$x = (m - a)\alpha + a \tag{3.71}$$

$$y = (n - c)\alpha + c \tag{3.72}$$

The product z of x and y becomes

$$z = xy = [(m - a)\alpha + a][(n - c)\alpha + c] \tag{3.73}$$

$$z = (m - a)(n - c)\alpha^2 + (m - a)\alpha\, c + a(n - c)\alpha + ac = f_1(\alpha) \tag{3.74}$$

If $ac \leq z \leq mn$, then the membership function of the fuzzy number $D = AB$ is an inverse of the function $f_1(\alpha)$, namely

$$D(z) = f_1^{-1}(z) \tag{3.75}$$

Similarly, consider the decreasing parts of the fuzzy numbers A and B:

$$x = (m - b)\alpha + b \tag{3.76}$$

$$y = (n - d)\alpha + d \tag{3.77}$$

$$z = xy = [(m - b)\alpha + b][(n - d)\alpha + d] \tag{3.78}$$

$$z = (m - b)(n - d)\alpha^2 + (m - b)\alpha\, d + b(n - d)\alpha + bd = f_2(\alpha) \tag{3.79}$$

As before, for any $mn \leq z \leq bd$ we have

$$D(z) = f_2^{-1}(z) \tag{3.80}$$

Note that in this case the fuzzy number D is does not have a triangular membership function, which means that multiplication of triangular fuzzy numbers does not preserve the original shape. Instead, multiplication of piecewise linear membership functions produces a quadratic form of the membership function of the resulting fuzzy number.

3.3.8 Division

Like multiplication, for the increasing parts of the membership functions

$$x = (m - a)\alpha + a \tag{3.81}$$

$$y = (n - c)\alpha + c \tag{3.82}$$

we compute the division $z = x/y$ which, after replacing x and y, is

$$z = \frac{x}{y} = \frac{(m - a)\alpha + a}{(n - c)\alpha + c} = g_1(\alpha) \tag{3.83}$$

so that, for $a/c \leq z \leq m/n$, the fuzzy number $E = A/B$ has the following membership function:

$$E(z) = g_1^{-1}(\alpha) \tag{3.84}$$

Analogously, for the decreasing parts of the membership functions

$$x = (m - b)\alpha + b \tag{3.85}$$

$$y = (n - d)\alpha + d \tag{3.86}$$

we obtain

$$z = \frac{x}{y} = \frac{(m - b)\alpha + b}{(n - d)\alpha + d} = g_2(\alpha) \tag{3.87}$$

Thus, for $m/n \le z \le b/d$, the membership function of $E = A/B$ is

$$E(z) = g_2^{-1}(\alpha) \tag{3.88}$$

Clearly, the membership function of E is a rational function. Hence, division, like multiplication, does not preserve the shape of the triangular membership functions.

3.4 Conclusions

We have discussed various approaches and algorithmic aspects of the design of fuzzy sets. The estimation of membership functions is a multifaceted problem and the choice of a suitable method relies on the choice of the available experimental data and domain knowledge. For the user-driven approaches, it is essential to evaluate and flag the consistency of the results. While some of the methods (the pairwise comparison) come with this essential feature, the results produced by the others have to be carefully inspected.

Transformation of fuzzy sets in the form of the extension principle and composition generalizes similar transformations found in set theory. They play an important role in providing further transformations through fuzzy relational equations, associative memories, and algebraic operations with fuzzy numbers.

Fuzzy numbers are convex and normal fuzzy sets defined on the set of real numbers. Operations with fuzzy numbers can be developed with the help of the extension principle. In particular, standard fuzzy arithmetic can be approached choosing the min t-norm. Several other choices are possible, but practice has shown that standard fuzzy arithmetic still is one of the highest applicability.

Exercises

Problem 3.1. In the horizontal mode of construction of a fuzzy set of *safe* speeds on a highway, the yes–no evaluations provided by a panel of nine experts are the following:

x	20	50	70	80	90	100	110	120	130	140	150	160
No. of yes responses	0	1	1	2	6	9	9	5	5	4	3	2

Determine the membership function and assess its quality by computing the corresponding confidence intervals. Interpret the results and identify the points of the universe of discourse that may require more attention.

Problem 3.2. In the vertical mode of membership function estimation, we are provided with the following experimental data:

α	0.3	0.4	0.5	0.6	0.7	0.8	0.9	1.0
Range of **X**	[−2,13]	[−1,12]	[0,11]	[1,10]	[2,9]	[3,8]	[4,7]	[5,6]

Plot the estimated membership function and suggest its analytical expression.

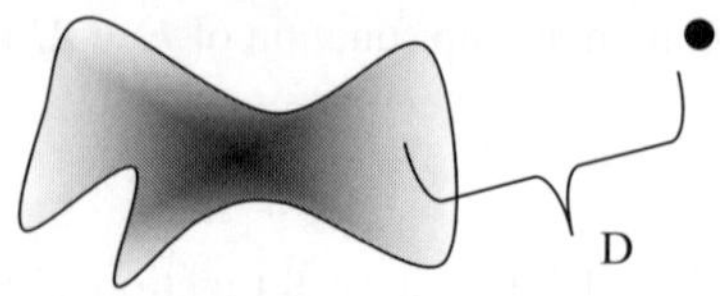

Figure 3.17 Forming a fuzzy set of distance between a geometric figure with fuzzy boundaries and a point.

Problem 3.3. In the calculations of the distance between a point and a certain geometric figure, we assumed that the boundaries of the figure are well defined. How can we proceed with a more general case when the boundaries are not clearly defined, that is, the figure itself is defined by some membership function, Figure 3.17? In other words, the figure is fully characterized by some membership function $R(\boldsymbol{x})$ where $\boldsymbol{x}$ is a vector of coordinates of x. If $R(\boldsymbol{x}) = 1$, the point fully belongs to the figure while lower values of $R(\boldsymbol{x})$ indicate that $\boldsymbol{x}$ is closer to the boundary of R.

Problem 3.4. Construct a fuzzy set describing the distance between the point of (5, 5) from the circle $x^2 + y^2 = 4$.

Problem 3.5. We maximize a function $f(x) = (x - 6)^4$ in the range of [3, 10]. Suggest a membership function describing a degree of membership of the optimal solution which minimizes $f(x)$. What conclusion could you draw based on the obtained form of the membership function?

Problem 3.6. The results of pairwise comparisons of four objects being realized on a scale of 1–5 are given in the following matrix form:

$$\begin{bmatrix} 1 & 5 & 2 & 4 \\ 1/5 & 1 & 3 & 1/3 \\ 1/2 & 1/3 & 1 & 1/5 \\ 1/4 & 3 & 5 & 1 \end{bmatrix}$$

What is the consistency of the findings? Evaluate the effect of the lack of transitivity. Determine the membership function of the corresponding fuzzy set.

Problem 3.7. In the method of pairwise comparisons, we use different scales involving various levels of evaluation, typically ranging from 5 to 9. What impact could the number of these levels have on the produced consistency of the results? Can you offer any guidelines on how to achieve a high consistency? What would be an associated tradeoff that one should take into consideration here?

Problem 3.8. Construct a fuzzy set of *large* numbers for the universe of discourse of integer numbers ranging from 1 to 10. It has been found that the experimental results of the pairwise comparison could be described in the form

$$r(x, y) = \begin{cases} x - y & \text{if} \quad x > y \\ 1 & \text{if} \quad x = y \end{cases}$$

(for $x < y$ we consider the reciprocal version of the above expression, that is, $1/(x - y)$).

Problem 3.9. In the fuzzy c-means (FCM) algorithm, the shape of the resulting membership function depends upon the value of the fuzzification coefficient (m). How does the mean value of the membership function relate to the values of "m". Run the FCM algorithm on the one-dimensional data set

$$\{1.3 \;\; 1.9 \;\; 2.0 \;\; 5.5 \;\; 4.9 \;\; 5.3 \;\; 4.5 - 1.3 \;\; 0.0 \;\; 0.3 \;\; 0.8 \;\; 5.1 \;\; 2.5 \;\; 2.4 \;\; 2.1 \;\; 1.7\}$$

considering that we have $c = 3$ clusters. Next, plot the relationship between the average of all membership grades and the associated fuzzification coefficient. For which values of "m" does the average of membership grades differ from 0.33 for less than δ? Consider several values of δ, say 0.2, 0.1, and 0.05. What can you say about the impact of "m" on the resulting average?

Problem 3.10. Consider a family of car makes, say $C_1, C_2, \ldots, C_n$. We are interested in forming fuzzy sets of economy, comfort, and safety, say A_{economy}, A_{comfort}, and A_{safety}. Use a method of pairwise comparison to build the corresponding fuzzy sets. Next, using the method of pairwise comparison, evaluate the car makes with respect to the overall quality (which involves economy, comfort, and safety). Given the already constructed fuzzy sets of the individual attributes and the overall quality A_{overall}, what relationship could you establish between them?

Problem 3.11. Consider a fuzzy set of a *safe* speed on an average highway, Figure 3.18.

How would this membership be affected when redefining this concept in the following settings of (a) an autobahn (note that on these German highways there is no speed limit) and (b) a snowy country road? Elaborate on the impact of various weather conditions on the corresponding membership function. From the standpoint of the elicitation of the membership function, how could you transform the original membership function to address the needs of

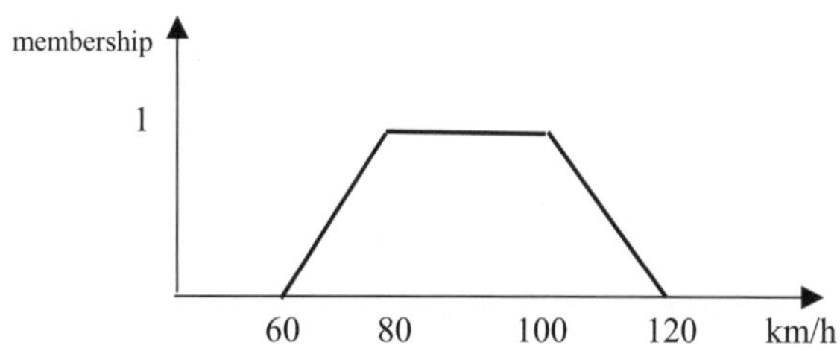

Figure 3.18 A fuzzy set of a *safe* speed on an average highway.

the specific context in which it is planned to be used?

$$G = \begin{bmatrix} 0.5 & 1.0 & 0.7 & 0.9 \\ 0.4 & 1.0 & 0.2 & 0.1 \\ 0.6 & 0.9 & 1.0 & 0.4 \end{bmatrix} \qquad W = \begin{bmatrix} 0.9 & 0.3 & 0.1 & 0.7 & 0.6 & 1.0 \\ 0.1 & 0.1 & 0.9 & 1.0 & 1.0 & 0.4 \\ 0.0 & 0.3 & 0.6 & 0.9 & 1.0 & 0.0 \\ 1.0 & 0.0 & 0.0 & 0.0 & 1.0 & 1.0 \end{bmatrix}$$

Problem 3.12. Consider $\mathbf{X} = \{1, 2, 3, 4\}$ and the fuzzy set $A = \{0.1/1, 0.2/2, 0.7/3, 1.0/4\}$ defined in this space. Also, let $\mathbf{Y} = \{1, 2, 3, 4, 5, 6\}$. Given a function f: $\mathbf{X} \to \mathbf{Y}$ such that $y = f(x) = x + 2$, show that $B = f(A) = \{0.1/3, 0.2/4, 0.7/5, 1.0/6\}$.

Problem 3.13. Determine the α-cuts of the fuzzy set A whose membership function is equal to

$$A(x) = \begin{cases} 2x - x^2 & \text{if} \quad 0 \le x \le 1 \\ 0 & \text{otherwise} \end{cases}$$

Let $f(x) = 2x - x^2$. Compute the image of the α-cuts of the fuzzy set A under "f". Sketch the transformations of the α-cuts.

Problem 3.14. Develop, analytically, the membership function of the fuzzy number F that is the subtraction of fuzzy numbers A and B, namely $F = A - B$.

Problem 3.15. Consider fuzzy numbers A and B whose membership functions are given in the form

$$A(x) = \begin{cases} \exp[-(x-m)^2/k] & a \le x \le b \\ 0 & \text{otherwise} \end{cases}$$

$$B(x) = \begin{cases} 0 & \text{if} \quad x \le a \\ \dfrac{x-a}{m-a} & \text{if} \quad x \in [a, m] \\ \dfrac{b-x}{b-m} & \text{if} \quad x \in [m, b] \\ 0 & \text{if} \quad x \ge b \end{cases}$$

Show that their α-cuts are given in the form

$$A_\alpha = \begin{cases} \left[m - \sqrt{\ln\left(\dfrac{1}{\alpha^k}\right)}, m - \sqrt{\ln\left(\dfrac{1}{\alpha^k}\right)}\right] & \text{if} \quad \alpha \ge \exp\left[-\left(\dfrac{-(a-m)^2}{k}\right)\right] \\ [a, b] & \text{if} \quad f\alpha < \exp\left[-\left(\dfrac{-(a-m)^2}{k}\right)\right] \end{cases}$$

and

$$B_\alpha = [(m-a)\alpha + a, (m-b)\alpha + b] \quad \forall \alpha \in [0, 1]$$

Sketch the membership functions of fuzzy sets of the addition, subtraction, multiplication, and division of A and B.

Problem 3.16. Are the parabolic fuzzy numbers $A, B, C\ldots$ whose membership functions come in the form

$$P(x, m, a) = \begin{cases} 1 - \left(\dfrac{x-m}{a}\right)^2 & \text{if} \quad x \in [m-a, m+a] \\ 0 & \text{otherwise} \end{cases}$$

closed under the addition operation? Justify your answer.

References

Bezdek, J. (1981) *Pattern Recognition with Fuzzy Objective Function Algorithms*, Plenum Press, New York.

Bortolan, G. and Pedrycz, W. (2002) An interactive framework for an analysis of ECG signals. *Artificial Intelligence in Medicine*, **24** (2), 109–132.

Chen, M. and Wang, S. (1999) Fuzzy clustering analysis for optimizing fuzzy membership functions. *Fuzzy Sets and Systems*, **103** (2), 239–254.

Chen, S.M. and Chen, J.H. (2009) Fuzzy risk analysis based on similarity measures between interval-valued fuzzy numbers and interval-valued fuzzy number arithmetic operators. *Expert Systems with Applications*, **36** (3), 6309–6317.

Civanlar, M. and Trussell, H. (1986) Constructing membership functions using statistical data. *Fuzzy Sets and Systems*, **18** (1), 1–13.

Dishkant, H. (1981) About membership functions estimation. *Fuzzy Sets and Systems*, **5** (2), 141–147.

Dombi, J. (1990) Membership function as an evaluation. *Fuzzy Sets and Systems*, **35** (1), 1–21.

Dubois, D. and Prade, H. (1980) *Fuzzy Sets and Systems: Theory and Applications*, Academic Press, New York.

Dubois, D. and Prade, H. (1985) A review of fuzzy set aggregation connectives. *Information Systems*, **36** (1–2), 85–121.

Dyckhoff, H. and Pedrycz, W. (1984) Generalized means as a model of compensative connectives. *Fuzzy Sets and Systems*, **14** (1), 143–154.

Hong, T. and Lee, C. (1996) Induction of fuzzy rules and membership functions from training examples. *Fuzzy Sets and Systems*, **84** (1), 389–404.

Klement, P. and Navara, M. (1999) A survey on different triangular norm-based fuzzy logics. *Fuzzy Sets and Systems*, **101** (2), 241–251.

Klement, P., Mesiar, R., and Pap, E. (2000) *Triangular Norms*, Kluwer, Dordrecht.

Klir, G. and Yuan, B. (1995) *Fuzzy Sets and Fuzzy Logic: Theory and Applications*, Prentice Hall, Upper Saddle River, NJ.

Kosko, B. (1992) *Neural Networks and Fuzzy Systems: A Dynamical Systems Approach to Machine Intelligence*, Prentice Hall, Upper Saddle River, NJ.

Kulak, O. and Kahraman, C. (2005) Fuzzy multi-attribute selection among transportation companies using axiomatic design and analytic hierarchy process. *Information Sciences*, **170** (2–4), 191–210.

Mares, M. (1997) Weak arithmetics of fuzzy numbers. *Fuzzy Sets and Systems*, **91** (2), 143–153.

Masson, M. and Denoeux, T. (2006) Inferring a possibility distribution from empirical data. *Fuzzy Sets and Systems*, **157** (3), 319–340.

Medaglia, A., Fang, S., Nuttle, H., and Wilson, J. (2002) An efficient and flexible mechanism for constructing membership functions. *European Journal of Operational Research*, **139** (1), 84–95.

Miller, G.A. (1956) The magical number seven plus or minus two: some limits of our capacity for processing information. *Psychological Review*, **63** (1), 81–97.

Nguyen, H. and Walker, E. (1999) *A First Course in Fuzzy Logic*, CRC Press, Boca Raton, FL.

Pedrycz, A., Dong, F., and Hirota, K. (2009) Finite α cut-based approximation of fuzzy sets and its evolutionary optimization. *Fuzzy Sets and Systems*, **160** (24), 3550–3564.

Pedrycz, W. and Gomide, F. (1998) *An Introduction to Fuzzy Sets: Analysis and Design*, MIT Press, Cambridge, MA.

Pedrycz, W. and Vukovich, G. (2002) On elicitation of membership functions. *IEEE Transactions on Systems, Man, and Cybernetics, Part A*, **32** (6), 761–767.

Saaty, T. (1980) *The Analytic Hierarchy Process*, McGraw-Hill, New York.

Saaty, T. (1986) Scaling the membership functions. *European Journal of Operational Research*, **25** (3), 320–329.

Saaty, T.L. (1986b) Axiomatic foundation of the analytic hierarchy process. *Management Science*, **32** (7), 841–855.

Schweizer, B. and Sklar, A. (1983) *Probabilistic Metric Spaces*, North-Holland, New York.

Stefanini, L. (2010) A generalization of Hukuhara difference and division for interval and fuzzy arithmetic. *Fuzzy Sets and Systems*, **161** (11), 1564–1584.

Stefanini, L., Sorin, L., and Guerra, M. (2006) Parametric representations of fuzzy numbers and application to fuzzy calculus. *Fuzzy Sets and Systems*, **157** (18), 2423–2455.

Turksen, I. (1991) Measurement of membership functions and their acquisition. *Fuzzy Sets and Systems*, **40** (1), 5–38.

van Laarhoven, P. and Pedrycz, W. (1983) A fuzzy extension of Saaty's priority theory. *Fuzzy Sets and Systems*, **11** (1–3), 199–227.

4

Continuous Models of Multicriteria Decision-Making and their Analysis

In this chapter, we concentrate on the construction, analysis, and application of continuous models of multicriteria decision-making (models of multiobjective decision-making). The basic definitions related to multicriteria decision-making are presented. The commonly utilized approaches to multiobjective decision-making are briefly described. A great deal of attention is given to the Bellman–Zadeh approach to decision-making in a fuzzy environment and its application to multicriteria problems. This approach can be regarded as a suitable conceptual and algorithmic method to develop harmonious solutions to this category of problems. The use of this approach is illustrated by solving problems coming from the multicriteria allocation of resources or their shortages as well as those problems encountered in power engineering.

4.1 Continuous Models ($\langle \mathbf{X}, \mathbf{M} \rangle$ Models) of Multicriteria Decision-Making

When solving continuous problems of multicriteria decision-making (problems of multiobjective decision-making) or analyzing $\langle \mathbf{X}, \mathbf{M} \rangle$ models, a set of objective functions $F(x) = \{F_1(x), F_2(x), \ldots, F_q(x)\}$ is considered while the problem itself calls for the simultaneous optimization of all objective functions, that is,

$$F_p(x) \rightarrow \underset{x \in \mathbf{L}}{\text{extr}}, \quad p = 1, 2, \ldots, q \tag{4.1}$$

where $q \geq 2$ and $\mathbf{L}$ is a set of feasible solutions in $\mathbf{R}^n$. Depending upon the nature of the problem under consideration, the term "extr" denotes the minimum or maximum.

As was stated in Section 1.2, in solving multicriteria problems we encounter the uncertainty of goals, which is difficult to overcome and handle because "we simply do not know what we want". This will be pondered upon in the following considerations.

From the point of view of the traditional (monocriteria) optimization, the problem formulated within the framework of the model (4.1), most likely, cannot be considered as being posed correctly. In particular, the notion of a complete optimal solution is considered by Sakawa (1993). More specifically, let us consider that all $F_p(x),\ p = 1, 2, \ldots,\ q$, are to be minimized. We say a point x^* is a complete optimal solution if and only if there exists $x^* \in \mathbf{L}$ such that $F_p(x^*) \leq F_p(x),\ p = 1, 2, \ldots,\ q$, for all $x \in \mathbf{L}$. Also, the terms *ideal* solution or *utopia* point are equivalent because, in general, a complete optimal solution that simultaneously minimizes (or maximizes) all objective functions does not exist if these objective functions conflict each other. Thus, in reality the scalar concept of an "optimum" cannot be applied directly and has to be redefined for multicriteria decision-making.

In terms of the analysis of the $\langle \mathbf{X}, \mathbf{M} \rangle$ models, the definition of optimality is not straightforward. The main difficulty comes from the presence of conflicting objective functions, where an improvement in the sense of one objective function may lead to a deterioration in other objective function(s). For example, the maximization of reliability of power supply can be reached by a satiation of electrical networks with switching devices. However, it leads to an increase of network costs, working against the objective to minimize costs. Tradeoffs exist between such conflicting objective functions, and the ultimate task is to find solutions that allow one to effectively balance these tradeoffs. Such a balance is achieved when a solution cannot improve any objective function without degrading one or more objective functions. These solutions are referred as nondominated solutions, efficient solutions, or Pareto-optimal solutions (Pareto, 1886), which were briefly discussed in Section 1.2 and are considered in more detail in the next section.

4.2 Pareto-Optimal Solutions

We provide some essential definitions (Hwang and Masud, 1979; Zeleny, 1982; Sakawa, 1993; Ehrgott, 2005) that are helpful when talking about the analysis of $\langle \mathbf{X}, \mathbf{M} \rangle$ models. To focus our considerations, we assume that the problem at hand requires minimization of the objective functions.

Domination: A solution x^1 dominates a solution x^2 if and only if:

- x^1 is not worse than x^2 in all objective functions, that is, $F_p(x^1) \leq F_p(x^2), \forall p = 1, 2, \ldots, q$;
- x^1 is strictly better than x^2 in at least one objective functions, that is, $\exists p = 1, 2, \ldots, q$: $F_p(x^1) < F_p(x^2)$.

Similarly, for the objective space, a solution $F^1(x)$ dominates another solution $F^2(x)$ if $F^1(x)$ is not worse than $F^2(x)$ in all values of objective functions, and $F^1(x)$ is better than $F^2(x)$ in at least one of the values of objective functions.

It is evident that that x^1 is better than x^2 (that is, x^1 dominates x^2), which happens when $F(x^1)$ dominates $F(x^2)$.

Weak domination: A solution x^1 weakly dominates a solution x^2 if and only if:

- x^1 is not worse than x^2 in all objective functions, that is, $F_p(x^1) \leq F_p(x^2), \forall p = 1, 2, \ldots, q$.

Pareto-optimal solution: A point x^* is a Pareto-optimal solution if there does not exist a solution $x \neq x^* \in \mathbf{L}$ that dominates it.

The Pareto-optimal solutions, as indicated above, are also named nondominated solutions and efficient solutions.

For the objective space, $F^*(x)$ is Pareto optimal if x is a Pareto-optimal solution.

All Pareto-optimal solutions form a Pareto-optimal solution set Ω^P (that has such a property that solutions $x \in \Omega^P$ cannot be simultaneously improved on all objective functions). The corresponding points in the objective space form a Pareto-optimal front Ω_F^P.

Weak Pareto-optimal solution: A point x^* is a weak Pareto-optimal solution if there does not exist a solution $x \neq x^* \in \mathbf{X}$ that weakly dominates it.

All weak Pareto-optimal solutions form a set of weak Pareto-optimal solutions Ω^{WP}. The corresponding points in the objective space form a Pareto-optimal frontier Ω_F^P.

Let Ω^{CO} denote a complete optimal solution set. Then, from the above definitions, we can construct the following relationships:

$$\Omega^{CO} \subseteq \Omega^P \subseteq \Omega^{WP} \tag{4.2}$$

In principle, the concept of the Pareto-optimal solution set is fundamental because a solution of a multicriteria decision-making problem must belong to this set. However, in general, its construction is usually a complicated and computationally cumbersome task. Diverse methods of building Pareto-optimal solution sets were discussed, for example, by Das and Dennis (1998), Deb (2001), Coelho, Van Veldhuizen, and Lamont (2002), and Statnikov and Matusov (2002).

In reality, the solution of multiobjective decision problems consists of several stages (Coelho, 2000). However, many researchers tend to concentrate on issues related to the construction of the Pareto-optimal solution sets, considering them as solutions to the multiobjective decision-making problems. Note that the Pareto-optimal solutions do not provide any insight into the process of decision-making itself (a DM still has to choose manually a final solution or a preferred solution), since they are really a useful generalization of a utility function under conditions of minimum information (that is, all objective functions are considered as having equal importance; in other words, a DM does not provide any preference for the objectives). Thus, the issue is how to incorporate DM preferences into the decision-making process.

Taking the above into account, it should be stressed that although the step of analyzing $\langle \mathbf{X}, \mathbf{M} \rangle$ models, associated with determining the Pareto-optimal solution set, is useful, it does not permit one to obtain unique solutions to real-world problems. It is necessary to choose a concrete Pareto-optimal solution through DMs' involvement in further information processing.

4.3 Approaches to the Use of DM Information

A possible way to classify approaches that help one to incorporate the information of a DM is based on the time (within the decision-making process) when this information is presented and applied. In accordance with this criterion, there exist three approaches: a priori, a posteriori, and an interactive one (for instance, Horn, 1997; Coelho, 2002).

If preferences are expressed a priori, a DM has to define them in advance (before actually realizing the decision process). In the procedures of *a priori* type, it is directly or indirectly assumed that all information, required to determine the most preferable solution, is incorporated into a formal model, that is, in the description of a set of alternatives and objective functions, and, consequently, can be extracted from the model by applying some transformations used in a constructive manner.

Procedures that are of *a posteriori* type are usually associated with the availability of some system of hypotheses or axioms, which are to be verified for each individual situation of decision-making. These hypotheses or axioms are considered as additional and are not included in the formal model. If the verification of the axioms leads to a positive result, it is possible to construct a convincing mode to choose the best alternative. This verification is associated with obtaining and applying additional information provided by a DM.

Although a priori and a posteriori decision-making approaches are common in the literature related to decision-making (Coelho, 2000), an *interactive* approach (it also uses additional information of a DM, however, as a progressive articulation of preferences) has been favored by researchers (Gardiner and Steuer, 1994) for several reasons, as discussed, for example, by Monarchi, Kisiel, and Duckstein (1973).

When applying the interactive approach, the procedures of successively improving the solution quality are realized as a transition from $x_\alpha^0 \in \Omega^P \subseteq \mathbf{L}$ to $x_{\alpha+1}^0 \in \Omega^P \subseteq \mathbf{L}$ by considering information I_α of a DM:

$$x_1^0, F(x_1^0) \xrightarrow{I_1} x_2^0 F(x_2^0) \xrightarrow{I_2} \cdots \xrightarrow{I_{\alpha-1}} x_\alpha^0, F(x_\alpha^0) \xrightarrow{I_\alpha} \cdots \xrightarrow{I_{\omega-1}} x_\omega^0, F(x_\omega^0) \tag{4.3}$$

The process (4.3) serves for two types of adaptation (taking this into account, the interactive approach itself is also called adaptive): computer to preferences of a DM and a DM to the problem. The first type of adaptation is based on information being received from a DM. The second type of adaptation is realized as a result of the steps $x_\alpha^0, F(x_\alpha^0) \xrightarrow{I_\alpha} x_{\alpha+1}^0, F(x_{\alpha+1}^0)$, which allow a DM to understand the relationships between his/her needs and the possibilities of their satisfaction by the model governed by (4.1). This explains the ability to construct sufficiently universal procedures of multiobjective decision-making. The types of such procedures are implied by the variety of applied forms of additional information representation. For instance, it is possible to distinguish the following forms:

- DM identifies significance of objective functions, that is, indicates proper assessment of weights of the criteria.
- DM identifies some desired levels of objective functions (goal values of objective functions, lower and/or higher admissible values of objective functions, admissible deviations from goal values of objective functions).
- DM compares sets of presented alternatives.
- Information provided by DM includes different combinations of the first three types of reports.

The taxonomy of the approaches that facilitates a way to incorporate information of a DM in the decision-making process is not complete and is relatively conditional. For instance, although many works in the field of multicriteria decision-making associate methods based on utility theory (Keeney and Raifa, 1976) with the a posteriori approach, some authors (Hwang and Masud, 1979; Lai and Hwang, 1996) link these methods to the a priori approach. Besides, there exist some decision-making procedures which cannot be uniquely related to one or another approach. Here we can refer to the mixed procedures (a priori and a posteriori, or a posteriori and interactive). Finally, the same method of the groups of multiobjective decision-making methods discussed in the next section can be realized within the framework of different approaches for incorporating information of a DM in the decision-making process. However, the construction of some taxonomy helps us to adequately represent the capabilities of diverse types of multicriteria decision-making methods as well as carefully identify their advantages and disadvantages.

4.4 Methods of Multiobjective Decision-Making

When formulating and solving multiobjective decision-making problems, it is necessary to develop answers to some specific questions. Among these questions, it is important to raise the following:

> **Normalization of objective functions**. In multiobjective decision-making problems, different objective functions may have different physical meaning and, consequently, are expressed in different units and their scales are not commensurable. Taking this into consideration, a comparison of the quality of obtained solutions for each objective function is impossible. The operation of unifying scales of objective functions to a unique scale is called normalization.
>
> **Choice of the principle of optimality**. In analyzing $\langle \mathbf{X}, \mathbf{M} \rangle$ models, the principle of optimality defines the properties of the optimal solution and answers in which sense the optimal solution excels all other possible solutions as well as offering guidelines on the search for optimal solutions. The principle of optimality is fundamental to multiobjective decision-making.
>
> **Consideration of priorities of objective functions**. Usually, considering the specificity of the problem, it becomes apparent that different objective functions have different importance, that is, one objective function has a higher priority relative to another one. It is intuitive to take this information into consideration in the choice of the principle of optimality, assigning higher priority to more important objective functions. In this regard, the following question arises: how do we define the formal description of the priority and the degree of its influence on the solution to the multiobjective problem?

The answers to the points posed above, and, subsequently, the development of multiobjective methods, were developed in different ways (Hwang and Masud, 1979; Zeleny 1982; Dubov, Travkin, and Yakimets, 1986; Lai and Hwang, 1996; Rao, 1996; Ehrgott, 2005). We identify those in common usage as:

- methods based on constructing convolutions;
- methods based on placing constraints on levels of objective functions, including a lexicographic method;
- methods of goal programming and of a global criterion.

The first group of attempts to solve multiobjective decision-making problems was based on their reduction to scalar (monocriteria) problems by constructing some types of convolutions (aggregations) (Kuhn and Tucker, 1951; Zadeh, 1963).

The simplest convolution reads as follows:

$$\Phi(x) = \sum_{p=1}^{q} f_p(x) \tag{4.4}$$

where $f_p(x) = F_p(x)$, $p = 1, 2, \ldots, q$, if all objective functions $F_p(x)$ are of a homogeneous character or exhibit the same semantics. However, if the objective functions have different meaning, it becomes impossible, as indicated above, to compare the quality of obtained solutions with respect to each objective function.

It is possible to identify several requirements related to the normalization of objective functions. However, the most important of them is the necessity to assign equal values to $f_p(x_p^0)$ or $f_p(x_p^{00})$ where $x_p^0 = \arg\min_{x\in\mathbf{L}} F_p(x)$ for minimized objective functions and $x_p^{00} = \arg\max_{x\in\mathbf{L}} F_p(x)$ for maximized objective functions. The inclusion of this requirement permits one to compare objective functions on the basis of their numeric values.

The above requirement is met by using the following normalization:

$$f_p(x) = \frac{\max\limits_{x\in\mathbf{L}} F_p(x) - F_p(x)}{\max\limits_{x\in\mathbf{L}} F_p(x) - \min\limits_{x\in\mathbf{L}} F_p(x)} \tag{4.5}$$

if the objective function $F_p(x)$ is to be minimized. If the objective function $F_p(x)$ is to be maximized, then we utilize the following version of normalization:

$$f_p(x) = \frac{F_p(x) - \min\limits_{x\in\mathbf{L}} F_p(x)}{\max\limits_{x\in\mathbf{L}} F_p(x) - \min\limits_{x\in\mathbf{L}} F_p(x)} \tag{4.6}$$

It becomes clear that in order to construct (4.5) or (4.6) for any objective function it is necessary to solve the following monocriteria problems:

$$F_p(x) \to \min_{x\in\mathbf{L}} \tag{4.7}$$

and

$$F_p(x) \to \max_{x\in\mathbf{L}} \tag{4.8}$$

Taking this into account, it is possible to apply another type of normalization

$$f_p(x) = \frac{\min\limits_{x\in\mathbf{L}} F_p(x)}{F_p(x)} \tag{4.9}$$

if the corresponding $F_p(x)$ is to be minimized, or to use the normalization

$$f_p(x) = \frac{F_p(x)}{\max\limits_{x\in\mathbf{L}} F_p(x)} \tag{4.10}$$

Table 4.1 Normalization of the objective function

$F_p(x)$	10	11	12	13	14	15	16	17	18	19	20
$f_p(x)$ (4.5)	1	0.9	0.8	0.7	0.6	0.5	0.4	0.3	0.2	0.1	0
$f_p(x)$ (4.6)	0	0.1	0.2	0.3	0.4	0.5	0.6	0.7	0.8	0.9	1
$f_p(x)$ (4.9)	1	0.91	0.83	0.77	0.71	0.67	0.63	0.59	0.56	0.53	0.50
$f_p(x)$ (4.10)	0.50	0.55	0.60	0.65	0.70	0.75	0.80	0.85	0.90	0.95	1

if the corresponding $F_p(x)$ is to be maximized.

The construction of (4.9) requires determining only the solution to the problem (4.7) while the construction of (4.10) requires finding the solution to the problem (4.8). However, although the construction of (4.9) and (4.10) is more rational from the computational point of view, the quality of normalized functions (4.9) and (4.10) may not always be acceptable. This is illustrated by the following example.

Example 4.1. Let us construct the normalization functions (4.5), (4.6), (4.9), and (4.10), when $\min_{x \in \mathbf{L}} F_p(x) = 10$ and $\max_{x \in \mathbf{L}} F_p(x) = 20$. The levels of $F_p(x)$ as well as the values of different $f_p(x)$ are shown in Table 4.1.

The data in Table 4.1 demonstrate that the use of (4.5) and (4.6) provides $0 \leq f_p(x) \leq 1$ and does not change the nature of $F_p(x)$. At the same time, the application of (4.9) and (4.10) leads to $0.5 \leq f_p(x) \leq 1$. Further, the use of (4.9) does not permit one to retain the essence of $F_p(x)$.

The use of Boldur's method (Roy, 1972) supports another way of normalization. In particular, we can assign "utilities" or "values" u'_p and u''_p to $F_p(x_p^0)$ and $F_p(x_p^{00})$, respectively, if $F_p(x)$ is to be minimized, or to $F_p(x_p^{00})$ and $F_p(x_p^0)$, respectively, if $F_p(x)$ is to be maximized. Then, if we accept linear interpolation, it is possible to construct the following aggregation:

$$\Phi(x) = \sum_{p=1}^{q} \left[\alpha_p F_p(x) + \beta_p\right] \tag{4.11}$$

where α_p and β_p can be determined by solving the following system of equations:

$$\begin{cases} \alpha_p F_p(x_p^0) + \beta_p = u'_p \\ \alpha_p F_p(x_p^{00}) + \beta_p = u''_p \end{cases} \tag{4.12}$$

if $F_p(x)$ is to be minimized, or solving the system of equations

$$\begin{cases} \alpha_p F_p(x_p^{00}) + \beta_p = u'_p \\ \alpha_p F_p(x_p^0) + \beta_p = u''_p \end{cases} \tag{4.13}$$

if $F_p(x)$ is to be maximized.

It is not difficult to verify that if the "utilities" or "values" $u'_p = 1$ and $u''_p = 0$, the aggregation (4.11) is reduced to the aggregation described by (4.5).

If we have to differentiate between the importance of different objective functions, it is possible to transform (4.4) into the form

$$\Phi(x) = \sum_{p=1}^{q} \lambda_p f_p(x) \tag{4.14}$$

where λ_p, $p = 1, 2, \ldots, q$, are weights, weighting factors, or importance factors, whose values reflect the relative importance of objective functions, which are to satisfy the following conditions:

$$\lambda_p \geq 0, \quad p = 1, 2, \ldots, q \tag{4.15}$$

and

$$\sum_{p=1}^{q} \lambda_p = 1 \tag{4.16}$$

The maximization of the convolution (4.14) is called the weighting function method (Lu *et al.*, 2008).

The application of the mappings (4.4) and (4.14), which have found wide practical applications, corresponds to the principle of uniform optimality (Lyapunov, 1972). At the same time, the application of aggregations of the type

$$\Phi(x) = \prod_{p=1}^{q} f_p(x) \tag{4.17}$$

as well as

$$\Phi(x) = \prod_{p=1}^{q} \lambda_p f_p(x) \tag{4.18}$$

or

$$\Phi(x) = \prod_{p=1}^{q} [f_p(x)]^{\lambda_p} \tag{4.19}$$

which have found some application, corresponds to the principle of just compromise.

The methods based on placing constraints at the level of objective functions are associated with specifying the desired levels for objective functions defined by the requirements of a DM (Benayoun *et al.*, 1971) and, then, by maximizing the convolutions, for example, those of the normalized objective functions (4.14).

The lexicographic method (Rao, 1996; Ehrgott, 2005) or a method of successive concessions (Podinovsky and Gavrilov, 1975) can be related to the group of placing constraints on the levels of objective functions.

In the lexicographic method, the objective functions are ranked in order of importance by a DM and are numbered according to this ranking. Taking this into account, assume that the most important objective function is $F_1(x)$ while $F_q(x)$ is the least important objective function. Further, for clarity of presentation, assume that all objective functions are to be minimized. Then, the original multiobjective problem can be replaced by a set of monocriteria problems. The first of them takes the following form:

$$F_1(x) \to \min_{x \in \mathbf{L}} \tag{4.20}$$

If x_1^0 is the solution to the problem (4.20), then the second problem

$$F_2(x) \to \min_{x \in \mathbf{L}} \tag{4.21}$$

is to be solved by taking into account the following additional constraint:

$$F_1(x_1^0) + \Delta F_1 \geq F_1(x) \tag{4.22}$$

where ΔF_1 is the concession on $F_1(x)$ to minimize $F_2(x)$.

If x_2^0 is the solution to the problem (4.21) and (4.22), then the concession ΔF_2 is set to minimize $F_3(x)$ and in this way obtain x_3^0. The process of setting the concessions is continued to obtain $x_4^0, x_5^0, \ldots, x_q^0$. Thus, the point x_p^0, $p = 1, 2, \ldots, q$, is considered as the solution of the original multiobjective problem.

The idea of goal programming was first presented by Charnes, Cooper, and Ferguson (1955), although the actual name first appears in Charnes and Cooper (1961). It was further developed by Lee (1972), Ignizio (1976), and Romero (1991), among others. An annotated bibliography on goal programming during the period 1990–2000 is presented in Jones and Tamiz (2002).

The goal programming method requires a DM to set goals for each objective that he/she wishes to attain. A preferred solution is then defined as the one that minimizes deviations from the set of goals. Thus, if some goals b_p, $p = 1, 2, \ldots, q$, are defined, then the problem of goal programming can be formulated as follows:

$$\text{minimize} \left[\sum_{p=1}^{q} \left(d_p^- + d_p^+ \right)^p \right]^{\frac{1}{p}}, \; p \geq 1 \tag{4.23}$$

subject to

$$x \in \mathbf{L} \tag{4.24}$$

$$F_p(x) + d_p^- - d_p^+ = b_p, \quad p = 1, 2, \ldots, q \tag{4.25}$$

$$d_p^- d_p^+ = 0, \quad p = 1, 2, \ldots, q \tag{4.26}$$

$$d_p^-, d_p^+ \geq 0, \quad p = 1, 2, \ldots, q \tag{4.27}$$

where b_p, $p = 1, 2, \ldots, q$, are the goals set by a DM for objectives (goals can be $F_p(x_p^0)$ for minimized objective functions or $F_p(x_p^{00})$ for maximized objective functions) and d_p^- and d_p^+ are the underachievement and overachievement of the pth goal, respectively. The value of p in (4.23) is based on the utility function chosen by a DM.

There exists a modification of goal programming, formulated within the framework of the model (4.23)–(4.27), called priority, pre-emptive, or lexicographic goal programming, when a DM, in addition to setting goals for objectives, is able to give an ordinal ranking of these objectives. It is evident that this modification should be used when there is a clear priority ordering among the goals to be achieved. If a DM is more interested in direct comparisons of the objectives then weighted goal programming can be used by introducing weights in (4.23).

In reality, it is possible to state that goal programming measures the distance to the goals by using the sum (the weighted sum) of absolute distances from given goals. Taking this into account, it is possible to mention the so-called global criteria method (Lai and Hwang, 1996), which differs from goal programming by the measurement of this distance using the Minkowski metric.

Without a comprehensive discussion of the strengths and weaknesses of the indicated group of methods (they are studied in detail by, for example, Hwang and Masud, 1979; Zeleny, 1982; Mashunin, 1986; Lai and Hwang, 1996; Ehrgott, 2005), it is necessary to indicate two fundamental weaknesses shared by all of them.

The first weakness is associated with the ability of methods based on placing constraints on levels of objective functions and methods of goal programming to produce solutions that are not Pareto optimal. This violates the basic concept of multicriteria decision-making. The second weakness is associated with the following considerations.

An important question in multicriteria decision-making is the quality of the solution itself. It is considered high if the levels of satisfying objectives are equal or close to each other (giving rise to so-called harmonious solutions) when the importance levels of the objective functions are equal (Ekel, 2001; Ekel and Galperin, 2003). It is not difficult to extend this concept for the case when the importance levels of the objective functions are different: the solutions are to be harmonious by taking into account the corresponding importance factors. From this point of view, the validity and advisability of the direction related to the principle of guaranteed results (Lyapunov, 1972) should be recorded. Other directions in multicriteria decision-making, in particular those indicated above, may lead to solutions with high levels of satisfaction of some criteria that are reached when assuring low levels of satisfaction of some other criteria. This situation could be completely unacceptable (for example, Ekel and Galperin, 2003; Canha *et al.*, 2007; Ekel, Menezes, and Schuffner Neto, 2007).

The lack of clarity on the concept of "optimal solution" arises from the basic methodological complexity when solving multicriteria problems. When applying the Bellman–Zadeh approach to decision-making in a fuzzy environment (Bellman and Zadeh, 1970) to solve multicriteria problems, this concept is defined with reasonable validity: the maximum degree of implementing goals serves as a criterion of optimality. This conforms to the principle of guaranteed results and provides constructive development lines in obtaining harmonious solutions. The Bellman–Zadeh approach helps one to realize an effective (from the computational standpoint) as well as rigorous (from the standpoint of obtaining solutions $x \in \Omega^P \subseteq \mathbf{L}$, at least, for convex Ω^P) method of analyzing multiobjective models. Finally, its use allows one to preserve a natural measure of uncertainty in decision-making and take into account the indices, criteria, and constraints of a qualitative character.

4.5 Bellman–Zadeh Approach and its Application to Multicriteria Decision-Making

When applying the Bellman–Zadeh approach to decision-making in a fuzzy environment for solving multicriteria problems, each objective function $F_p(x)$ can be replaced by a fuzzy objective function or a fuzzy set A_p. A fuzzy solution D with the setting up of fuzzy sets A_p is obtained as a result of the intersection $D = \cap_{p=1}^{q} A_p$ with a membership function

$$D(x) = \mathop{\wedge}_{p=1}^{q} A_p(x) = \min_{p=1,2,\ldots,q} A_p(x), \quad x \in \mathbf{L} \tag{4.28}$$

Its use allows us to obtain a solution providing the maximum degree of belongingness to the fuzzy solution D

$$\max D(x) = \max_{x \in \mathbf{L}} \min_{p=1,2,\ldots,q} A_p(x) \tag{4.29}$$

and reduces the problem (4.1) to a search for

$$x^0 = \arg \max_{x \in \mathbf{L}} \min_{p=1,2,\ldots,q} A_p(x) \tag{4.30}$$

To illustrate the use of (4.28)–(4.30), let us consider the following simple example.

Example 4.2. The membership functions of the three fuzzy objective functions $A_1(x)$, $A_2(x)$, and $A_3(x)$ are presented in Table 4.2.

Applying (4.28), we construct the fuzzy solution $D(x)$ presented in Table 4.3, which according to (4.30) gives rise to the solution $x^0 = 5$.

If we consider multiobjective problems, to obtain (4.30) it is necessary to construct the membership functions $A_p(x)$, $p = 1, 2, \ldots, q$, reflecting a degree of achieving their "own" optima by $F_p(x)$, $x \in \mathbf{L}$, $p = 1, 2, \ldots, q$. Taking into account the relationships (4.5) and (4.6), we apply the membership functions

$$A_p(x) = \left[\frac{\max_{x \in \mathbf{L}} F_p(x) - F_p(x)}{\max_{x \in \mathbf{L}} F_p(x) - \min_{x \in \mathbf{L}} F_p(x)} \right]^{\lambda_p} \tag{4.31}$$

Table 4.2 Membership functions of fuzzy objective functions

x	1	2	3	4	5	6	7	8	9	10
$A_1(x)$	0.1	0.2	0.8	1.0	0.9	0.7	0.5	0.3	0.2	0.1
$A_2(x)$	0.1	0.2	0.4	0.6	0.8	1.0	0.6	0.3	0.1	0.9
$A_3(x)$	0.4	0.6	1.0	0.9	0.7	0.6	0.5	0.4	0.3	0.2

Table 4.3 Membership function of a fuzzy solution

x	1	2	3	4	5	6	7	8	9	10
$D(x)$	0.1	0.2	0.4	0.6	0.7	0.6	0.5	0.3	0.1	0.1

for the minimized objective functions, or the membership functions

$$A_p(x) = \left[\frac{F_p(x) - \min_{x\in\mathbf{L}} F_p(x)}{\max_{x\in\mathbf{L}} F_p(x) - \min_{x\in\mathbf{L}} F_p(x)} \right]^{\lambda_p} \tag{4.32}$$

for the maximized objective functions.

Thus, the solution to problem (4.1) requires an analysis of $2q + 1$ monocriteria problems (4.7), (4.8), and (4.29), respectively.

Since the solution x^0 has to belong to $\Omega^P \subseteq \mathbf{L}$, it is necessary to construct

$$\bar{D}(x) = \bigwedge_{p=1}^{q} A_p(x) \wedge P(x) = \min\left\{ \min_{p=1,2,\ldots,q} A_p(x),\ P(x) \right\} \tag{4.33}$$

where $P(x) = \begin{cases} 1 & \text{if} \quad x \in \Omega^P \\ 0 & \text{if} \quad x \notin \Omega^P \end{cases}$.

The procedures for solving the problem (4.29), discussed in the next section, provide a way of obtaining $x^0 \in \Omega^P \subseteq \mathbf{L}$ according to (4.33). Thus, this says something about the equivalence of $\bar{D}(x)$ and $D(x)$. This line of thinking helps us to dispense with the need to implement a cumbersome procedure for building $\Omega^P \subseteq \mathbf{L}$.

Finally, the existence of additional conditions (indices, criteria, and/or constraints) of a qualitative character, defined by linguistic variables, reduces (4.30) to

$$x^0 = \arg\max_{X\in\mathbf{L}} \min_{p=1,2,\ldots,q+s} A_p(x) \tag{4.34}$$

where $A_p(x), x \in \mathbf{L}, p = q+1, q+2, \ldots, s$, are membership functions of fuzzy values of linguistic variables, which reflect the nature of these additional conditions.

There is some theoretical justification behind the validity of applying the min operator in (4.27)–(4.29); see, for example, Bellman and Giertz (1974). Taking this into account, it is necessary to note that there exist many families of aggregation operators, which may be used in place of the min operator. Consequently, it is possible to generalize (4.28) as follows:

$$D(x) = \text{agg}(A_1(x), A_2(x), \ldots, A_q(x)), \quad x \in \mathbf{L} \tag{4.35}$$

Despite the fact that some properties of these aggregation operators are well established, there is no clear and intuitively appealing interpretation of these properties. Likewise, there is a lack of a unifying interpretation of the operators themselves (Beliakov and Warren, 2001). An important question emerges: among the many types of aggregation operators, how do we select the one that is adequate for the particular problem at hand? Although some selection criteria were suggested in Zimmermann (1996), the majority of investigations is focused on choosing the operators on the basis of some available experimental evidence. Thus, it is possible to assert that the selection of the operators, to a significant extent, is experience-based. Considering this, we discuss below experiments not only showing the use of the min operator, but also involving the product operator (which can be considered as one of a family of t-norm operators, see for instance Yager (1988), and which has found a quite visible position in decision-making problems). Using the product, we reduce (4.28) to the following form:

$$D(x) = \prod_{p=1}^{q} A_p(x) \tag{4.36}$$

This leads to the expression

$$\max D(x) = \max_{x \in \mathbf{L}} \prod_{p=1}^{q} A_p(x) \tag{4.37}$$

from which we obtain

$$x^0 = \arg\max_{x \in \mathbf{L}} \prod_{p=1}^{q} A_p(x) \tag{4.38}$$

4.6 Multicriteria Resource Allocation

The problem of multicriteria allocation of resources or their shortages (these problems are equivalent from the mathematical and conceptual points of view) among consumers (departments, regions, projects, etc.) brings about the possibility to use diverse types of objective functions (linear, fractional, quadratic, etc., Ekel *et al.*, 1998) in the optimization problem expressed by (4.1) and defined in a feasible region

$$\mathbf{L} = \left\{ \boldsymbol{x} \in \mathbf{R}^n \,|\, 0 \le x_i \le B_i,\ \sum_{i=1}^{n} x_i = B \right\} \tag{4.39}$$

where $\boldsymbol{x} = [\,x_1\ x_2\ \dots\ x_n\,]$ is a vector of limitations (for the sake of our considerations) for consumers, B_i is a permissible value of limitation for the ith consumer, while B is a total value of limitations for all consumers considered in this planning or control problem.

To describe a general scheme for solving the problem formalized by the model (4.1) and (4.39), it is advisable to introduce a linguistic variable Q – *limitation for consumer* to provide a DM with the possibility of considering conditions that are difficult to formalize. Thus, the general scheme assumes the availability of a procedure for building a term set $T(Q)$ of the linguistic variable and membership functions for its fuzzy values. In addition, if the solution x_α^0 with the values $A_p(x_\alpha^0)$, $p = 1, 2, \dots, q$, is not satisfactory, a DM has to have the possibility to correct it, passing to $x^0_{\alpha+1}$ by changing the importance of one or more objective functions. Thus, in the general scheme we also assume the availability of the procedure for constructing and correcting a vector of importance factors $\Lambda = [\lambda_1\ \lambda_2\ \ldots\ \lambda_n]$.

The general scheme for solving the problem described by (4.1) and (4.39), which has been used for implementing an adaptive interactive decision-making system (AIDMS1) (to be described in the next section) is associated with the following sequence of steps:

1. Solution of the problems (4.7) and (4.8) in order to obtain x^0_p, $p = 1, 2, \ldots, q$, and x^{00}_p, $p = 1, 2, \ldots, q$, respectively.
2. Construction of the membership functions expressed by (4.31) or (4.32).
3. Construction of an initial vector of the importance factors $\Lambda = [\lambda_1\ \lambda_2\ \ldots\ \lambda_n]$.
4. Analysis of the availability of initial conditions defined by the linguistic variable. If these conditions are not available, then go to Step 8; otherwise, go to Step 5.
5. Verification of compatibility of the initial conditions and, if necessary, their correction.
6. Solving the problem (4.29) with the goal to obtain x^0_α defined by (4.34).
7. Analysis of the current solution x^0_α. If a DM is satisfied with this solution, then go to Step 10; otherwise, go to Step 8, taking $\alpha := \alpha + 1$.
8. Correction of the vector of the importance factors.
9. Insertion of additional conditions defined by the linguistic variables; then go to Step 5.
10. Completion of computing – the solution x^0 has been obtained.

The main functions of the calculating kernel of the AIDMS1 are associated with obtaining x^0_p, $p = 1, 2, \ldots, q$, and x^{00}_p, $p = 1, 2, \ldots, q$, which are produced by solving the problems (4.7) and (4.8) and by obtaining x^0 according to (4.34). The solution of (4.7) and (4.8) is rather straightforward. The maximization of (4.28) can be based on a nonlocal search that comes as a modification of the Gelfand and Tsetlin "long valley" method (Raskin, 1976).

Experimental evidence shows that variables in (4.28) can be divided into inessential and essential ones. The change of inessential variables leads to essential variations of (4.28). The change of essential variables leads to inessential variations of (4.28). Thus, a structure of (4.28) may be considered as a multidimensional "long valley". If we use direct search methods (Rao, 1996), this circumstance requires the ascent from different initial points x^0_p (Pareto points), if we minimize $F_p(x)$, or x^{00}_p (Pareto points), if we maximize $F_p(x)$, to find the most convincing solution x^0. This explains the use of a nonlocal search. The procedure for this search can be outlined as follows:

1. The sequence $\{x^{(l)}\}, l = 1, 2, \ldots, q$, is built by starting from points x^0_p, if we minimize $F_p(x)$, or x^{00}_p, if we maximize $F_p(x)$, obtained as a result of execution of Step 1 of the general scheme. This sequence satisfies the property $\min_{1\le p\le q} A_p(x^{(l)}) \geq \min_{1\le p\le q} A_p(x^{(l+1)})$, $l = 1, 2, \ldots, q-1$.
2. The local search for x^0 is carried out from $x^{(1)} (l = 1)$. As a result of this search, we obtain a point $x^{(1)0}$ with the corresponding $A_p(x^{(1)0})$, $p = 1, 2, \ldots, q$.
3. The local search for x^0 is carried out starting from $x^{(l+1)}$. As a result, we obtain a point $x^{(l+1)0}$ with the corresponding $A_p(x^{(l+1)0})$, $p = 1, 2, \ldots, q$.

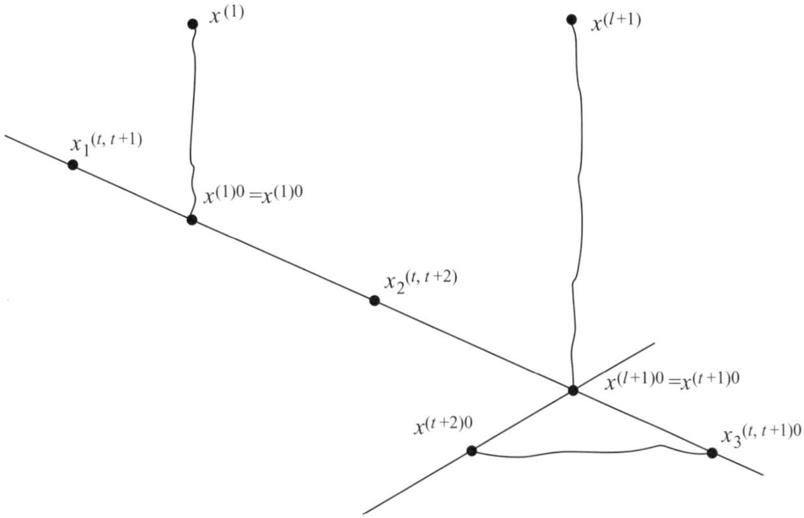

Figure 4.1 Nonlocal search for x^0.

4. The following examination is executed:
   (a) if $x^{(1)0} \neq x^{(l+1)0}$, then go to Operation 5;
   (b) if $x^{(1)0} = x^{(l+1)0}$ for $l \neq q - 1$, then go to Operation 3, by incrementing $l := l + 1$;
   (c) if $x^{(1)0} = x^{(l+1)0} = x^{(q)0}$, then go to Operation 8, by setting $x^0 = x^{(1)0}$.
5. A line between points $x^{(t)0}$ and $x^{(t+1)0}$ is formed to generate points $x_s^{(t,t+1)}$, $s = 1,\ 2,\ 3$ (see Figure 4.1). Among them (if they are acceptable from the point of view of the constraints (4.39)), a point $x^{(t,t+1)0} = \arg\max_t \min_{1\le p\le q} A_p(x_s^{(t,t+1)})$ is selected to define a direction for future search.
6. The next local search for x^0 is carried out starting from $x^{(t,t+1)0}$. As a result of this search, we obtain a point $x^{(t+2)0}$ (see Figure 4.1).
7. We execute the analysis: if three "last" points $x^{(t)0}$, $x^{(t+1)0}$, and $x^{(t+2)0}$ differ on $\min_{1\le p\le q} A_p(x^{(t)0})$, $\min_{1\le p\le q} A_p(x^{(t+1)0})$, and $\min_{1\le p\le q} A_p(x^{(t+2)0})$ less than the desired accuracy level, then go to Operation 8, taking $x^0 = \arg\max[\min_{1\le p\le q} A_p(x^{(t+2)})$, $\mathrm{minmin}_{1\le p\le q} A_p(x^{(t+2)})$, $\min_{1\le p\le q} A_p(x^{(t+2)})]$, otherwise go to Operation 5, taking $x^{(t)0} := x^{(t,t+1)0}$ and $x^{(t+1)0} := x^{(t+2)0}$.
8. Calculations are completed and the solution $x^0 \in \Omega^P \subseteq \mathbf{L}$ has been obtained.

The computing realized in Operations 2, 3, and 6 of the algorithm is possible by making use of any search method (in particular, within the framework of the AIDMS1, we used a modification of the univariate method; Rao, 1996). If $x^{(m)}$ is a current point, the transition to $x^{(m+1)}$ is expedient if

$$(\forall p = 1, 2, \ldots, q): A_p(x^{(m+1)}) \geq \min_{1\le p\le q} A_p(x^{(m)}) \tag{4.40}$$

In contrast, if

$$(\exists p = 1, 2, \ldots, q): A_p(x^{(m+1)}) < \min_{1 \le p \le q} A_p(x^{(m)}) \tag{4.41}$$

the transition to $x^{(m+1)}$ is not expedient from the point of view of maximizing (4.28). This way of evaluating the expediency of the transition to the next point $x^{(m+1)}$ leads to the solution (4.30) that is Pareto, if all inexpedient transitions are rejected.

The AIDMS1 includes a procedure for constructing and correcting the term set $T(Q)$ and membership functions for fuzzy values of the linguistic variable Q – *Limitation for consumer*. The initial term set available for a DM comes in the form $T(Q) = \langle$*near*, *approximately*, *slightly less*, *considerably less*, *slightly more*, and *considerably more*$\rangle$. The corresponding membership functions are defined in the form

$$A_p(x_i) = \exp[-k(R_i - x_i)^2] \tag{4.42}$$

$$A_p(x_i) = \begin{cases} 1 - \exp[-k(R_i - x_i)^2] & x_i \le R_i \\ 0 & x_i > R_i \end{cases} \tag{4.43}$$

$$A_p(x_i) = \begin{cases} 1 - \exp[-k(R_i - x_i)^2] & x_i \ge R_i \\ 0 & x_i < R_i \end{cases} \tag{4.44}$$

where k is a coefficient defined by the given accuracy of solution; R_i is a "specific value" which is related to the condition (for example, *considerably less* than R_i) that is to be taken into account.

The membership function described by (4.42) corresponds to the terms *near* and *approximately*, (4.43) to *slightly less* and *considerably less*, and (4.44) to *slightly more* and *considerably more*.

Furthermore, the AIDMS1 includes several procedures for forming and correcting the vector $\Lambda = [\lambda_1\ \lambda_2 \ldots \lambda_n]$ of importance factors. One of them is based on Saaty's pairwise comparison approach (Saaty, 1980) as discussed in Chapter 3.

4.7 Adaptive Interactive Decision-Making System for Multicriteria Resource Allocation

The AIDMS1 is used for solving the problem (4.1) and (4.39) for the case of linear objective functions. It has been developed in the C++ programming language. The software is run in a graphical environment of the Microsoft Windows operating system. Below, we present several typical windows that appear in the process of initial data preparation as well as a few windows that appear in the process of multicriteria resource shortage allocation.

An initial window (see Figure 4.2) is used to invoke the system by clicking on the Database option.

The Database window, Figure 4.3, is used to load information available in the database (by clicking on the Load button) or for preparing and storing input information (by clicking on the Save button). In the second case, Number of Functions (q), Number of Variables

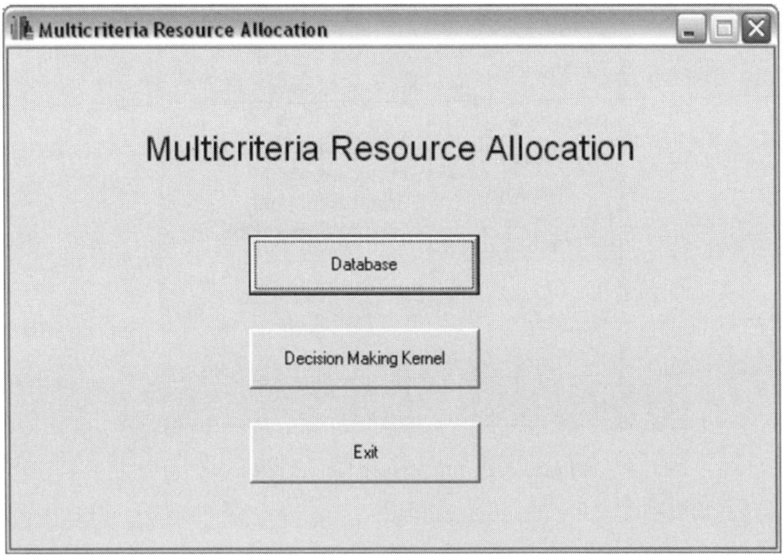

Figure 4.2 Initial window.

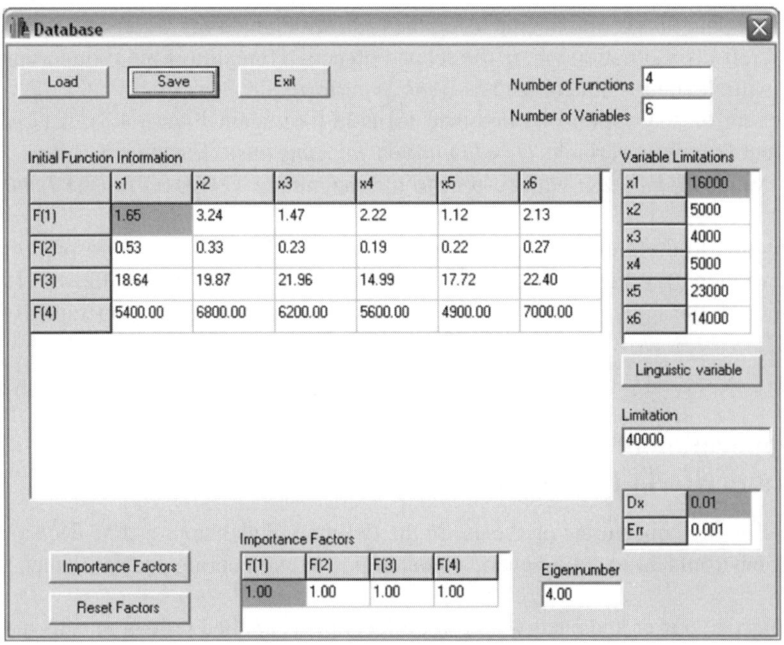

Figure 4.3 Database interface.

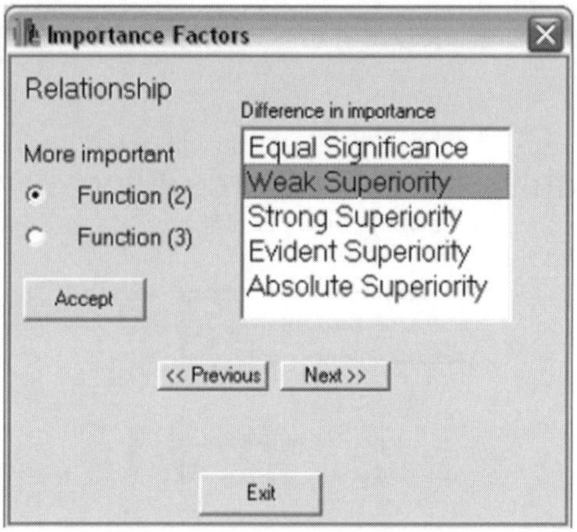

Figure 4.4 Importance factors interface.

(n), Initial Function Information $(c_{pi},\ p = 1, 2, \dots, q,\ i = 1, 2, \dots, n)$, Variable Limitations $(B_i,\ i = 1, 2, \dots, n)$, and Limitation (B) are defined. Further, the variable increment (D_X) and the desired accuracy (Err) are defined as well. The screenshot given in Figure 4.3 shows the input required in the example of multicriteria resource shortage allocation.

By clicking on the Importance Factors button (see Figure 4.4), one constructs or corrects the vector of importance factors by indicating which of two goals is more important and estimating the corresponding distinction degree using the rank scale given in Chapter 3. As an example, Figure 4.4 reflects a situation where the second objective function is more important then the third one with the rank characterized as *Weak Superiority*.

By clicking on the option of Linguistic variable (see again Figure 4.3), it is possible to consider the linguistic variable *Q – Limitation for consumer*. Figure 4.5 shows the use of the fuzzy value *slightly less* with respect to the magnitude 12 000.00 for the *Limitation for consumer 1*.

The solution to the problem of multicriteria resource shortage allocation presented in the Database window (see Figure 4.3), when using the min operator, is visualized in Figure 4.6. The solution to the same problem obtained with the use of the product operator is shown in Figure 4.7.

4.8 Application of the Bellman–Zadeh Approach to Multicriteria Problems

In this section, we concentrate on the use of the Bellman–Zadeh approach to decision-making in a fuzzy environment to solve several essential power engineering problems such as:

- multiobjective power and energy shortage allocation as applied to load management;
- multicriteria power system operation to realize dispatch on several objectives;
- multicriteria optimization of network configuration in distribution systems;
- energetically effective (bicriteria) voltage control in distribution systems.

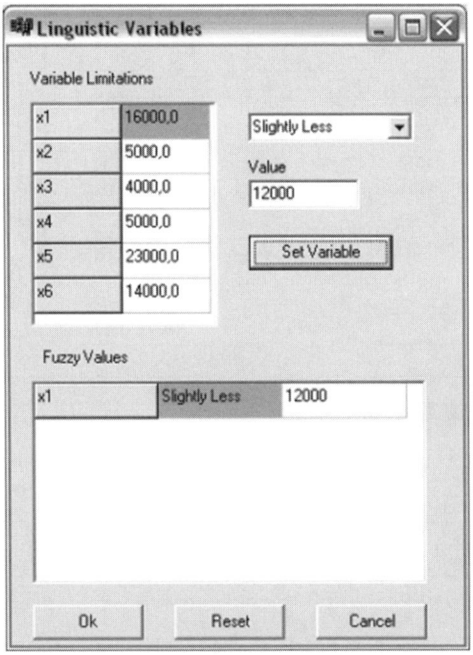

Figure 4.5 Linguistic variables interface.

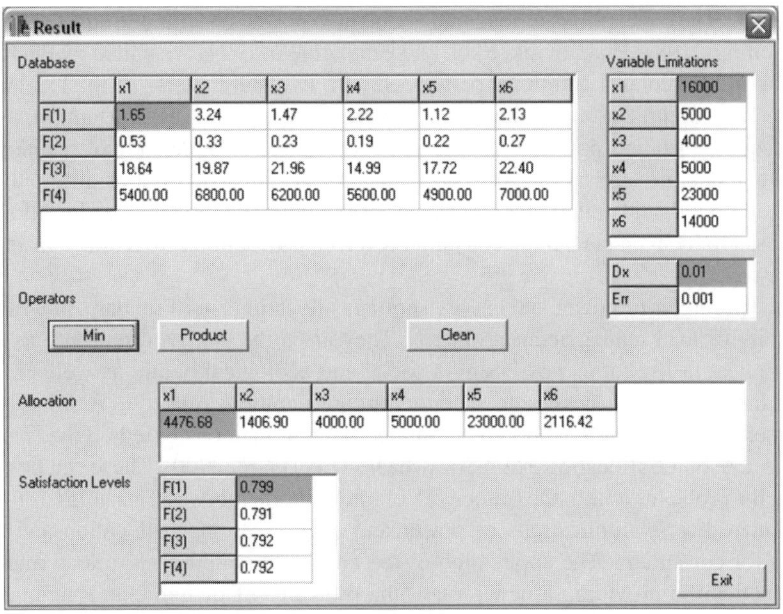

Figure 4.6 Execution interface when applying the min operator.

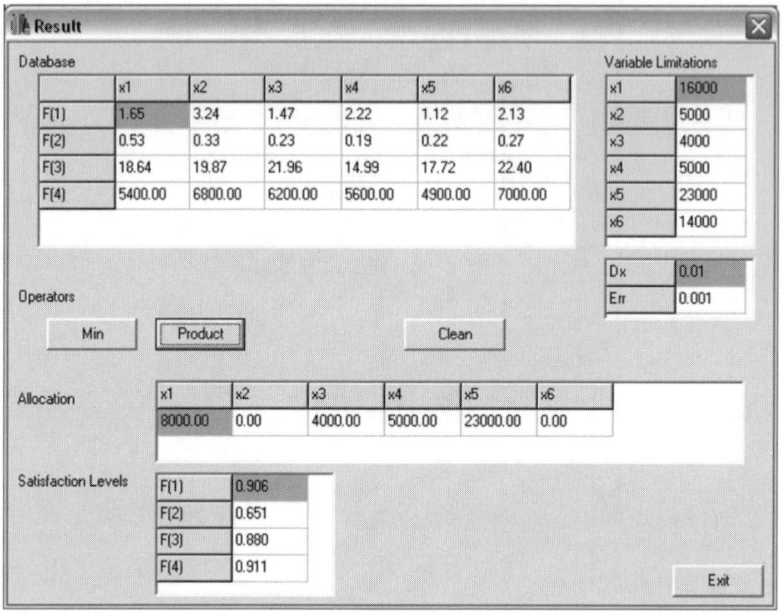

Figure 4.7 Execution interface when applying the product operator.

Example 4.3. The conceptions of load management in power systems and subsystems (Talukdar and Gellings, 1987; Prakhovnik, Ekel, and Bondarenko, 1994) are united by the following. The elaboration of control actions is performed on a two-stage basis. At the level of energy control centers, optimization of allocating power and energy shortages (natural, associated with inadequate installed power of generating sources and/or deficiency of primary energy, or with the economic advisability of load management) is carried out at different levels of territorial, temporal, and situational hierarchies of planning and operation. This allows one to draw up tasks for consumers. At the consumer level, control actions are realized in accordance with these tasks.

Thus, the problems of power and energy shortage allocation are of fundamental importance in the family of load management problems. They are to be analyzed not only as technical and economical tasks, but as possessing a social and ecological nature as well. In addition, when solving them, it is necessary to account for considerations related to forming incentives to consumers. Consequently, it should be pointed out that methods based on the fundamental principles of resource allocation exhibit drawbacks (Ekel *et al.*, 2000). These can be overcome by casting the problems within the framework of multicriteria models. This helps us to consider and minimize diverse implications of power and energy shortage allocation and to create incentives for consumers. The application of the multicriteria approach to load management is also beneficial in providing a new look at the problems of planning and the operation of electricity markets (Stoft, 2002). In particular, all market participants aspire to maximize their benefits (economical, technological, social, political, etc.). The goals of market participants, as a rule, come into conflict, which may be resolved by searching for a corresponding compromise.

Its objective is the formation of mutually advantageous and harmonious relations between the market participants.

The substantial analysis of problems of power and energy shortage allocation, systems of economic management, including taxation policy, as well as readily available reported information, has led to the construction of a general set of goals to solve these problems in a multicriteria statement. The complete list includes 17 types of goals, some of which are given below:

1. Primary limitation of consumers with a lower cost of production or given services on consumed 1 kWh of energy (achievement of a minimal drop in total production and/or given services).

15. Primary limitation of consumers with a lower value of the demand coefficient (primary limitation of consumers with greater possibilities of production outside of the peak time).

16. Primary limitation of consumers with a lower duration of using maximum load in 24 hours (primary limitation of consumers with greater possibilities in transferring maximum load in the daily interval).

17. Primary limitation of consumers with a lower duration of using maximum load in month (quarter, year) (primary limitation of consumers with greater possibilities in transferring maximum load in the month (quarter, year) interval).

The general set of goals is sufficiently complete because it is directed at decreasing diverse negative consequences for consumers and creating incentives for them. This set of goals is universal because it can serve as the basis for building models at different levels of the load management hierarchy by aggregating information and posterior decomposition of the problems in accordance with different indices. The corresponding list of goals can be defined for every case by a DM, who can be an individual or a group (for example, the DM may be the leading organizations of a country or state, a council of directors of enterprises, and so on, whose decision regarding the list of goals can be considered as the legislative one at the corresponding level)

Consider the solution to the problems of power shortage allocation formalized within the framework of the model (4.1) and (4.39) for six consumers for $B^1 = 40\,000$ kW and $B^2 = 60\,000$ kW when using the min operator. For comparison, we also use the product operator as well as Boldur's method considered in Section 4.4. The goals listed above are taken into consideration and described by the linear objective functions

$$F_{p1}(x) = \sum_{i=1}^{6} c_{pi} x_i, \quad p = 1, 15, 16, 17 \tag{4.45}$$

that are to be minimized. Here x_i, $i = 1, 2, \ldots, 6$, are limitations of power supply for consumers. The coefficients c_{pi}, $p = 1, 15, 16, 17$, $i = 1, 2, \ldots, 6$, are determined by specific characteristics of consumers. Table 4.4 provides the initial information for $B^1 = 40\,000$ kW as well as for $B^2 = 60\,000$ kW. This information corresponds to input information given in Figure 4.3.

Table 4.4 Initial information

I	1	2	3	4	5	6
$c_{1,i}$, monetary units (kWh)	1.65	3.24	1.47	2.22	1.12	2.13
$c_{15,i}$	0.53	0.33	0.23	0.19	0.22	0.27
$c_{16,i}$ (h)	18.64	19.87	21.96	14.99	17.72	22.40
$c_{17,i}$ (h)	5400	6800	6200	5600	4900	7000
A_i (kW)	16 000	5000	4000	5000	23 000	14 000

The obtained results for the min operator (x^0), product operator (x^{00}) as well as Boldur's method (x^{000}) are presented in Table 4.5 and Table 4.6. The solutions obtained when using the min and product operators for $B^1 = 40\,000$ kW are also presented in Figures 4.6 and 4.7.

To reflect the quality of solutions obtained on the basis of different approaches, Table 4.7 includes the mean magnitudes of absolute values $\Delta(x)$ of deviations of membership function levels (satisfaction levels) $A_p(x)$ from their mean values $\hat{A}_p(x)$ which are calculated as follows:

$$\Delta(x) = \frac{1}{4}\sum_{i=1}^{4}\left|A_p(x) - \hat{A}_p(x)\right| \tag{4.46}$$

where

$$\hat{A}_p(x) = \frac{1}{4}\sum_{i=1}^{4} A_p(x) \tag{4.47}$$

The results shown in Table 4.7 demonstrate that $x^0 \succ x^{00}$ and $x^0 \succ x^{000}$. The high quality of the solutions x^0 is also confirmed by inequalities $\min_p A_p(x^0) > \min_p A_p(x^{00})$ and $\min_p A_p(x^0) > \min_p A_p(x^{000})$, which are observed for both cases. This confirms the validity of the use of the principle of guaranteed results based on applying the min operator with the most noncompensatory behavior (Yager, 1988).

To show the possibility of correcting solutions as a result of changing the importance of the objective functions, let us assume, for example, that the second objective function ($p = 15$) has the level "*weak superiority*" relative to other objective functions (note that other objective functions have the level "*identical significance*" relative to each other). These

Table 4.5 Power shortage allocation

I	1	2	3	4	5	6
$x^{1,0}$	4476.68	1406.90	4000.00	5000.00	23 000.00	2116.00
$x^{1,00}$	8000.00	0	4000.00	5000.00	23 000.00	0
$x^{1,000}$	8000.00	0	4000.00	5000.00	23 000.00	0
$x^{2,0}$	12 868.07	2763.63	4000.00	5000.00	23 000.00	12 368.30
$x^{2,00}$	15 510.77	0	4000.00	5000.00	23 000.00	12 489.23
$x^{2,000}$	16 000.00	0	4000.00	5000.00	23 000.00	12 000.00

Table 4.6 Levels of membership functions

P	1	15	16	17
$A_p(x^{1,0})$	0.799	0.791	0.792	0.792
$A_p(x^{1,00})$	0.906	0.651	0.880	0.911
$A_p(x^{1,000})$	0.906	0.651	0.880	0.911
$A_p(x^{2,0})$	0.629	0.624	0.624	0.629
$A_p(x^{2,00})$	0.968	0.400	0.688	0.879
$A_p(x^{2,000})$	0.986	0.446	0.727	0.932

pairwise comparisons lead to the results $\lambda_1 = 0.67$, $\lambda_2 = 2.00$, $\lambda_3 = 0.67$, and $\lambda_4 = 0.67$. The corresponding solution for $B^1 = 40\,000$ kW is $x_1^{1,0} = 2196.11$ kW, $x_2^{1,0} = 1262.74$ kW, $x_3^{1,0} = 4000.00$ kW, $x_4^{1,0} = 5000.00$ kW, $x_5^{1,0} = 23\,000.00$ kW, and $x_6^{1,0} = 4541.15$ kW with $A_2(x^{1,0}) = 0.944$ and $A_1(x^{1,0}) = 0.675$, $A_3(x^{1,0}) = 0.602$, and $A_4(x^{1,0}) = 0.599$. It is possible to increase (to a higher extent) the importance of the second objective function ($p = 15$) by utilizing, for example, the level "*Evident Significance*" relative to other objective functions. In this case, we obtain $\lambda_1 = 0.40$, $\lambda_2 = 2.80$, $\lambda_3 = 0.40$, and $\lambda_4 = 0.40$, and the solution is $x_1^{1,0} = 1036.58$ kW, $x_2^{1,0} = 1079.98$ kW, $x_3^{1,0} = 4000.00$ kW, $x_4^{1,0} = 5000.00$ kW, $x_5^{1,0} = 23\,000.00$ kW, and $x_6^{1,0} = 5883.44$ kW, with $A_2(x^{1,0}) = 0.980$ and $A_1(x^{1,0}) = 0.499$, $A_3(x^{1,0}) = 0.364$, $A_4(x^{1,0}) = 0.364$.

Let us consider the influence of the linguistic variable Q – *Limitation for consumer*. For example, the introduction of the condition "*considerably less* than 5000 kW" for the fourth consumer leads to a change of the solution shown in Table 4.5 as follows: $x_4^{1,0} = 1936.10$ kW and $x_1^{1,0} = 5313.92$ kW, $x_2^{1,0} = 3312.00$ kW, $x_3^{1,0} = 4000.00$ kW, $x_5^{1,0} = 23\,000.00$ kW, $x_6^{1,0} = 1999.98$ kW. At the same time, the introduction of the condition "*slightly less* than 5000 kW" for the fourth consumer leads to a change of the solution given in Table 4.5 as follows: $x_4^{1,0} = 3257.52$ kW and $x_1^{1,0} = 4888.11$ kW, $x_2^{1,0} = 3750.00$ kW, $x_3^{1,0} = 4000.00$ kW, $x_5^{1,0} = 23\,000.00$ kW, $x_6^{1,0} = 1104.57$ kW.

Example 4.4. The use of the results described above stipulates that we can apply the multicriteria approach to power system operation to realize dispatch with several objectives involved (say, minimum fuel cost, minimum losses, maximum degree of security, minimum environmental impact, etc.). This is illustrated by a case study of the standard IEEE 30-bus system presented in Figure 4.8 (bus 1 is a slack bus) when considering the objectives of minimizing losses $L(x)$, reducing sulfur oxide emissions $E_{SOx}(x)$, and reducing nitrogen oxide emissions $E_{NOx}(x)$.

Table 4.7 Mean deviations

Δ	B^1	B^2
$\Delta(x^0)$	0.003	0.003
$\Delta(x^{00})$	0.093	0.195
$\Delta(x^{000})$	0.093	0.186

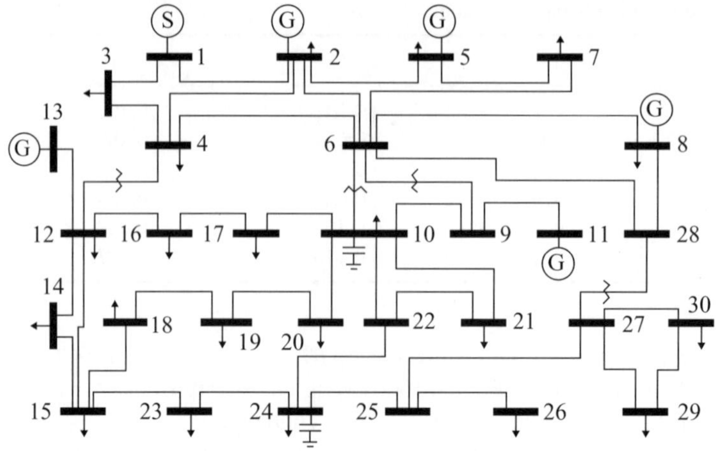

Figure 4.8 System diagram.

The details of the characteristics of the generators are listed in Table 4.8. It includes the coefficients (Nimura *et al.*, 2001) for estimating levels of SOx and NOx emissions on the basis of the following relationships:

$$E_{SOx,i}(x_i) = a_{SOx,i}x_i^2 + b_{SOx,i}x_i + c_{SOx,i} \tag{4.48}$$

and

$$E_{NOx,i}(x_i) = a_{NOx,i}x_i^2 + b_{NOx,i}x + c_{NOx,i} \tag{4.49}$$

The initial state is the following: $x_1^{(0)} = 85.50\,\text{MW}$, $x_2^{(0)} = 66.92\,\text{MW}$, $x_3^{(0)} = 66.76\,\text{MW}$, $x_4^{(0)} = 58.91\,\text{MW}$, and $x_5^{(0)} = 48.67\,\text{MW}$, with $L(x^{(0)}) = 9.04\,\text{MW}$, $E_{SOx}(x^{(0)}) = 220.72\,\text{kg/h}$, and $E_{NOx}(x^{(0)}) = 284.52\,\text{kg/h}$.

Consideration of (4.48) and (4.49) creates no difficulties at all. At the same time, the presentation of the function $L(x)$ in an explicit form gives rise to some difficulties. One way around this problem is the application of the procedures for sequential multicriteria optimization using sensitivity models reflecting the loss change occurring at each optimization step.

Table 4.8 Generator characteristics

i	x_1	x_2	x_3	x_4	x_5
Bus	2	5	8	11	13
Fuel	Hydro	Gas	Oil	Coal	Hydro
$a_{SOx,i}$	0	0	0.010	0.015	0
$b_{SOx,i}$	0	0	0.800	1.200	0
$a_{NOx,i}$	0	0.010	0.015	0.030	0
$b_{NOx,i}$	0	0.200	0.300	0.600	0

Table 4.9 Results obtained in successive steps of multicriteria optimization

Step	$x_1^{(m)}$	$x_2^{(m)}$	$x_3^{(m)}$	$x_4^{(m)}$	$x_5^{(m)}$
0	85.50	66.92	66.76	58.91	48.67
1	85.86	70.26	67.25	56.67	46.71
2	84.34	73.78	66.36	53.91	48.38
3	82.44	77.47	65.20	51.31	50.35
4	83.56	81.35	64.94	48.74	48.18

It is legitimate to construct the sensitivity models (Ekel *et al.*, 2002) on the basis of experimental design techniques (Box, Hunter, and Hunter, 1978; Jain, 1991). In particular, by varying the generation magnitudes for the initial state at levels $x_i^{(0)} - \delta x_i^{(0)}$ and $x_i^{(0)} + \delta x_i^{(0)}$, $i = 1, 2, \ldots, 5$, and applying 2^{5-2} fractional design (Box, Hunter, and Hunter, 1978; Jain, 1991), we construct the sensitivity model in the following form:

$$\bar{L}(x^{(0)}) = 0x_1 - 0.0814x_2 - 0.0494x_3 - 0.0475x_4 + 0x_5 \tag{4.50}$$

The linearized objective function (4.50) has been constructed for $\delta = 0.05$.

In the general case, the linearized objective function

$$\bar{L}(x^{(m)}) = c_1^{(m)}x_1 + c_2^{(m)}x_2 + c_3^{(m)}x_3 + c_4^{(m)}x_4 + c_5^{(m)}x_5 \tag{4.51}$$

together with the objective functions (4.48) and (4.49) and the constraints for generators

$$\max\{0, x_i^{(m)} - \delta x_i^{(m)}\} \leq x_i \leq \min\{x_i^{(m)} + \delta x_i^{(m)}, x_i^{\max}\}, i = 1, 2, \ldots, 5 \tag{4.52}$$

form the multicriteria problem for the *m*th step.

Table 4.9 and Table 4.10 include the results produced at successive steps of the sequential multicriteria optimization. Table 4.11 contains the values of the coefficients of the linearized function $\bar{L}(x^{(m)})$ obtained in successive steps.

Example 4.5. The problems of optimizing network configuration (network reconfiguration) in distribution systems are associated with altering the topologies of the distribution systems by changing the state of their switches (in other words, by changing the locations of their disconnections). These problems are solved in long- and short-term planning and operation,

Table 4.10 Levels of objective functions

Step	$L(x^{(m)})$	$E_{SOx}(x^{(m)})$	$E_{NOx}(x^{(m)})$
0	9.04	220.72	284.52
1	8.91	215.23	281.81
2	8.78	205.41	274.69
3	8.60	195.73	268.60
4	8.55	188.25	265.70

Table 4.11 Coefficients of the linearized function $\bar{L}(x^{(m)})$

Step	$c_1^{(m)}$	$c_2^{(m)}$	$c_3^{(m)}$	$c_4^{(m)}$	$c_5^{(m)}$
0	0	−0.0814	−0.0494	−0.0475	0
1	0	−0.0801	−0.0494	−0.0481	−0.0326
2	0	−0.0786	−0.0497	−0.0482	−0.0310
3	0	−0.0765	−0.0502	−0.0492	0
4	0	−0.0756	−0.0508	−0.0492	0

and can be applied to design studies. The increased interest in them is associated with the wide automation of distribution systems whose switches are remotely monitored and controlled. This makes it possible to solve them online in real time. Many studies, using different approaches, have focused on developing solutions to these problems on the basis of diverse approaches. At the same time, the majority of existing studies are focused on solving monocriteria problems (usually, power losses or energy losses are minimized). Further, network reconfiguration problems are inherently multicriteria in nature because they have an impact on reliability, service quality, and economical feasibility of power supply.

Taking the above observations into account, the developed computing system, named DNOS, designed to deal with multicriteria optimization of network configuration in distribution systems, is useful for considering and minimizing the objective functions of power losses, energy losses, the system average interruption frequency index (SAIFI) (IEEE, 2004), the system average interruption duration index (SAIDI) (IEEE, 2004), undersupply of energy, poor energy quality consumption (consumption of energy outside of permissible limits), and integrated overload of network elements in diverse combinations.

The solutions to monocriteria as well as multicriteria problems of optimizing the network configuration are based on the use of the univariate method (Rao, 1996) and its modifications, which are flexible and easily adaptable to different practical solution strategies as well as to the technology of representing information on a network topology. It is worth noting that the solution to the multicriteria problems is associated with analyzing the conditions (4.40) and (4.41) at each optimization step to make a decision on the transition from solution $x^{(m)}$ to solution $x^{(m+1)}$.

Let us consider the results coming from solving a simple problem of optimizing the configuration of a 13.8 kV network which includes 77 buses and 87 branches. As the minimized objective functions, power losses ΔP, energy losses ΔW, and poor energy consumption ΔN have been considered.

The results of monocriteria optimization on ΔP, ΔW, and ΔN are presented in Table 4.12, Table 4.13, and Table 4.14, respectively. Table 4.15 contains the results of the multicriteria optimization. To better understand the advantages of applying the multicriteria approach, Table 4.16 shows comparative results of the monocriteria and multicriteria optimization. These results show that the application of the multicriteria approach leads to a harmonious solution with small deviations from locally optimal solutions obtained for each criterion. This point is confirmed by considering the results (see Table 4.17) of solving other simple problems of monocriteria and multicriteria optimizing configurations of a 13.8 kV network, which includes 24 buses and 29 branches, on the objective functions of SAIFI, SAIDI, and undersupply energy ΔE, which reflect power supply reliability.

Table 4.12 Levels of objective functions (monocriteria optimization on ΔP)

Objective function	Initial state	Optimal state	Objective function reduction (%)
ΔP (kW)	99.15	64.10	35.35
ΔW (kWh)	1191.32	849.09	28.73
ΔN (kWh)	502.42	1086.09	–116.17

Table 4.13 Levels of objective functions (monocriteria optimization on ΔW)

Objective function	Initial state	Optimal state	Objective function reduction (%)
ΔP (kW)	99.15	64.56	34.89
ΔW (kWh)	1191.32	835.49	29.87
ΔN (kWh)	502.42	349.97	30.34

Table 4.14 Levels of objective functions (monocriteria optimization on ΔN)

Objective function	Initial state	Optimal state	Objective function reduction (%)
ΔP (kW)	99.15	84.88	14.39
ΔW (kWh)	1191.32	1146.49	3.76
ΔN (kWh)	502.42	145.92	70.96

Table 4.15 Levels of objective functions (multicriteria optimization)

Objective function	Initial state	Optimal state	Objective function reduction (%)
ΔP (kW)	99.15	66.98	32.45
ΔW (kWh)	1191.32	854.66	28.26
ΔN (kWh)	502.42	188.28	52.53

Table 4.16 Levels of objective functions (monocriteria and multicriteria optimization)

Objective function	Optimization on ΔP	Optimization on ΔW	Optimization on ΔN	Multicriteria optimization
ΔP (kW)	64.10	64.56	84.88	66.98
ΔW (kWh)	849.09	835.48	1146.49	854.66
ΔN (kWh)	1086.09	349.97	145.92	188.28

Table 4.17 Levels of objective functions (monocriteria and multicriteria optimization)

Objective function	Initial state	Optimization on SAIFI	Optimization on SAIDI	Optimization on ΔE	Multicriteria optimization
SAIFI (interruptions/year)	18.59	14.02	14.62	17.89	14.47
SAIDI (hours/year)	45.00	51.77	25.73	40.01	31.18
ΔE (kWh)	771.79	636.11	576.34	382.43	506.85

Example 4.6. The problem of optimizing modes for the operation of basic means of voltage control in distribution systems is associated with choosing off-load taps for distribution transformers and conditions for operating tap changing under load transformers or voltage regulators at feeding substations. The techniques for optimizing voltage control implemented within the framework of the developed computing system, named VCOS, are directed at minimizing poor energy quality consumption on the basis of applying a synthesis of the integral criteria of voltage quality and calculations of permissible voltage levels.

In particular, the energy-weighted average voltage drops from buses at feeding substations to the centers of loads of low-voltage networks of distribution transformers are used to choose their off-load taps.

The choice of conditions for operating tap changing under load transformers at feeding substations is considered as a stage that follows the choice of off-load taps for the distribution transformer. However, if it is impossible to change the off-load taps, this stage is considered as independent.

The optimal conditions for operating tap changing under load transformers at a feeding substation may be obtained if it is possible to provide the voltage addition at the bus of this substation that is equal in magnitude and opposite in sign to the power-weighted voltage levels at the centers of loads of low-voltage networks of all distribution transformers for each step of loads curves. Thus, it is possible to construct a relationship $E_\omega = f_E(\omega)$ where $\omega = 1, 2, \dots, \Omega$ is a step of load curves.

Some regulation laws can be considered, say (1) voltage stabilization ($E_S = a_S$), (2) voltage control with current correction ($E_I = a_I + b_I I$), and (3) voltage control with active power correction ($E_P = a_P + b_P P$).

Knowing, for example, the active load curve for tap changing under a load transformer at a feeding substation $P_\omega = f_P(\omega)$, $\omega = 1, 2, \dots, \Omega$, and $E_\omega = f_E(\omega)$, $\omega = 1, 2, \dots, \Omega$, and eliminating the parameter ω, we obtain a set of points in the system of coordinates $P - E_\omega$. By applying the method of least squares to this set of points, the regulation law $E_P = a_P + b_P P$ can be constructed.

The approach described above serves for minimizing poor energy quality consumption. However, in accordance with a situational hierarchy, the need can arise for energetically effective voltage control, considering the static load characteristics (in particular, the results of De Steese, Merrick, and Kennedy (1990) demonstrate the possibility of essentially reducing peak load and energy consumption as a result of reducing the voltage). Thus, it is beneficial to look at the second problem statement dealing with the minimization of poor energy quality consumption and reducing peak load and/or energy consumption. Its solution is one of the functions of the VCOS system implemented on the basis of the results described above.

Let us consider the results obtained when solving the problem of optimizing voltage control in a 13.8 kV network which includes 6 feeders with 2629 distribution transformers. The voltage control E_P and a reducing peak load consumption law E_P^C are presented in Figure 4.9. In this example, where the criteria have equal significance, the peak load is reduced by 6.68%; however, poor energy quality consumption is increased by 24.8%. Figure 4.10 presents the solution where the importance of the criterion of poor energy quality consumption has *Absolute Superiority* in relation to reducing peak load consumption. The E_P^C law of Figure 4.10 leads to a reduction of peak load by 2.8% and an increase in poor energy quality consumption by 6.5%.

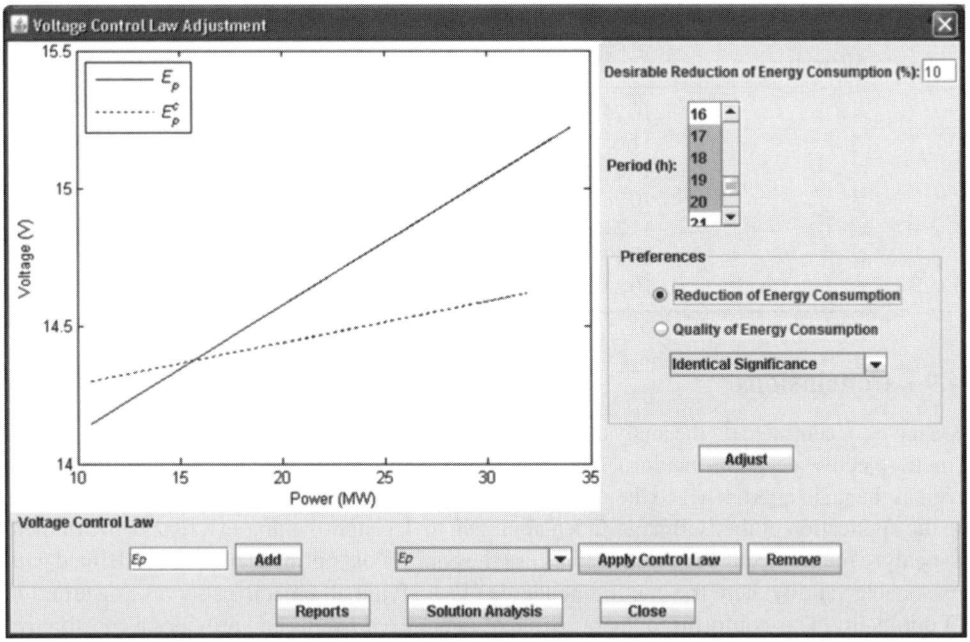

Figure 4.9 Voltage control laws with identical levels of significance of criteria.

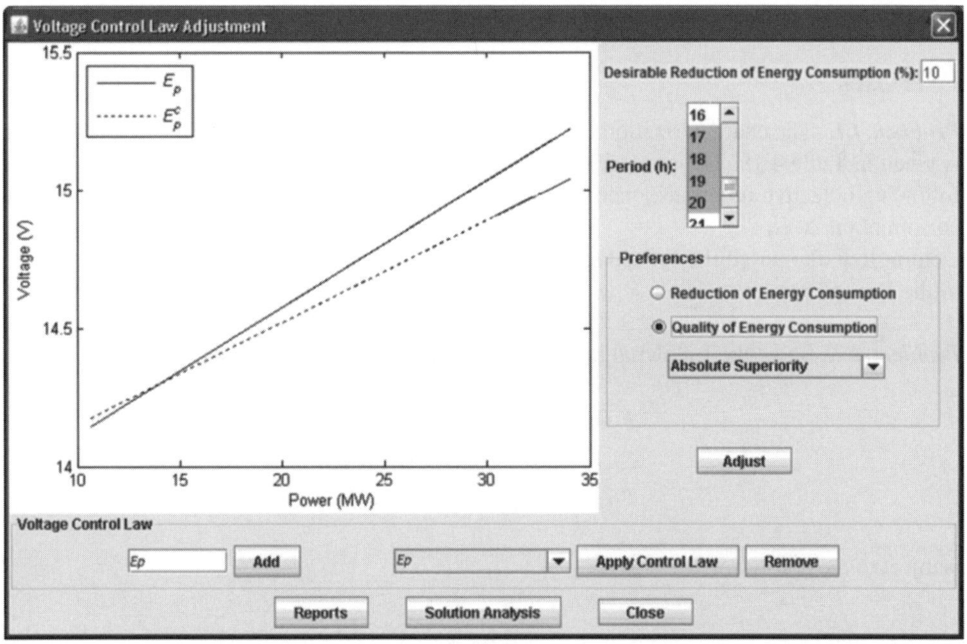

Figure 4.10 Voltage control laws with different levels of significance of criteria.

Table 4.18 Levels of objective functions

Alternative	ΔW	ΔE	ΔN
x_1	$10\,241 \times 10^3$	312×10^3	1027×10^3
x_2	$11\,635 \times 10^3$	300×10^3	1234×10^3
x_3	$10\,210 \times 10^3$	314×10^3	1116×10^3
x_4	$10\,284 \times 10^3$	316×10^3	1211×10^3
x_5	$11\,243 \times 10^3$	324×10^3	1190×10^3
x_6	$10\,493 \times 10^3$	331×10^3	1017×10^3
x_7	$10\,233 \times 10^3$	318×10^3	1098×10^3

4.9 Conclusions

We have concentrated on the analysis of continuous problems of multicriteria decision-making (multiobjective decision-making). The main concepts of multiobjective decision-making as well as the main approaches to their solutions have been studied. Much attention has been paid to the application of the Bellman–Zadeh approach to decision-making in a fuzzy environment to analyze multiobjective problems. In its use the concept of "optimal solution" is defined with reasonable validity: here the maximum degree of satisfying all objectives serves as a criterion of optimality. This conforms to the principle of guaranteed results and provides a constructive way to obtain harmonious solutions on the basis of analyzing associated *max–min* models. The application of the presented results has been illustrated through practical examples arising in the realm of power engineering.

Exercises

Problem 4.1. The characterization of seven alternatives for choosing a power supply scheme is given in Table 4.18. The comparison of alternatives is to be carried out on the basis of the following objective functions: energy losses ΔW, undersupply energy ΔE, and poor energy consumption ΔN.
Demonstrate the possibility of reducing the number of alternatives on the basis of the notion of the Pareto-optimal solution.

Problem 4.2. Solve the following problem of multiobjective optimization:

$$\begin{aligned} F_1(x) &= 3x_1 + 13x_2 \rightarrow min \\ F_2(x) &= 8x_1 + 6x_2 \rightarrow min \\ F_3(x) &= 4x_1 + 5x_2 \rightarrow min \end{aligned}$$

subject to the constraints

$$\begin{aligned} 0 \leq x_1 \leq 20 \\ 0 \leq x_2 \leq 10 \\ x_1 + x_2 = 20 \end{aligned}$$

Apply the convolution (4.4) where the normalization of the objective function is realized on the basis of (4.5).

Problem 4.3. Solve the problem presented in Problem 4.2 by applying Boldur's method.

Problem 4.4. Solve the same problem (Problem 4.2) by applying the method based on placing constraint levels of objective functions to improve the quality of the solution (so that we can obtain a more harmonious solution).

Problem 4.5. Consider the problem (Borisov, Krumberg, and Fedorov, 1990) of selecting a manager for an enterprise. There are five candidates for this position. They are evaluated on the basis of the following indicators: professional skills (C_1), organizational capabilities (C_2), work experience (C_3), authority (C_4), ability to work with people (C_5), and age (C_6). The results of evaluating the candidates are as follows:

$$C_1 = \{0.9/x_1, 0.9/x_2, \ 0.6/x_3, \ 0.8/x_4, \ 0.5/x_5\}$$
$$C_2 = \{0.8/x_1, 0.9/x_2, \ 0.5/x_3, \ 0.7/x_4, \ 0.6/x_5\}$$
$$C_3 = \{0.7/x_1, 0.9/x_2, \ 0.8/x_3, \ 0.5/x_4, \ 0.3/x_5\}$$
$$C_4 = \{0.9/x_1, 0.8/x_2, \ 0.5/x_3, \ 0.6/x_4, \ 0.5/x_5\}$$
$$C_5 = \{0.9/x_1, 0.9/x_2, \ 0.4/x_3, \ 0.7/x_4, \ 0.6/x_5\}$$
$$C_6 = \{0.9/x_1, 0.4/x_2, \ 0.8/x_3, \ 0.7/x_4, \ 0.5/x_5\}$$

Try to select the best candidate assuming that all indicators are of the same importance.

Problem 4.6. Consider the problem (Borisov, Krumberg, and Fedorov, 1990) of selecting a location for a building of an enterprise. There are four alternatives. They are evaluated on the basis of the following indicators: proximity to a consumer (C_1), proximity to sources of raw materials (C_2), and availability of labor force (C_3). The results of evaluating the alternatives are as follows:

$$C_1 = \{0.5/x_1, 0.7/x_2, \ 0.3/x_3, \ 0.6/x_4\}$$
$$C_2 = \{0.5/x_1, 0.4/x_2, \ 0.8/x_3, \ 0.4/x_4\}$$
$$C_3 = \{0.2/x_1, 0.1/x_2, \ 0.6/x_3, \ 0.9/x_4\}$$

Apply the Bellman–Zadeh approach for the two cases:

(a) the first two indicators have identical importance and are two times more important than the third criterion;
(b) the first indicator is two times more important than the last two indicators which have identical importance.

Problem 4.7. Solve the following problem of multiobjective optimization:

$$F_1(x) = 5x_1 + 9x_2 \rightarrow \min$$
$$F_2(x) = 8x_1 + 3x_2 \rightarrow \min$$

subject to the constraints

$$x_1 + x_2 = 10$$
$$x_1, x_2 \geq 0$$

Apply the Bellman–Zadeh approach and consider two scenarios:

(a) the objective functions have identical importance;
(b) the first objective function is two times more important than the second objective function.

Problem 4.8. Solve the following problem of multiobjective optimization:

$$F_1(x) = 4x_1 + 9x_2 \rightarrow \min$$
$$F_2(x) = 3x_1 + 6x_2 \rightarrow \max$$

subject to the constraints

$$x_1 + x_2 = 10$$
$$x_1, x_2 \geq 0$$

Apply the Bellman–Zadeh approach and study two cases:

(a) the objective functions have identical importance;
(b) the first objective function is two times less important than the second objective function.

References

Beliakov, G. and Warren, J. (2001) Appropriate choice of aggregation operators in fuzzy decision support systems. *IEEE Transactions on Fuzzy Systems*, **9** (6), 773–784.

Bellman, R. and Giertz, M. (1974) On the analytic formalism of the theory of fuzzy sets. *Information Sciences*, **5** (2), 149–157.

Bellman, R.E. and Zadeh, L.A. (1970) Decision-making in a fuzzy environment. *Management Science*, **17** (1), 141–164.

Benayoun, R., de Montgolfier, J., Tergny, J., and Larichev, O.I. (1971) Linear programming with multiple objective functions: STEP method (STEM). *Mathematical Programming*, **1** (3), 366–375.

Borisov, A.N., Krumberg, O.A., and Fedorov, I.P. (1990) *Decision-Making on the Basis of Fuzzy Models*, Zinatne, Riga (in Russian).

Box, G.E.P., Hunter, W.G., and Hunter, J.S. (1978) *Statistics for Experiments: An Introduction to Design, Data Analysis and Model Building*, John Wiley & Sons, Inc., New York.

Canha, L., Ekel, P., Queiroz, J., and Schuffner Neto, F. (2007) Models and methods of decision-making in fuzzy environment and their applications to power engineering problems. *Numerical Linear Algebra with Applications*, **14** (3), 369–390.

Charnes, A. and Cooper, W.W. (1961) *Management Models and Industrial Applications of Linear Programming*, John Wiley & Sons, Inc., New York.

Charnes, A., Cooper, W.W., and Ferguson, R. (1955) Optimal estimation of executive compensation by linear programming. *Management Science*, **1** (2), 138–151.

Coelho, C.A.C. (2000) Handling preferences in evolutionary multiobjective optimization: a survey. Proceedings of the 2000 Congress on Evolutionary Computation, San Diego, pp. 30–37.

Coelho, C.A.C. (2002) Evolutionary multi-objective optimization: Critical Review, in *Evolutionary Optimization* (eds R. Sarker, M. Mohammadian, and X. Yao), Kluwer, Boston, pp. 117–146.

Coelho, C.A.C., Van Veldhuizen, D. A., and Lamont, G.B. (2002) *Evolutionary Algorithms for Solving Multi-Objective Problems*, Kluwer, New York.

Das, I. and Dennis, J.E. (1998) Normal-boundary intersection: a new method for generating the Pareto surface in nonlinear multicriteria optimization problems. *SIAM Journal of Optimization*, **8** (4), 631–657.

De Steese, J.G., Merrick, S.B., and Kennedy, B.W. (1990) Estimating methodology for a large regional application of conservation voltage reduction. *IEEE Transactions on Power Systems*, **5** (3), 862–870.

Deb, K. (2001) *Multi-Objective Optimization Using Evolutionary Algorithms*, John Wiley & Sons, Ltd, Chichester.

Dubov, Ya.A., Travkin, C.I., and Yakimets, V.N. (1986) *Multicriteria Models for Forming and Choosing System Alternatives*, Nauka, Moscow (in Russian).

Ehrgott, M. (2005) *Multicriteria Optimization*, Springer-Verlag, Berlin.

Ekel, P., Junges, M., Morra, J., and Paletta, F. (2002) Fuzzy logic based approach to voltage and reactive power control in power systems. *International Journal of Computer Research*, **11** (2), 159–170.

Ekel, P.Ya. (2001) Methods of decision-making in fuzzy environment and their applications. *Nonlinear Analysis: Theory, Methods and Applications*, **47** (5), 979–990.

Ekel, P.Ya. and Galperin, E.A. (2003) Box-triangular multiobjective linear programs for resource allocation with application to load management and energy market problems. *Mathematical and Computer Modelling*, **37** (1), 1–17.

Ekel, P.Ya., Menezes, M., and Schuffner Neto, F. (2007) decision-making in fuzzy environment and its application to power engineering problems. *Nonlinear Analysis: Hybrid Systems*, **1** (4), 527–536.

Ekel, P.Ya., Terra, L.D.B., Junges, M.F.D., *et al.* (1998) Multicriteria power and energy shortage allocation using fuzzy set theory. Proceedings of the International Symposium on Bulk Power Systems Dynamics and Control IV: Restructuring, Santorini, pp. 311–318.

Ekel, P.Ya., Terra, L.D.B., Junges, M.F.D., *et al.* (2000) Multicriteria load management in power systems. Proceedings of the IEEE and IEE International Conference on Electric Utility Deregulation and Restructuring and Power Technologies, London, pp. 167–172.

Gardiner, L.R. and Steuer, R.E. (1994) Unified interactive multiple objective programming: an open architecture for accommodating new procedures. *Journal of the Operational Research Society*, **45** (12), 1456–1466.

Horn, J. (1997) Multicriteria decision-making, in *Handbook of Evolutionary Computation*, IOP Publishing and Oxford University Press, Bristol, pp. F1.9:1–F1.9:15.

Hwang, C.L. and Masud, A.S.M. (1979) *Multiple Objective Decision Makinh: Methods and Applications*, Springer-Verlag, Heidelberg.

IEEE (2004) *IEEE Guide for Electric Power Distribution Reliability Indices: IEEE Standard 1366TM-2003*, IEEE, New York.

Ignizio, J.P. (1976) *Goal Programming and Extensions*, Lexington Books, Lanham, MD.

Jain, R. (1991) *The Art of Computer Systems Performance Analysis: Techniques for Experimental Design, Measurement, Simulation, and Modeling*, John Wiley & Sons, Inc., New York.

Jones, D.F. and Tamiz, M. (2002) Goal programming in the period 1990-2000, in *Multiple Criteria Optimization: State of the Art Annotated Bibliographic Surveys* (eds M. Ehrgott and X. Gandibleux), Kluwer, Boston, pp. 129–170.

Keeney, R.L. and Raiffa, H. (1976) *Decision with Multiple Objectives: Preferences and Value Tradeoffs*, John Wiley & Sons, Inc., New York.

Kuhn, H.W. and Tucker, A.W. (1951) Nonlinear programming. Proceedings of the Second Berkeley Symposium on Mathematical Statistics and Probability, Berkeley, pp. 481–492.

Lai, Y.J. and Hwang, C.L. (1996) *Fuzzy Multiple Objective Decision-Making: Methods and Applications*, Springer-Verlag, Berlin.

Lee, S.M. (1972) *Goal Programming for Decision Analysis*, Auerback, Philadelphia.

Lu, J., Zhang, G., Ruan, D., and Wu, F. (2008) *Multi-Objective Group Decision-Making: Methods, Software and Applications with Fuzzy Set Techniques*, Imperial College Press, London.

Lyapunov, A.A. (ed.) (1972) *Operational Research: Methodological Aspects*, Nauka, Moscow (in Russian).

Mashunin, Yu.K. (1986) *Methods and Models of Vector Optimization*, Nauka, Moscow (in Russian).

Monarchi, D.E., Kisiel, C.C., and Duckstein, L. (1973) Interactive multiobjective programming in water resources: a case study. *Water Resources Research*, **9** (4), 837–850.

Nimura, T., Moreira, F.A., Nakashima, T., *et al.* (2001) Multiple attribute performance evaluation of independent power producers in a deregulated power system based on fuzzy sets. Proceedings of the International Conference on Intelligent System Application to Power Systems, Rio de Janeiro, pp. 122–126.

Pareto, V. (1886) *Cours d'économie politique*, Lousanne Rouge, Lousanne.

Podinovsky, V.V. and Gavrilov, V.M. (1975) *Optimization on Sequentially Utilized Criteria*, Sovetskoe Radio, Moscow (in Russian).

Prakhovnik, A.V., Ekel, P.Ya., and Bondarenko, A.F. (1994) *Models and Methods of Optimizing and Controlling Modes of Operation of Electric Power Supply Systems*, ISDO, Kiev (in Ukrainian).

Rao, S. (1996) *Engineering Optimization*, John Wiley & Sons, Inc., New York.

Raskin, L.G. (1976) *Analysis of Complex Systems and Elements of Optimal Control Theory*, Sovetskoe Radio, Moscow (in Russian).

Romero, C. (1991) *Handbook of Critical Issues in Goal Programming*, Pergamon Press, Oxford.

Roy, B. (1972) Décisions avec critères multiples: problèmes et méthodes. *Merta International*, **11** (2), 121–151.

Saaty, T. (1980) *The Analytic Hierarchy Process*, McGraw-Hill, New York.

Sakawa, M. (1993) *Fuzzy Sets and Interactive Multiobjective Optimization*, Plenum Press, New York.

Statnikov, R.B. and Matusov, J.B. (2002) *Multicriteria Analysis in Engineering*, Kluwer, Dordrecht.

Stoft, S. (2002) *Power System Economics: Designing Markets for Electricity*, John Wiley & Sons, Inc., New York.

Talukdar, S. and Gellings, C. (1987) *Load Management*, IEEE Press, New York.

Yager, R.R. (1988) On ordered weighted averaging operators in multicriteria decision-making. *IEEE Transactions on Systems, Man, and Cybernetics*, **18** (3), 183–190.

Zadeh, L.A. (1963) Optimality and nonscalar-valued performance criteria. *IEEE Transactions on Automatic Control*, **8** (1), 59–60.

Zeleny, M. (1982) *Multiple Criteria Decision-Making*, McGraw-Hill, New York.

Zimmermann, H.J. (1996) *Fuzzy Set Theory and Its Application*, Kluwer, Boston.

5

Introduction to Preference Modeling with Binary Fuzzy Relations

In this chapter, we present an introduction to preference modeling realized in terms of binary fuzzy relations and address certain difficulties that arise in the extension of the classical or Boolean preference structures of binary relations (which is a well-established research area) to the fuzzy environment. Particularly, the extension of the classical structures to their fuzzy counterparts requires the selection of a De Morgan triplet and of adequate functions to construct suitable binary fuzzy relations of strict preference, indifference, and incomparability. Unfortunately, it is not that simple to implement this extension (we must be prudent when it comes to the selection of t-norms). In this context, the current chapter recalls some concepts related to binary fuzzy relations and some specific t-norms, t-conorms, and negation operators, which will play an important role. Then, we define preference structures of binary fuzzy relations (or fuzzy preference structures) and outline a method for constructing these fuzzy structures, without losing important characteristics of the classical preference structures of binary relations.

5.1 Binary Fuzzy Relations and their Fundamental Properties

As stated in Chapter 2, the binary fuzzy relation (BFR) consists of a fuzzy set with bidimensional (that is, defined for two arguments) membership function $\boldsymbol{R} : \mathbf{X} \times \mathbf{X} \rightarrow [0, 1]$. In essence, such relations associate with each ordered pair of elements (X_k, X_l), where $X_k, X_l \in \mathbf{X}$, a number $\boldsymbol{R}(X_k, X_l)$ coming from the unit interval that reflects a degree to which elements X_k and X_l are in relation $\boldsymbol{R}$.

Particularly, when we deal with discrete (finite) sets of alternatives in preference modeling, the BFRs are commonly represented in two different ways:

- a square matrix $\boldsymbol{R}$ of dimension equal to the number of elements of $\mathbf{X}$, whose each entry r_{kl} corresponds to $\boldsymbol{R}(X_k, X_l)$;

Fuzzy Multicriteria Decision-Making: Models, Methods and Applications Witold Pedrycz, Petr Ekel and Roberta Parreiras

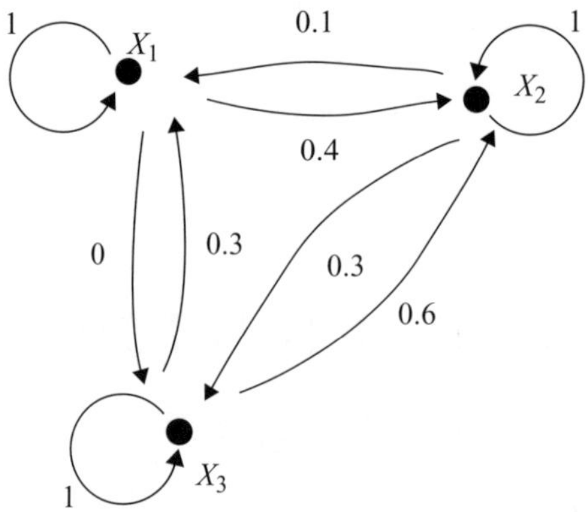

Figure 5.1 Relation R of Example 4.1 represented as a graph.

- a weighted graph where each element from $\mathbf{X}$ is symbolized by a node and the relations between the elements are represented as weighted arcs, in such a way that $\boldsymbol{R}(X_k, X_l)$ corresponds to an arc oriented from X_k toward X_l.

Example 5.1. Consider a set $\mathbf{X} = \{X_1, X_2, X_3\}$, where all possible pairs of elements are interrelated as follows: $\boldsymbol{R}(X_1, X_1) = \boldsymbol{R}(X_2, X_2) = \boldsymbol{R}(X_3, X_3) = 1$, $\boldsymbol{R}(X_1, X_2) = 0.4$, $\boldsymbol{R}(X_1, X_3) = 0$, $\boldsymbol{R}(X_2, X_1) = 1$, $\boldsymbol{R}(X_2, X_3) = 0.7$, $\boldsymbol{R}(X_3, X_1) = 0.3$, and $\boldsymbol{R}(X_3, X_2) = 0.6$. The fuzzy binary relation $\boldsymbol{R}$ can be equivalently represented as the directed graph shown in Figure 5.1 or as a square matrix with the entries

$$\boldsymbol{R} = \begin{bmatrix} 1 & 0.4 & 0 \\ 0.1 & 1 & 0.3 \\ 0.3 & 0.6 & 1 \end{bmatrix} \tag{5.1}$$

Given a generic BFR $\boldsymbol{R}$, its inverse (or transpose) relation $\boldsymbol{R}^{-1}$, its complementary relation $\overline{\boldsymbol{R}}$, and its dual relation $\boldsymbol{R}^d$ are, respectively, defined as follows (Fodor and Roubens, 1994b):

$$\boldsymbol{R}^{-1}(X_k, X_l) = \boldsymbol{R}(X_l, X_k) \tag{5.2}$$

$$\overline{\boldsymbol{R}}(X_k, X_l) = 1 - \boldsymbol{R}(X_k, X_l) \tag{5.3}$$

$$\boldsymbol{R}^d(X_k, X_l) = 1 - \boldsymbol{R}(X_l, X_k) = (\boldsymbol{R}^{-1}(X_k, X_l))^c \tag{5.4}$$

Example 5.2. By applying (5.2)–(5.4) to matrix $\boldsymbol{R}$ of Example 5.1, the following matrices are obtained:

$$\boldsymbol{R}^{-1} = \begin{bmatrix} 1 & 0.1 & 0.3 \\ 0.4 & 1 & 0.6 \\ 0 & 0.3 & 1 \end{bmatrix} \tag{5.5}$$

$$\overline{\boldsymbol{R}} = \begin{bmatrix} 0 & 0.6 & 1 \\ 0.9 & 0 & 0.7 \\ 0.7 & 0.4 & 0 \end{bmatrix} \tag{5.6}$$

$$\boldsymbol{R}^d = \begin{bmatrix} 0 & 0.9 & 0.7 \\ 0.6 & 0 & 0.4 \\ 1 & 0.7 & 0 \end{bmatrix} \tag{5.7}$$

Now, let us recall from Chapter 2 some basic concepts related to fuzzy relations, which will be utilized here. These are the operations of intersection, union, and complement of fuzzy relations, as well as some properties utilized to characterize binary fuzzy relations.

It should be noted that in this chapter, for the benefit of increased readability, the operations of intersection, union, and complement of fuzzy relations are represented using a specific notation that indicates which particular t-norm, t-conorm, and negation is utilized to implement these operations. In this way, given two generic BFRs defined in $\mathbf{X} \times \mathbf{X}$, namely $\boldsymbol{Q}$ and $\boldsymbol{R}$, and a pair of alternatives, $(X_k, X_l) \in \mathbf{X} \times \mathbf{X}$, we have:

- $\boldsymbol{Q} \cap_T \boldsymbol{R}$ corresponds to the intersection of $\boldsymbol{Q}$ and $\boldsymbol{R}$, implemented with a t-norm T: $T(\boldsymbol{Q}(X_k, X_l), \boldsymbol{R}(X_k, X_l))$;
- $\boldsymbol{Q} \cup_S \boldsymbol{R}$ corresponds to the union of $\boldsymbol{Q}$ and $\boldsymbol{R}$, implemented with a t-conorm S: $S(\boldsymbol{Q}(X_k, X_l), \boldsymbol{R}(X_k, X_l))$;
- $N(\boldsymbol{R})$ corresponds to the complement of $\boldsymbol{R}$, implemented with the negation operator N.

In particular, a specific type of De Morgan triplet, based on the Lukasiewicz triplet, will prove to be important for the development of fuzzy preference structures. But, before presenting this triplet, it is necessary to introduce some basic concepts such as automorphism ϕ and the ϕ-transform of a t-norm and of a t-conorm. These concepts are defined in conformity with the main references (Fodor and Roubens, 1994a; De Baets and Van De Walle, 1997; Bufardi, 1998).

Any continuous and strictly increasing function $\phi : [0, 1] \rightarrow [0, 1]$ is called an automorphism of the unity interval, if it satisfies the boundary conditions $\phi(0) = 0$ and $\phi(1) = 1$.

Considering that $x, y \in [0, 1]$, a ϕ-transform of a t-norm T with the automorphism ϕ is the t-norm

$$T^{\phi}(x, y) = \phi^{-1}(T(\phi(x), \phi(y))) \tag{5.8}$$

Similarly, a ϕ-transform of a t-conorm S with the automorphism ϕ is the t-conorm

$$S^{\phi}(x, y) = \phi^{-1}(S(\phi(x), \phi(y))) \tag{5.9}$$

The Archimedean t-norm is defined as the t-norm satisfying the condition

$$T(x, x) < x \quad \forall x \in [0, 1] \tag{5.10}$$

A t-norm T is said to have zero divisors if it satisfies the relationship

$$T(x, y) = 0 \quad \text{for some positive } x \text{ and } y \tag{5.11}$$

The Lukasiewicz t-norm is a commonly encountered example of a continuous Archimedean t-norm with zero divisors. An important result demonstrated in Ovchinnikov and Roubens (1991) is that any continuous Archimedean t-norm T with zero divisors can be represented as a ϕ-transform of the Lukasiewicz t-norm as follows:

$$T_\phi(x, y) = \phi^{-1}(\max(\phi(x) + \phi(y) - 1, 0)) \tag{5.12}$$

where $x, y \in [0, 1]$.

If the same automorphism ϕ is applied to obtain the ϕ-transform of the complement N, the ϕ-transform of the Lukasiewicz t-norm T and the ϕ-transform of the Lukasiewicz t-conorm S, a Lukasiewicz-like De Morgan triplet (T_ϕ, S_ϕ, N_ϕ) is obtained, where T_ϕ is given by (5.12), N_ϕ is given by

$$N_\phi(x) = \phi^{-1}(1 - \phi(x)) \tag{5.13}$$

and S_ϕ can be determined as

$$S_\phi(x, y) = \phi^{-1}(\min(\phi(x) + \phi(y), 1)) \tag{5.14}$$

Furthermore, we consider several properties to characterize the binary fuzzy relations, which are of interest in the context of preference modeling (Öztürk, Tsoukiàs, and Vincke, 2005). Next, we present a list of properties of binary fuzzy relations, which is more extensive than the one presented in Chapter 2:

$$\text{Reflexivity}: \boldsymbol{R}(X_k, X_k) = 1 \quad \forall X_k \in \mathbf{X} \tag{5.15}$$
$$\text{Irreflexivity}: \boldsymbol{R}(X_k, X_k) = 0 \quad \forall X_k \in \mathbf{X} \tag{5.16}$$
$$\text{Symmetry}: \boldsymbol{R}(X_k, X_l) = \boldsymbol{R}(X_l, X_k) \quad \forall X_k, X_l \in \mathbf{X} \tag{5.17}$$
$$T\text{-asymmetry}: T(\boldsymbol{R}(X_k, X_l), \boldsymbol{R}(X_l, X_k)) = 0 \quad \forall X_k, X_l \in \mathbf{X} \tag{5.18}$$
$$T\text{-antisymmetry}: T(\boldsymbol{R}(X_k, X_l), \boldsymbol{R}(X_l, X_k)) = 0 \quad \forall X_k, X_l \in \mathbf{X} | X_k \neq X_l \tag{5.19}$$
$$T\text{-transitivity}: \boldsymbol{R}(X_k, X_n) \geq T(\boldsymbol{R}(X_k, X_l), \boldsymbol{R}(X_l, X_n)) \quad \forall X_k, X_l, X_n \in \mathbf{X} \tag{5.20}$$
$$S\text{-completeness}: S(\boldsymbol{R}(X_k, X_l), \boldsymbol{R}(X_l, X_k)) = 1 \quad \forall X_k, X_l \in \mathbf{X} | X_k \neq X_l \tag{5.21}$$
$$\text{Additive reciprocity}: \boldsymbol{R}(X_k, X_l) + \boldsymbol{R}(X_l, X_k) = 1 \quad \forall X_k, X_l \in \mathbf{X} \tag{5.22}$$

As can be seen, some of these properties, namely T-asymmetry, T-antisymmetry, T-transitivity, and S-completeness, require the selection of t-norms and t-conorms to implement the logical connectives AND and OR as operations on the unit interval. One can note that the selection of different operators to implement the logical connectives results in different definitions for these properties.

In particular, the family of T-transitivities deserves special attention here. As indicated in Chapter 2, the basic idea behind transitivity is that the strength of the direct relationship between two elements should not be weaker than their indirect relationship involving other elements. In the case of preference modeling, transitivity has been adopted as a consistency condition in the sense that it can be associated with the following idea: if someone says that X_k is better than X_j and that X_j is better than X_l, then it is expected that this person prefers X_k to X_l at least until a minimum strength and it is not expected that they prefer X_l to X_k (however, it is important to indicate that there are counterexamples of the validity of this assumption, as will be seen in Section 5.5). From a practical viewpoint, depending on the selected transitivity property, the corresponding consistency condition may become more rigorous or could be somewhat relaxed. The family of T-transitivities includes:

- the min-transitivity condition for fuzzy preference relations

$$\boldsymbol{R}(X_k, X_n) \geq \min(\boldsymbol{R}(X_k, X_l), \boldsymbol{R}(X_l, X_n)) \tag{5.23}$$

- the product-transitivity condition

$$\boldsymbol{R}(X_k, X_n) \geq \boldsymbol{R}(X_k, X_l).\boldsymbol{R}(X_l, X_n) \tag{5.24}$$

- the Lukasiewicz-transitivity condition

$$\boldsymbol{R}(X_k, X_n) \geq \max(\boldsymbol{R}(X_k, X_l) + \boldsymbol{R}(X_l, X_n) - 1, 0) \tag{5.25}$$

Example 5.3. Given three alternatives, $X_1, X_2, X_3 \in \mathbf{X}$, consider that their relationships are captured in the form of a symmetric BFR $\boldsymbol{R}$ in such a way that $\boldsymbol{R}(X_1, X_2) = \boldsymbol{R}(X_2, X_1) = 0.5$, $\boldsymbol{R}(X_2, X_3) = \boldsymbol{R}(X_3, X_2) = 0.2$. Suppose that the missing value of the relation $\boldsymbol{R}(X_1, X_3)$ (and $\boldsymbol{R}(X_3, X_1)$) can be convincingly estimated by admitting that $\boldsymbol{R}$ satisfies:

(a) the Lukasiewicz-transitivity condition

$$\boldsymbol{R}(X_1, X_3) \geq \max(0.5 + 0.2 - 1, 0) \tag{5.26}$$

$$\boldsymbol{R}(X_1, X_3) \geq 0.3 \tag{5.27}$$

(b) the min-transitivity condition

$$\boldsymbol{R}(X_1, X_3) \geq \min(0.5, 0.2) \tag{5.28}$$

$$\boldsymbol{R}(X_1, X_3) \geq 0.2 \tag{5.29}$$

5.2 Preference Modeling with Binary Fuzzy Relations

Let us consider that a DM is asked to compare two alternatives $X_k, X_l \in \mathbf{X}$ and determine which one of these two he/she prefers. One of the following answers is expected:

- X_k and X_l are indifferent;
- X_k is strictly better than X_l;

- X_l is strictly better than X_k;
- X_k and X_l are incomparable (the DM may not be able to compare the alternatives).

Accordingly, in order to realistically characterize this comparison between two alternatives, three main types of judgments can be distinguished, namely indifference, strict preference, and incomparability. These judgments can be modeled by means of BFRs, in such a way that the membership function of each BFR reflects the credibility or the intensity of the observed judgment, being quantified in the interval [0,1]. The coherence between the model and the corresponding judgment is assured by requiring that each BFR should have some basic properties, in accordance with the nature of the judgment that it is supposed to reflect.

Next, we define and characterize the indifference, strict preference, and incomparability judgments as BFRs. Let us start with the judgment of indifference, which is utilized in situations where a DM thinks that both alternatives satisfy equally well his/her interests. Indifference can be modeled as a BFR I with the following properties:

- reflexivity: a DM is always indifferent to X_k and X_k;
- symmetry: statement "X_k is indifferent to X_l" is equivalent to statement "X_l is indifferent to X_k".

Strict preference is the judgment utilized when a DM can identify which one is the best of two alternatives. Strict preference can be modeled as a BFR $\boldsymbol{P}$, which is supposed to satisfy the conditions:

- irreflexivity: a DM cannot strictly prefer X_k to X_k;
- T-asymmetry: a DM cannot strictly prefer X_k to X_l and X_l to X_k, at the same time.

Incomparability is utilized when a DM cannot express his/her opinion, that is, in those situations where a DM is asked about his/her preference and the answer is "I do not know", due to missing or uncertain information or as a consequence of the existence of conflicting information. The incomparability judgment is reflected by means of the BFR J, which must possess the following properties:

- irreflexivity: a DM cannot say that X_k is incomparable to X_k;
- symmetry: statement "X_k is incomparable to X_l" is equivalent to statement "X_l is incomparable to X_k".

When we focus on fuzzy models of preference expressed in terms of BFRs, it is also important to consider the fuzzy nonstrict preference relation $\boldsymbol{R}$ (also named or fuzzy large preference relation or fuzzy weak preference relation). The relation $\boldsymbol{R}(X_k, X_l)$ reflects the degree to which X_k is at least as good as (or is not worse than; or weakly dominates) X_l. In a somewhat loose sense (Kulshreshtha and Shekar, 2000) $\boldsymbol{R}(X_k, X_l)$ also represents the degree of truth of the statement "X_k is at least as good as X_l". Accordingly, $\boldsymbol{R}$ is a reflexive relation, which can be defined as the union of strict preference and indifference:

$$\boldsymbol{R} = \boldsymbol{P} \cup_S \boldsymbol{I} \tag{5.30}$$

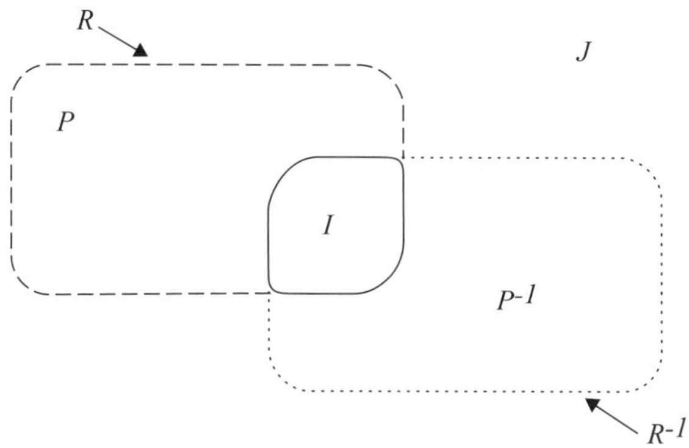

Figure 5.2 Partition of $\mathbf{X} \times \mathbf{X}$ and relationships between $\boldsymbol{P}$, $\boldsymbol{I}$, $\boldsymbol{J}$, $\boldsymbol{R}$, and $\boldsymbol{R}^{-1}$.

Equivalently, in a less intuitive way, this relation can be stated in the form

$$\boldsymbol{R}^d = \boldsymbol{P} \cup_S \boldsymbol{J} \tag{5.31}$$

With the use of the operations on fuzzy sets, it is possible to define $\boldsymbol{I}$, $\boldsymbol{P}$, and $\boldsymbol{J}$ exclusively in terms of $\boldsymbol{R}$ (Figure 5.2 shows the partition of $\mathbf{X} \times \mathbf{X}$ and the existing interactions among $\boldsymbol{I}$, $\boldsymbol{P}$, and $\boldsymbol{J}$ (Fodor and Roubens, 1993)). By considering that $\boldsymbol{I}$ corresponds to all pairs of alternatives that simultaneously satisfy $\boldsymbol{R}(X_k, X_l)$ and $\boldsymbol{R}(X_l, X_k)$, the indifference relation can be stated as

$$\boldsymbol{I} = \boldsymbol{R} \cap_T \boldsymbol{R}^{-1} \tag{5.32}$$

Similarly, as $\boldsymbol{P}(X_k, X_l)$ implies $\boldsymbol{R}(X_k, X_l)$ and $N(\boldsymbol{R}(X_l, X_k))$, the strict preference can be specified as

$$\boldsymbol{P} = \boldsymbol{R} \cap_T \boldsymbol{R}^d, \boldsymbol{P} \cap_T \boldsymbol{P}^{-1} = \phi \tag{5.33}$$

Finally, as the relation $\boldsymbol{J}(X_k, X_l)$ implies $N(\boldsymbol{R}(X_k, X_l))$ and $N(\boldsymbol{R}(X_l, X_k))$, the incomparability relation is given by

$$\boldsymbol{J} = \overline{\boldsymbol{R}} \cap_T \boldsymbol{R}^d, \boldsymbol{J} = \boldsymbol{J}^{-1} \tag{5.34}$$

Therefore, once we have at hand the values of $\boldsymbol{R}(X_k, X_l)$ and $\boldsymbol{R}(X_l, X_k)$, the estimation of $\boldsymbol{I}$, $\boldsymbol{P}$, and $\boldsymbol{J}$ is realized with the use of (5.32), (5.33), and (5.34), respectively. In applications, when the cardinality of $\mathbf{X}$ is low, a DM is usually asked to pick the values in the unit interval that reflect the level of credibility or just the strength of his/her nonstrict preference for one alternative over another. At this point, we should pay attention to the fact that there are several ways of constructing the fuzzy nonstrict preference relations. Particularly, they can be directly assessed by a DM or indirectly determined by a DM, who is supposed to express his/her preferences using other preference formats. This subject will be further discussed in Chapter 6.

Regardless of the method utilized to construct a fuzzy nonstrict preference relation, the fuzzy strict preference and indifference relations can be derived from that relation in a number of ways. The most popular way is also the one that happened to be introduced the earliest. In his original work (Orlovsky, 1978), Orlovsky proposed to derive the fuzzy strict preference and fuzzy indifference relations from the fuzzy nonstrict preference relation as follows:

$$\boldsymbol{P}(X_k, X_l) = \max(\boldsymbol{R}(X_k, X_l) - \boldsymbol{R}(X_l, X_k), 0) \tag{5.35}$$

$$\boldsymbol{I}(X_k, X_l) = \min(\boldsymbol{R}(X_k, X_l), \boldsymbol{R}(X_l, X_k)) \tag{5.36}$$

Example 5.4. Consider that $\boldsymbol{R}$ corresponds to the fuzzy nonstrict preference relation provided by an expert who has been asked to compare three alternatives by using values coming from the unit interval that reflect the degree of nonstrict preference of one alternative over another:

$$\boldsymbol{R} = \begin{bmatrix} 1 & 0 & 0.4 \\ 1 & 1 & 1 \\ 1 & 0.8 & 1 \end{bmatrix} \tag{5.37}$$

By applying (5.35) and (5.36) to (5.37), the following relations are derived:

$$\boldsymbol{P} = \begin{bmatrix} 0 & 0 & 0 \\ 1 & 0 & 0.2 \\ 0.6 & 0 & 0 \end{bmatrix} \tag{5.38}$$

$$\boldsymbol{I} = \begin{bmatrix} 1 & 0 & 0.4 \\ 0 & 1 & 0.8 \\ 0.4 & 0.8 & 1 \end{bmatrix} \tag{5.39}$$

The reader can easily note in Example 5.4 that the resulting relations, $\boldsymbol{P}$ and $\boldsymbol{I}$, satisfy their respective properties, that is, $\boldsymbol{P}$ is irreflexive and min asymmetric (a proof of the min asymmetry of $\boldsymbol{P}$ can be found in Fodor and Roubens, 1994a), and $\boldsymbol{I}$ is reflexive and symmetric. It is interesting to observe that $\boldsymbol{I}$ has entries distributed symmetrically along the main diagonal, as a natural outcome of the symmetry of the indifference relation. Furthermore, whereas the diagonal of $\boldsymbol{I}$ is filled with "ones", being a consequence of the reflexivity of $\boldsymbol{I}$, the diagonal of $\boldsymbol{P}$ is filled with zeros, as a result of the irreflexivity of $\boldsymbol{P}$.

After the pioneering contribution of Orlovsky, several researchers developed other definitions of fuzzy preference relations. Among the most relevant contributions, we can refer to Ovchinikov (1981), Roubens (1989), and Ovchinikov and Roubens (1991). They defined the fuzzy strict preference relation, as summarized in Table 5.1, and relation $\boldsymbol{I}$ in the same way as Orlovsky did (Fodor and Roubens, 1993).

An overview of these works, as well as other similar studies, can be found in De Baets and Fodor (1997). However, it is important to indicate that these works correspond to independent efforts to define fuzzy preference relations (De Baets and Fodor, 1997). A more formal class of results can be also found in the literature. They correspond to axiomatic methods for extending the classical or Boolean preference models to the fuzzy environment, in an attempt to derive the fuzzy strict preference, indifference, and incomparability relations from a fuzzy

Table 5.1 Fuzzy strict preference relations

Reference	Expressions
Ovchinikov (1981)	$\boldsymbol{P}(X_k, X_l) = \begin{cases} \boldsymbol{R}(X_k, X_l) & \text{if } \boldsymbol{R}(X_k, X_l) > \boldsymbol{R}(X_l, X_k) \\ 0 \end{cases}$
Roubens (1989)	$\boldsymbol{P}(X_k, X_l) = \boldsymbol{T}(R(X_k, X_l),\ 1 - \boldsymbol{R}(X_l, X_k))$
Ovchinikov and Roubens (1991)	$\boldsymbol{P}(X_k, X_l) = T_\phi(\boldsymbol{R}(X_k, X_l), N_\phi(\boldsymbol{R}(X_l, X_k)))$

large preference relation $\boldsymbol{R}$, without losing the fuzzy counterparts of the Boolean preference structures. In what follows, we will address the supporting concepts related to the construction of fuzzy preference structures.

5.3 Preference Structure of Binary Fuzzy Preference Relations

In general, given the De Morgan triplet (T, S, N), a fuzzy preference structure (FPS) of BFRs can be defined as a collection of fuzzy binary relations $\boldsymbol{P}$, $\boldsymbol{I}$, and $\boldsymbol{J}$ that satisfies some requirements. It is instructive to start with a full list of conditions to be satisfied by these relations, in order to avoid losing some practical aspects of the structure. However, it should be indicated that this list is not the minimal one, that is, some of the conditions shown there are redundant (De Baets and Van De Walle, 1997).

For every $(X_k, X_l) \in \mathbf{X} \times \mathbf{X}$, it is required that (De Baets and Van De Walle, 1997):

$$\boldsymbol{I} \text{ is reflexive and symmetric} \tag{5.40}$$

$$\boldsymbol{P} \text{ is irreflexive and } T\text{-asymmetric} \tag{5.41}$$

$$\boldsymbol{J} \text{ is irreflexive and symmetric} \tag{5.42}$$

$$\boldsymbol{P} \cap_T \boldsymbol{I} = \emptyset \tag{5.43}$$

$$\boldsymbol{P} \cap_T J = \emptyset \tag{5.44}$$

$$\boldsymbol{J} \cap_T \boldsymbol{I} = \emptyset \tag{5.45}$$

$$\boldsymbol{P} \cup_S \boldsymbol{P}^{-1} \cup_S \boldsymbol{I} \cup_S \boldsymbol{J} = \mathbf{X} \times \mathbf{X} \tag{5.46}$$

Conditions (5.40)–(5.42) refer to the set of properties that relations $\boldsymbol{P}$, $\boldsymbol{I}$, and $\boldsymbol{J}$ must have in order to guarantee that the observed characteristics of the judgments provided by a DM are properly reflected by the model (as discussed in the previous section). Conditions (5.43)–(5.45) preserve the mutually exclusive character of these judgments. Furthermore, conditions (5.43)–(5.45), in conjunction with the completeness condition (5.46), assert that each couple of alternatives necessarily belongs exactly to one of relations $\boldsymbol{P}$, $\boldsymbol{P}^{-1}$, $\boldsymbol{I}$, or $\boldsymbol{J}$.

At this point, it should be clear to the reader that the definition of a FPS requires the selection of a De Morgan triplet and that the selection of different triplets obviously results in different FPSs. However, a negative result demonstrated by Alsina (1985) indicates that if a De Morgan triplet (T, S, N) is utilized to represent the complement, intersection, and union of BFRs, then the equality

$$\boldsymbol{Z} = (\boldsymbol{Z} \cap_T \boldsymbol{W}) \cup_S (\boldsymbol{Z} \cap_T \overline{\boldsymbol{W}}) \tag{5.47}$$

is not satisfied for any BFRs $\boldsymbol{Z}$ and $\boldsymbol{W}$. If we consider Z as being $\boldsymbol{R}$ and $\boldsymbol{W}$ as $\boldsymbol{R}^d$, then this implies that the relationships $\boldsymbol{P} = \boldsymbol{R} \cap_T \boldsymbol{R}^d, \boldsymbol{I} = \boldsymbol{R} \cap_T \boldsymbol{R}^{-1}$, and $\boldsymbol{R} = \boldsymbol{P} \cup_S \boldsymbol{I}$ are inconsistent for any reflexive fuzzy binary relation $\boldsymbol{R}$. For this reason, the extension (fuzzification) of the classical preference structures is not as straightforward as we might have expected (Bufardi, 1999). In the next section, we will see which t-norms can be considered in the definition of a FPS satisfying properties (5.40)–(5.46).

5.4 A Method for Constructing a Fuzzy Preference Structure

The nonintuitive result demonstrated by Alsina (1985) motivated several researchers to work on the development of axiomatic ways of constructing a FPS (see, for instance, Ovchinnikov and Roubens, 1992; Fodor and Roubens, 1994a; Bufardi, 1998; Llamazares, 2003; Fodor and Rudas, 2006; Fodor and De Baets, 2008). This section presents the main results associated with the method introduced by Fodor and Roubens (1994a), which starts by considering the following axioms:

- Independence of irrelevant alternatives: Given two alternatives X_k and X_l, the value of $\boldsymbol{P}(X_k, X_l)$, $\boldsymbol{I}(X_k, X_l)$, and $\boldsymbol{J}(X_k, X_l)$ must depend only on the values of $\boldsymbol{R}(X_k, X_l)$ and $\boldsymbol{R}(X_l, X_k)$. In this way, there are mappings from $[0, 1] \times [0, 1] \rightarrow [0, 1]$, namely $P(x, y)$, $I(x, y)$, and $J(x, y)$, that satisfy the following conditions:

$$\boldsymbol{P}(X_k, X_l) = P(\boldsymbol{R}(X_k, X_l), N(\boldsymbol{R}(X_l, X_k))) \tag{5.48}$$

$$\boldsymbol{I}(X_k, X_l) = I(\boldsymbol{R}(X_k, X_l), \boldsymbol{R}(X_l, X_k)) \tag{5.49}$$

$$\boldsymbol{J}(X_k, X_l) = J(N(\boldsymbol{R}(X_k, X_l)), N(\boldsymbol{R}(X_l, X_k))) \tag{5.50}$$

  Each function, say $P(x, y)$, $I(x, y)$, and $J(x, y)$, works as a generator function of a strict preference relation, an indifference relation, and an incomparability relation, respectively. They must be defined in such a way that the *positive association principle* and the *symmetry property* are satisfied:
- Positive association principle: Functions $P(x, y)$, $I(x, y)$, and $J(x, y)$ must be nondecreasing with respect to the first argument and nonincreasing with respect to the second argument. The monotonicity of $P(x, y)$, $I(x, y)$, and $J(x, y)$ with respect to its respective arguments is coherent with the data shown in Table 5.2, which presents the Boolean (two-valued) conditions for the classical binary relations $\boldsymbol{P}$, $\boldsymbol{I}$, and $\boldsymbol{J}$, for different states of $\boldsymbol{R}(X_k, X_l)$ and $\boldsymbol{R}(X_l, X_k)$ (Fodor and Roubens, 1994a; Fodor and Roubens, 1994b).

Table 5.2 Monotonicity of $\boldsymbol{P}$, $\boldsymbol{I}$, and $\boldsymbol{J}$

$\boldsymbol{R}(X_k, X_l)$	$\boldsymbol{R}(X_l, X_k)$	$\boldsymbol{P}(X_k, X_l)$	$\boldsymbol{I}(X_k, X_l)$	$\boldsymbol{J}(X_k, X_l)$
0	0	0	0	1
0	1	0	0	0
1	0	1	0	0
1	1	0	1	0

The positive association principle can be justified by the following observations: in the case of $\boldsymbol{P}$, considering the first argument of $P(x, y)$, if the level of nonstrict preference of X_k over X_l increases, it is naturally expected that the level of strict preference of X_k over X_l does not decrease. On the other hand, if we consider the second argument of $P(x, y)$, if the level in which X_l is at least as good as X_k is increasing, then we expect that the level of strict preference of X_k over X_l does not increase. Similar observations justify the principle of positive association for $I(x, y)$ and $J(x, y)$.

- Symmetry: As $\boldsymbol{I}$ and $\boldsymbol{J}$ are symmetric relations, functions $I(x, x)$ and $J(x, x)$ must be symmetric concerning their respective arguments.

Starting from these axioms, the construction of a fuzzy structure consists of defining six continuous functions defined in $[0, 1] \times [0, 1] \rightarrow [0, 1]$. They are the generator functions $P(x, y)$, $I(x, y)$, and $J(x, y)$ and the triplet of De Morgan operators (T, S, N), which can be determined by solving the system of equations for any pair $(X_k, X_l) \in \mathbf{X} \times \mathbf{X}$,

$$\begin{cases} S(\boldsymbol{P}(X_k, X_l), \boldsymbol{I}(X_k, X_l)) = \boldsymbol{R}(X_k, X_l) \\ S(\boldsymbol{P}(X_k, X_l), \boldsymbol{J}(X_k, X_l)) = \boldsymbol{R}^d(X_k, X_l) \end{cases} \tag{5.51}$$

where $\boldsymbol{P}$, $\boldsymbol{I}$, and $\boldsymbol{R}$ are given by expressions (5.48)–(5.50) respectively, which are supposed to satisfy the axioms of the positive association principle and that of symmetry. Obviously, solving (5.51) is not a simple task. Fortunately, the results presented by Fodor and Roubens (1994a) allow us to characterize the complete set of solutions for this system of equations.

It has been proved that if a particular set of continuous functions (T, S, N, P, I, J) is a solution of (5.51), then there is an automorphism ϕ of the unit interval such that T is the ϕ-transform of the Lukasiewicz t-norm as defined by (5.12), N is the ϕ-transform of the standard negation, as given by (5.13), and S is the ϕ-transform of the Lukasiewicz t-conorm as given by (5.14). In other words, this means that T must be Archimedean with zero divisors. Van de Walle, De Baets, and Kerre (1998) demonstrated that when T is non-Archimedean, having zero divisors, the degree of strict preference, indifference, and incomparability in [0,1[is bounded by a value strictly lower than one. This violates the assignment principle, which is the most important requirement for a fuzzy preference structure (Fodor and De Baets, 2008):

- Assignment principle: For any pair of alternatives $(X_k, X_l) \in \mathbf{X} \times \mathbf{X}$, a DM can pick from the unit interval any value to assign at least one of the degrees $\boldsymbol{P}(X_k, X_l)$, $\boldsymbol{P}(X_l, X_k)$, $\boldsymbol{I}(X_k, X_l)$, and $\boldsymbol{J}(X_k, X_l)$.

Furthermore, it is also demonstrated that the solution to (5.51) necessarily satisfies the following inequalities (Fodor and Roubens, 1994a):

$$T_\phi(\boldsymbol{R}(X_k, X_l), N_\phi(\boldsymbol{R}(X_l, X_k))) \leq \boldsymbol{P}(X_k, X_l) \leq \min(\boldsymbol{R}(X_k, X_l), N_\phi(\boldsymbol{R}(X_l, X_k))) \tag{5.52}$$

$$T_\phi(\boldsymbol{R}(X_k, X_l), \boldsymbol{R}(X_l, X_k)) \leq \boldsymbol{I}(X_k, X_l) \leq \min(\boldsymbol{R}(X_k, X_l), \boldsymbol{R}(X_l, X_k)) \tag{5.53}$$

$$T_\phi(N_\phi(\boldsymbol{R}(X_k, X_l)), N_\phi(\boldsymbol{R}(X_l, X_k))) \leq \boldsymbol{J}(X_k, X_l) \leq \min(N_\phi(\boldsymbol{R}(X_k, X_l)), N_\phi(\boldsymbol{R}(X_l, X_k))) \tag{5.54}$$

Among possible solutions of (5.51), the one given by

$$\begin{cases} \boldsymbol{P} = \max(\boldsymbol{R}(X_k, X_l) - \boldsymbol{R}(X_l, X_k), 0) \\ \boldsymbol{I} = \min(\boldsymbol{R}(X_k, X_l), \boldsymbol{R}^{-1}(X_k, X_l)) \\ \boldsymbol{J} = \min(1 - \boldsymbol{R}(X_k, X_l), 1 - \boldsymbol{R}(X_l, X_k)) \end{cases} \tag{5.55}$$

deserves special attention. The reader should note that the expression for $\boldsymbol{P}$ in (5.55) can be obtained by substituting the generator function $P(x, y)$ in (5.48) by the Lukasiewicz t-norm, as follows:

$$\boldsymbol{P}(X_k, X_l) = \max(\boldsymbol{R}(X_k, X_l) + N(\boldsymbol{R}(X_l, X_k)) - 1, 0). \tag{5.56}$$

Given that $N(\boldsymbol{R}(X_l, X_k)) = 1 - \boldsymbol{R}(X_l, X_k)$, it is easy to see that (5.56) and (5.55) are equivalent. It is also worth noting that this corresponds to an extreme solution of (5.51), as the formula of $\boldsymbol{P}$ in (5.55) returns the lower limit of (5.52) and, at the same time, both $\boldsymbol{I}$ and $\boldsymbol{J}$ (defined as in (5.55)) correspond to the upper limits of (5.53) and of (5.54), respectively.

Further, the preference structure given by (5.55) has an interesting property: if the fuzzy nonstrict preference relation satisfies the min-transitivity condition, then both $\boldsymbol{I}$ and $\boldsymbol{P}$ certainly satisfy the min-transitivity, as demonstrated by Burfardi (1998). From a practical point of view, this property is rather meaningful, given that min-transitivity has been traditionally taken as a consistency condition for pairwise judgments (Zimmermann, 1996). However, as will be discussed in the next section and in Chapter 6, the current literature still lacks a fair consistency condition for the pairwise judgments expressed in terms of fuzzy preference relations.

Finally, it is also important to indicate that, as one can see by comparing the definitions of $\boldsymbol{P}$ and $\boldsymbol{I}$ in (5.55) to the expressions proposed by Orlovsky for fuzzy strict preference and fuzzy indifference relations (see (5.35) and (5.36)), they are equivalent. Indeed, the preference model proposed by Orlovsky can be viewed as a particular case of (5.55), where the incomparability relation corresponds to an empty set and $\boldsymbol{R}$ is complete, in the sense that it satisfies the max-completeness property

$$\max(\boldsymbol{R}(X_k, X_l), \boldsymbol{R}(X_l, X_k)) = 1 \tag{5.57}$$

Figures 5.3–5.6 offer graphical representations of relations $\boldsymbol{P}$, $\boldsymbol{I}$, and $\boldsymbol{J}$ being expressed in terms of $\boldsymbol{R}$ and $\boldsymbol{R}^{-1}$. As can be seen, when condition (5.57) is satisfied, the incomparability relation $\boldsymbol{J}$ corresponds to an empty relation.

5.5 Consistency of Fuzzy Preference Relations

In the Boolean context, the concept of consistency has traditionally been defined in terms of acyclicity, that is, the absence of cycles or sequential judgments such as

$$\boldsymbol{R}(X_1, X_2), \boldsymbol{R}(X_2, X_3), \ldots, \boldsymbol{R}(X_k, X_{k+1}), \boldsymbol{R}(X_{k+1}, X_1) \tag{5.58}$$

In the context of fuzzy sets, the traditional requirement to characterize consistency has been enriched by also considering the intensity of preferences. Among the most utilized consistency

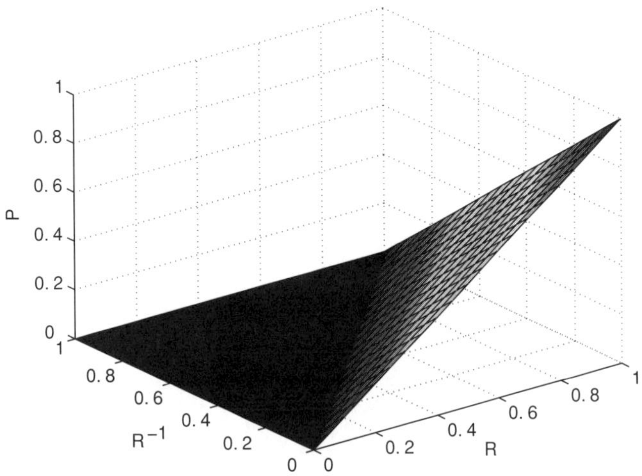

Figure 5.3 Fuzzy strict preference relation $\boldsymbol{P}$ defined by (5.55).

conditions for fuzzy preference relations, we can name the min-transitivity, which is given by (5.23), and the weak-transitivity condition, which is given by the following:

$$\begin{aligned}&\text{If } \boldsymbol{R}(X_k, X_j) \geq \boldsymbol{R}(X_j, X_k) \text{ and } \boldsymbol{R}(X_j, X_l) \geq \boldsymbol{R}(X_l, X_j),\\ &\text{then } \boldsymbol{R}(X_k, \boldsymbol{R}_l) \geq \boldsymbol{R}(X_l, X_k) \quad \forall X_k, X_j, X_l \in \mathbf{X}\end{aligned} \tag{5.59}$$

The interpretation of (5.59) leads to the minimum requirement that should be satisfied by the preference judgments supplied by a rational person. This implies that if someone says that X_k is preferred to X_j and that X_j is preferred to X_l, then X_k is preferred to X_l, without considering the strength of the preferences.

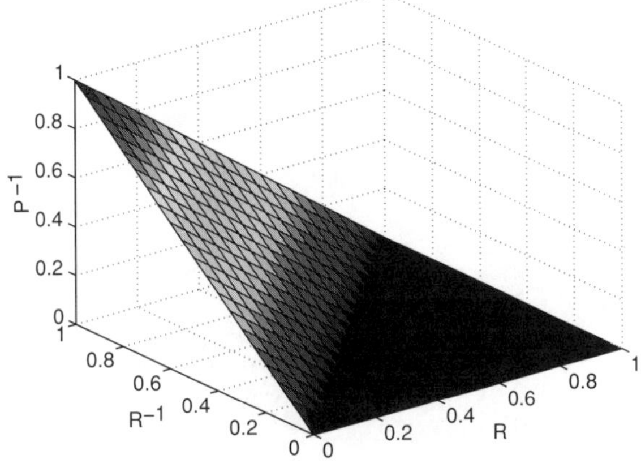

Figure 5.4 The inverse $\boldsymbol{P}^{-1}$ of the fuzzy strict preference relation defined by (5.55).

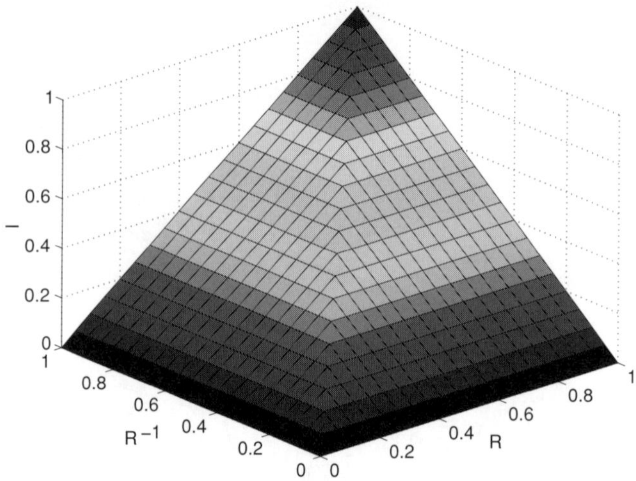

Figure 5.5 Fuzzy indifference relation $\boldsymbol{I}$ defined by (5.55).

The min-transitivity is a stronger consistency condition than the weak transitivity, as it enhances the weak transitivity by also requiring that the strength of preference of X_k over X_l should be higher than the strength of preference observed between X_k and X_j, as well as between X_j and X_l. However, as indicated by Herrera-Viedma *et al.* (2004), min-transitivity may be considered excessively rigorous as even a fuzzy preference relation, perfectly consistent for practical purposes, may not satisfy the min-transitivity.

In reality, the problem concerning the consistency of fuzzy preference relations lies in the fact that it is usually difficult to guarantee any level of consistency within the pairwise comparisons provided by the DMs in the decision-making process. Under this scenario, the

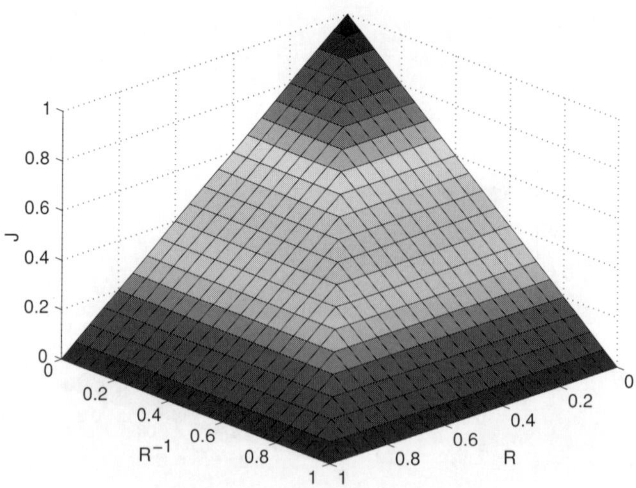

Figure 5.6 Fuzzy incomparability relation $\boldsymbol{J}$ defined by (5.55).

definition of a reasonable consistency condition for fuzzy preference relations has been a topic of debate in the fuzzy community (Herrera-Viedma *et al.*, 2004). This debate has been motivated mainly by the following aspects:

- in real applications, a DM is usually incapable of providing perfectly consistent judgments, in particular when the set of alternatives is large;
- the lack of consistency within the judgments can lead to incompatible conclusions;
- some works in the literature recognize that rational judgments may not be always transitive.

One instructive example of intransitive preferences is related to the comparison of three chess players, X_k, X_j, and X_l, exhibiting very different strategies (Hwang and Yoon, 1981). Whereas it is expected that X_k triumphs over X_j and that X_j triumphs over X_l, due to the peculiarities of their respective strategies, when X_k challenges X_l, one can expect that X_l will be the winner, because of a particular capacity of X_l to take advantage of the weakness of X_k.

A classical example that illustrates the fact that indifference is not transitive refers to the amount of sugar in a cup of coffee. One can judge that adding a lump of sugar or no sugar to the coffee is indifferent, because it is almost impossible to perceive a change in the taste of the coffee; for the same reason, one can judge that adding a lump or two is indifferent; adding two lumps or three is indifferent; and so forth. If it is assumed that indifference is transitive, we can make the erroneous conclusion that it does not matter if one or ten lumps of sugar are dropped in the coffee, which obviously is not true. In reality, situations where a series of subsequent indifferences results in a preference are not rare.

Taking all of these observations into consideration, usually, the T-transitivity property has not been imposed as a requirement to be specified by the fuzzy nonstrict preference, the fuzzy strict preference, or the fuzzy indifference relations. However, the fact that inconsistencies within pairwise judgments may lead to incoherent results for decision-making problems has motivated several authors to investigate whether certain fuzzy preference relations satisfy the T-transitivity. In practice, this finding is rather valuable for the selection of a method of analysis of $\langle \mathbf{X}, \boldsymbol{R} \rangle$ models adequate for dealing with cyclic preferences, whenever it is not possible to guarantee the T-transitivity of the fuzzy preference relations of the $\langle \mathbf{X}, \boldsymbol{R} \rangle$ model under consideration.

5.6 Conclusions

In this chapter, we have considered some selected issues related to preference modeling through preference structures of binary fuzzy relations, including the definition of fuzzy preference structures, a method for constructing a fuzzy preference structure, and the insertion of the particular results of the pioneering and seminal paper of Orlovsky (Orlovsky, 1978), into an axiomatic framework.

As indicated in the text, extending the Boolean definition of preference structure to fuzzy settings requires the adherence to certain conditions, related to the selection of a De Morgan triplet to implement the AND, OR, and NOT operators on the unit interval. Fortunately, with the advance of the research on this topic, it has been possible to construct reasonable preference models, with large applicability in the areas of multicriteria analysis and decision-making. The reader can refer to Fodor and De Baets (2008) for more recent results on the development of fuzzy preference structures.

The fact that in real-world situations the transitivity may not be inherent to the fuzzy preference relations has led the research community to consider the development of rather flexible structures. Usually, the T-transitivity property has not been imposed as a requirement to be verified by the fuzzy nonstrict preference, the fuzzy strict preference, and the fuzzy indifference relations. However, the fact that inconsistent fuzzy preference relations may lead to incoherent results for decision-making problems has motivated several authors to investigate whether certain fuzzy preference relations satisfy the T-transitivity property. A further discussion on the consistency of preference relations is presented in the next chapter.

Exercises

Problem 5.1. Verify whether the fuzzy relation

$$\boldsymbol{R} = \begin{bmatrix} 1 & 0.5 & 0.5 & 0.5 \\ 0.5 & 1 & 0.5 & 0.5 \\ 0.6 & 0.6 & 1 & 0.6 \\ 0.7 & 0.5 & 0.5 & 1 \end{bmatrix}$$

satisfies (a) reflexivity, (b) min-asymmetry, (c) max-completeness, (d) min-transitivity.

Problem 5.2. Derive the fuzzy strict preference relation and the fuzzy indifference relation from the fuzzy nonstrict preference relation of Problem 5.1, by using (5.35) and (5.36), respectively.

Problem 5.3. Verify whether the resultant fuzzy indifference relation from Problem 5.2 is (a) reflexive, (b) symmetric, and (c) min-transitive.

Problem 5.4. Verify whether the fuzzy strict preference relation obtained in Problem 5.2 is (a) irreflexive, (b) min-asymmetric, and (c) min-transitive.

Problem 5.5. Verify whether the fuzzy relations

$$\text{(a)} \quad \boldsymbol{R} \begin{bmatrix} 1 & 0.9 & 0.8 \\ 0.01 & 1 & 0.94 \\ 0.11 & 1 & 1 \end{bmatrix} \qquad \text{(b)} \quad \boldsymbol{R} \begin{bmatrix} 1 & 0 & 0 \\ 0 & 1 & 0 \\ 1 & 1 & 1 \end{bmatrix}$$

satisfy max-completeness. Derive the incomparability relation from $\boldsymbol{R}$ with the use of (5.55). Relate the obtained result to the max-completeness property.

References

Alsina, C. (1985) On a family of connectives for fuzzy sets. *Fuzzy Sets and Systems*, **16** (3), 231–235.

Burfardi, A. (1998) On the construction of fuzzy preference structures. *Journal of Multi-criteria Decision Analysis*, **7** (3), 169–175.

Bufardi, A. (1999) On the fuzzification of the classical definition of preference structure. *Fuzzy Sets and Systems*, **104** (2), 323–332.

De Baets, B. and Fodor, J.C. (1997) Twenty years of fuzzy preference structures. *Rivista di Matematica per le Scienze Economiche e Sociali*, **20** (1), 45–66.

De Baets, B. and Van de Walle, B. (1997) Minimal definitions of classical and fuzzy preference structures. Proceedings of the Annual Conference of the North American Fuzzy Information Processing Society, Syracuse, pp. 299–304.

Fodor, J.C. and De Baets, B. (2008) Fuzzy preference modeling: fundamentals and recent advances, in *Fuzzy Sets and Their Extensions: Representation, Aggregation and Models* (eds H. Bustince, F. Herrera, and J. Montero), Springer-Verlag, New York, pp. 207–217.

Fodor, J.C. and Roubens, M. (1993) Fuzzy strict preference relations in decision making. Proceedings of the Second IEEE International Conference on Fuzzy Systems, San Francisco, pp. 1145–1149.

Fodor, J.C. and Roubens, M. (1994a) Valued preference structures. *European Journal of Operational Research*, **79** (2), 277–286.

Fodor, J.C. and Roubens, M. (1994b) *Fuzzy Preference Modelling and Multicriteria Decision Support*, Kluwer, Boston, MA.

Fodor, J.C. and Rudas, I. (2006) Two functional equations in fuzzy preference modeling, INES'06. Proceedings of the International Conference on Intelligent Engineering Systems, Cidade, pp. 17–23.

Herrera-Viedma, E., Herrera, F., Chiclana, F., and Luque, M. (2004) Some issues on consistency of fuzzy preference relations. *European Journal of Operational Research*, **154** (1), 98–109.

Hwang, C.L. and Yoon, K. (1981) *Multiple Attribute Decision Making – Methods and Applications: A State of the Art Survey*, Springer-Verlag, New York.

Kulshreshtha, P. and Shekar, B. (2000) Interrelations among fuzzy preference-based choice functions and significance of rationality conditions: a taxonomic and intuitive perspective. *Fuzzy Sets and Systems*, **109** (3), 429–445.

Llamazares, B. (2003) Characterization of fuzzy preference structures through Lukasiewicz triplets. *Fuzzy Sets and Systems*, **136** (2), 217–235.

Orlovsky, S.A. (1978) Decision making with a fuzzy preference relation. *Fuzzy Sets and Systems*, **1** (3), 155–167.

Ovchinnikov, S. (1981) Structure of fuzzy binary relations, *Fuzzy Sets and Systems*, **6** (2), 169–195.

Ovchinnikov, S. and Roubens, M. (1991) On strict preference relations. *Fuzzy Sets and Systems*, **43** (3), 319–326.

Ovchinnikov, S. and Roubens, M. (1992) On fuzzy strict preference, indifference and incomparability relations. *Fuzzy Sets and Systems*, **49** (1), 15–20.

Öztürk, M., Tsoukiàs, A., and Vincke, P. (2005) Preference modelling, in *Multiple Criteria Decision Analysis: State of the Art Surveys* (eds J. Figueira, S. Greco, and M. Ehrgott), Springer-Verlag, New York, pp. 265–292.

Roubens, M. (1989) Some properties of choice functions based on valued binary relations. *European Journal of Operational Research*, **40** (3), 309–321.

Van de Walle, B., De Baets, B., and Kerre, E.E. (1998) Characterizable fuzzy preference structures. *Annals of Operations Research*, **80** (0), 105–136.

Zimmermann, H.-J. (1996) *Fuzzy Set Theory and Its Applications*, Kluwer, Boston.

6

Construction of Fuzzy Preference Relations

In dealing with $\langle \mathbf{X}, \boldsymbol{R} \rangle$ models, which are constructs for multiattribute decision-making, based on fuzzy preference relations, a fundamental question arises on how one can construct fuzzy preference relations to reflect the preferences of a DM. In practice, a DM can directly quantify the fuzzy preference relation. Alternatively, a DM can use other ways to express preferences, which can be then converted into fuzzy preference relations with the use of adequate transformations. Taking this into consideration, in this chapter we present five main preference formats which cover a majority of real-world situations when preparing information for decision-making problems. These formats are the ordering of alternatives, utility values, fuzzy estimates, multiplicative preference relations, and fuzzy preference relations. The chapter also presents transformation functions used for converting different preference formats into fuzzy preference relations. Taking into account the importance of collecting consistent judgments, some questions of repairing inconsistencies in the estimates of judgments of DMs/experts are also discussed.

6.1 Preference Formats

In real-world applications, every professional involved in any decision process has their own perception of the problem, a different way of thinking, and usually access to different sources of information. As a consequence, it is quite natural to envision circumstances where every DM selects a different way to express their preferences. Furthermore, several factors may lead a DM to select a different way for expressing preferences about each criterion. Among these factors we can list the following:

- Each criterion comes with its significance (a fundamental feature which provides significance on the difference between two degrees evaluated for this criterion). Depending on whether this significance has a qualitative or quantitative character, the use of certain preference formats can make the preference elicitation process easier and also more reliable.

Fuzzy Multicriteria Decision-Making: Models, Methods and Applications Witold Pedrycz, Petr Ekel and Roberta Parreiras

- Each criterion is associated with information arising from different sources and with information having different levels of uncertainty.
- A DM may find that his/her preference strengths can be better reflected or quantified by a specific preference format.
- The fact that a DM may possess previous knowledge or experience in expressing a specific preference format can motivate him/her to choose it again.

In the current literature on decision-making, we can distinguish eight preference formats that can be utilized to establish preferences for the alternatives under consideration (Zhang, Chen, and Chong, 2004; Zhang, Wang, and Yang, 2007). With the availability of different formats, a DM can select the one that makes him/her feel more comfortable in articulating his/her own preferences. Next, we describe five preference formats, which cover most of the real-world situations. They include:

1. The ordering of the alternatives
2. Utility values
3. Fuzzy estimates
4. Multiplicative preference relations
5. Fuzzy preference relations.

6.1.1 Ordering of Alternatives

When a DM encounters difficulties in assessing quantitatively the strength of preferences, it is advantageous to use information of purely ordinal character. By asking a DM to provide a complete ranking of the alternatives in accordance with his/her preferences, the DM is released from having to quantify the difference in his/her preference strengths between any two alternatives. In this way, the chances of deriving recommendations based on incorrect information are reduced.

The ordering of alternatives from best to worst can be represented as an array $\boldsymbol{O} = [O(X_1)\ldots O(X_n)]$, with $O(X_k)$ being a permutation function which returns the position of alternative X_k among the integer values $\{1, 2, \ldots, n\}$ (Chiclana, Herrera, and Herrera-Viedma, 1998).

Example 6.1. Consider that a DM is asked to order five alternatives $\mathbf{X} = \{X_1, X_2, \ldots, X_5\}$, from best (first position) to worst (last position) for a given criterion. Table 6.1 shows the DM's judgments and the corresponding ordered array. For instance, no. 3 in vector $\boldsymbol{O}$ reflects the fact that X_5 is the third best alternative, according to the DM preferences.

Table 6.1 Ordering of alternatives

Alternatives	X_1	X_2	X_3	X_4	X_5
Positions	Second	First	Fourth	Fifth	Third
Array	$\boldsymbol{O} = [2\ 1\ 4\ 5\ 3]$				

6.1.2 Utility Values

The terms "utility function" and "value function" are utilized to refer to two types of preference models. Utility theory deals with preference models for risky decisions, that is, decisions involving alternatives whose consequences are uncertain and as a consequence involve risks. The value theory is considered as a simplification of utility theory for dealing with decisions under certainty. Here, we intend to focus on preference models based on utility functions and value functions. However, we describe only the preference elicitation process of value functions and indicate references (Keeney and Raiffa, 1976; Von Winterfeldt and Edwards, 1986) for preference elicitation procedures associated with the construction of utility functions. Although the elicitation procedure for each type of model is definitely different, there is no need to make a distinction among them from the point of view of the use of transformation functions used for converting them to fuzzy preference relations. Thus, for simplicity of presentation, from now on, we use the term "utility" to refer to both types of models and, thereby, the text is coherently maintained with some relevant references about such transformation functions (Chiclana, Herrera, and Herrera-Viedma, 1998; Chiclana, Herrera, and Herrera-Viedma, 2001; Zhang, Wang, and Yang, 2007).

Now, let us focus on the representation of the preferences of a DM with the use of the preference function called the utility function $U(x) : \mathbf{X} \to [0, 1]$ (by convention, the highest value of the utility function is equal to one and its lowest value is equal to zero). In the literature, we can distinguish between two main types of utility functions: the ordinal and the cardinal ones. The ordinal utility function is related to the ordering of the alternatives rather than expressing the preference strength of one alternative over another. In real-world applications, the ordinal utility function is usually modeled by a maximizing profit function or a minimizing cost function, defined over the significance axis of the criterion being studied (Dyer, 2005).

The ordinal utility function $U(x) : \mathbf{X} \to [0, 1]$ is supposed to preserve the preference ordering of the alternatives in such a way that

$$\text{if } U(X_k) > U(X_l), \text{ then } X_k \text{ is preferred to } X_l \tag{6.1}$$

$$\text{if } U(X_k) = U(X_l), \text{ then } X_k \text{ is indifferent to } X_l \tag{6.2}$$

Since in ordinal utility functions the ranking of the numbers is all that matters, any monotonic transformation of this function is considered equivalent to it. Indeed, the main weakness of ordinal utility functions lies in the fact that different ordinal utility functions can be utilized to reflect the same ordering of the alternatives. However, in the aggregation across the criteria in the multicriteria analysis, each one of these admissible functions may lead to a different outcome. Nevertheless, such types of ambiguity can be diminished by using the measurable or cardinal utility function (rather than the ordinal utility function) to capture the strength of preferences. Indeed, the use of the cardinal utility function can contribute significantly to the effectiveness of decision-making by resolving ambiguities that may emerge in the process of multicriteria analysis.

A very important type of cardinal utility function (Farquhar and Keller, 1989; Belton, 1999; Dyer, 2005) is founded on differences in preference strengths, in such a way that, for given a measurable utility function $U(x) : \mathbf{X} \to [0, 1]$, if we have

$$U(X_k) - U(X_l) > U(X_j) - U(X_k) \tag{6.3}$$

then we can infer that the difference in the preference between X_k and X_l is greater than the difference in preference between X_j and X_k.

It is worth noting that the ratio between two preference degrees expressed as cardinal utilities based on interval scales is not meaningful, only their differences or the ratio between their differences are significant. The ratio between preference strengths makes sense only when utilities are measured on a ratio scale. This is the most rigorous type of preference measure, being admissible only when the preference strengths are measured on an appropriate scale with an absolute zero. Under such circumstances, one can state the preference for any alternative X_k over another alternative X_l as a ratio $U(X_k)/U(X_l)$ between their respective utilities, that is, it is possible to determine how many times alternative X_k is better (or worse) than the other alternative X_l by means of the ratio $U(X_k)/U(X_l)$.

The interval-scale cardinal utility functions can be determined in several ways. Here we consider two of them: the direct rating and the bisection techniques. In both of them, a DM is supposed to compare differences in preference strengths in order to determine the utility of each alternative. It should be mentioned that these procedures do not accommodate intransitivity in any form. In this way, a DM is required to reconsider all judgments from which intransitivity arises. Next, we begin by describing the direct rating. It consists of determining utility values, taking as reference only two anchor points, in such a way that it is not necessary to define a scale for characterizing other performances than the alternatives being evaluated (Winterfeldt and Edwards, 1986; Belton, 1999).

6.1.2.1 Direct Rating Technique

Step 1. A DM identifies two anchors, which correspond to the worst evaluated alternative and the best evaluated alternative, considering the criterion being studied. The values 0 and 1, respectively, are assigned to them.

Step 2. A DM rates the remaining alternatives in between the extreme points of the scale, in such a way that the spacing between the alternatives reflects the strength of preferences of one alternative over another.

Step 3. A DM reviews the assessments and, whenever necessary, updates them (the process stops only if the DM is in perfect accordance with the elicited utility values). If the criterion under consideration can be captured by an attribute which can be measured on a numerical scale, it is possible to plot the assessed points and draw a smooth curve passing through them.

Example 6.2. Consider that four apartments for rent, namely X_1, X_2, X_3, and X_4, are to be compared by a DM, by taking into account their respective layouts. When analyzing this decision criterion, the DM is particularly concerned about how the living area of each apartment is distributed among its rooms.

In Step 1, when the DM involved in decision-making is asked to indicate the anchors, the apartments X_1 and X_3, respectively, are identified as the alternatives having the best and the worst layouts: whereas X_1 has a large living area satisfactorily distributed among eight rooms, X_3 is very small and has four rooms, which is not a sufficient number of room in his opinion.

In Step 2, the DM thinks that the difference in preference of X_1 over X_2 is much lower than the difference in preference of X_2 over X_3, as X_2 also has a large living area, although

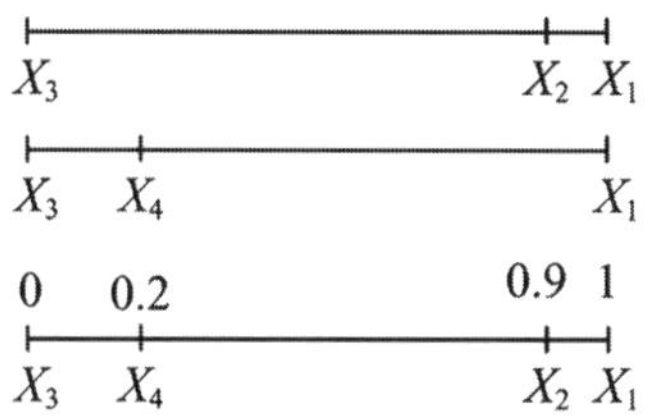

Figure 6.1 Utilities elicited through the direct rating technique.

it only has five rooms. Besides, the DM thinks that the difference in preference of X_4 over X_3 is much lower than the difference in preference of X_1 over X_4, as X_4 is also very small, being a little larger than X_3 and having just one more room than it. Finally, the DM judges that X_2 is preferred to X_4, which is coherent with the previously given explanations; the DM defined the utility values as shown in Figure 6.1. In Step 3, those assessments are confirmed to be coherent with the preferences of the DM.

It should be mentioned that in the construction of an underlying scale for a criterion, a DM can think on a local or global scale. For the local scale, the extreme points are defined by taking as reference only the elements of set **X**. A global scale is defined by taking as reference a wider set of possibilities, being the extreme points of the scale set as the ideal and the worst conceivable performances, or the best and worst performances that in reality could be observed for this criterion (Belton, 1999). The advantage of a global scale lies in the fact that once it has been defined, there is no need to modify it if **X** changes. On the other hand, it is definitely more difficult to construct a global scale than a local one.

With the use of the mid-value splitting or bisection procedure, having defined a measurable attribute that captures the criterion under study, a DM is required to seek mid-value points as described below.

6.1.2.2 Bisection Procedure

Step 1. A DM identifies two anchors, namely x_0 and x_1, which correspond to the worst and the best evaluated alternatives for this criterion. The values 0 and 1, respectively, are assigned to them.

Step 2. A DM determines an object $x_{0.5}$ that is positioned in the middle between the two extreme anchors in such a way that, in the DM's opinion, exchanging x_0 for $x_{0.5}$ is as attractive as exchanging $x_{0.5}$ for x_1. After assessing $x_{0.5}$, the procedure is continued to determine $x_{0.75}$ and $x_{0.25}$ (and so forth), which corresponds to the mid-values of intervals $[x_{0.5}, x_1]$ and $[x_0, x_{0.5}]$. In general, as is indicated by Winterfeldt and Edwards (1986), three points cautiously assessed by a DM may provide enough information to approximate the curve of the utility function.

Step 3. The assessed points are plotted and a smooth curve is drawn passing through them. If a DM considers that the curve is a satisfactory representation of his/her preferences, then the process is terminated.

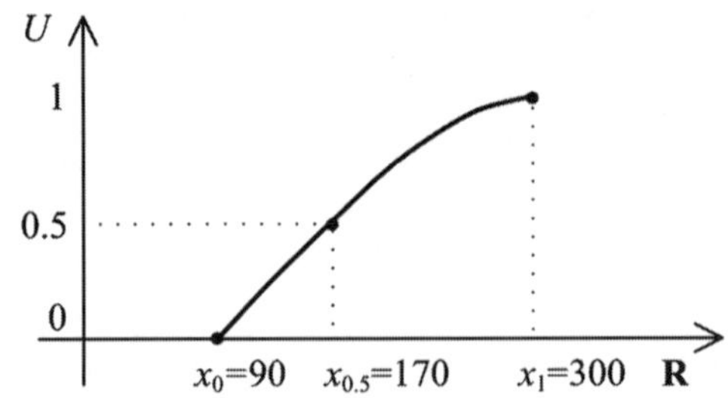

Figure 6.2 Smooth utility function elicited by bisection procedure.

Example 6.3. Consider that four apartments for rent are to be compared by a DM, by taking into account only their respective living areas. The DM is interested in a very large apartment (the preference increases with the construction size) and the available alternatives, namely X_1, X_2, X_3, and X_4, respectively, have $250\,\text{m}^2$, $220\,\text{m}^2$, $110\,\text{m}^2$, and $120\,\text{m}^2$ of living area.

In Step 1, the DM defined a global scale, by indicating $90\,\text{m}^2$ as the worst admissible area and $300\,\text{m}^2$ as the ideal area for the desired apartment (more than $300\,\text{m}^2$ is useless and hard to keep clean in the DM's opinion).

In Step 2, the DM identifies the middle point $x_{0.5} = 170\,\text{m}^2$ between the anchors. This means that, in the DM's opinion, the exchange of $90\,\text{m}^2$ for $170\,\text{m}^2$ is equally attractive as the exchange of $170\,\text{m}^2$ for $300\,\text{m}^2$. The fitted curve obtained in Step 3 is given in Figure 6.2 and reflects the DM preferences in a satisfactory manner, as the DM thinks that the improvement in the utility function associated with the rise of the living area is more significant when the living area is near to x_0, and less significant when the constructed area is near to x_1.

6.1.3 *Fuzzy Estimates*

The elements of $\mathbf{X}$ can be evaluated with the use of fuzzy estimates $\mathbf{L} = \{l(X_1), \ldots, l(X_n)\}$, with $l(X_k)$ being the fuzzy estimate associated with alternative X_k from the point of view of a given criterion F. The fuzzy estimate $l(X_k)$ refers to a fuzzy number that can be directly specified by a DM or indirectly expressed by means of linguistic terms from some set S such as, for instance, $S(F) = \{$*low quality*, *average quality*, *high quality*$\}$. In the latter case, the linguistic terms must be converted into fuzzy estimates, as they are required to perform the analysis of the problem. Although one can find that the use of linguistic terms makes the preference elicitation process more intuitive, it is important to indicate that the effectiveness of the elicitation can be diminished due to the existence of differences between numerical interpretations of the linguistic terms in the expert's mind and their numerical representation in the model being utilized. In this context, the techniques discussed in Chapter 3 for constructing and equalizing fuzzy sets may be helpful for reducing this type of elicitation error.

Example 6.4. A DM utilized linguistic terms from a set $S(F) = \{$*very poor*, *poor*, *average*, *good*, *very good*$\}$ to evaluate the layout of each apartment discussed in the previous example. The set of linguistic terms along with their respective representation through fuzzy sets are shown in Figure 6.3. The elicited preferences are included in Table 6.2.

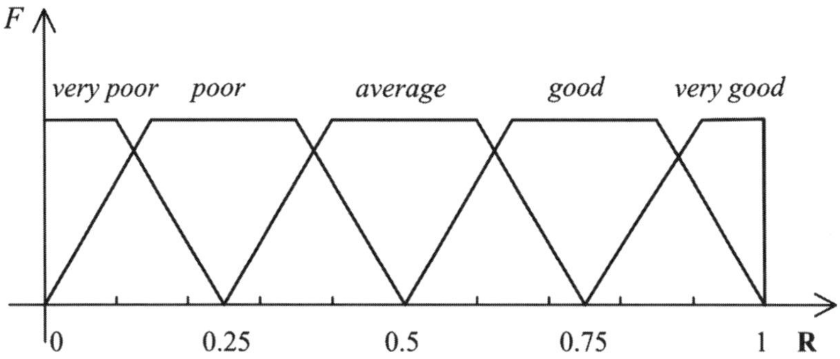

Figure 6.3 Set of linguistic terms.

6.1.4 Multiplicative Preference Relations

The multiplicative preference relation can be represented as an $n \times n$ positive reciprocal matrix $\boldsymbol{M}$ reflecting the preference intensity ratio between the alternatives in accordance with the AHP approach (Saaty, 1980), as discussed in Chapter 3. Each entry $\boldsymbol{M}(X_k, X_l)$ of this reciprocal matrix represents a preference intensity ratio and can be interpreted as "X_k is $\boldsymbol{M}(X_k, X_l)$ times more dominant than X_l" (Saaty, 1980) or as "X_k is $\boldsymbol{M}(X_k, X_l)$ times as good as X_l" (Chiclana, Herrera, and Herrera-Viedma, 2001).

The elicitation process realized by the AHP allows a DM to express preferences verbally, through the use of several linguistic terms, or numerically on the basis of different ratio scales. If a DM uses linguistic terms, these judgments are later converted into numbers, in order to proceed with the analysis of the decision-making problem. In this way, regardless of how the elicitation process has been carried out, it is necessary to define an adequate ratio scale. The selection of a proper ratio scale should be done by considering the entire set of objects about which ratio comparisons are to be performed (Harker and Vargas, 1987; Salo and Hämäläinen, 1997).

Under the condition of multiplicative reciprocity, once a DM provides $\boldsymbol{M}(X_k, X_l)$, the value of $\boldsymbol{M}(X_l, X_k)$ is automatically inferred as $\boldsymbol{M}(X_l, X_k) = 1/\boldsymbol{M}(X_k, X_l)$. Although multiplicative preference relations accommodate intransitivity, it is desirable to collect judgments that are as consistent as possible, since the methods for further analysis of such relations usually require them to be transitive in order to guarantee results of high quality. As already mentioned

Table 6.2 Evaluation of alternatives by means of linguistic terms

Apartment	Evaluation
X_1	*very good*
X_2	*good*
X_3	*very poor*
X_4	*poor*

in Chapter 3, the perfect consistency of a multiplicative preference relation is expressed as the satisfaction of a multiplicative transitivity property, that is,

$$\boldsymbol{M}(X_k, X_j) = \boldsymbol{M}(X_k, X_l) \cdot \boldsymbol{M}(X_l, X_j) \quad \forall j, k, l \in \{1, 2, \ldots, n\} \tag{6.4}$$

In the elicitation process of multiplicative preference relations, it becomes necessary to collect $n(n-1)/2$ pairwise comparisons or else, by enforcing multiplicative transitivity, it is possible to collect just $(n-1)$ pairwise comparisons and estimate the missing ones with the help of (6.4). As described in Chapter 3, when a DM provides all pairwise comparisons, if they do not satisfy (6.4) it is possible to identify and highlight the inconsistent pairwise comparisons so that the DM can review them. It is also possible to make use of some automated method to improve the consistency of the provided multiplicative preference relations (Zeshui and Cuiping, 1999).

Example 6.5. Let us consider the preference elicitation process realized by means of multiplicative preference relations based on the 1–9 scale (refer to Table 3.1) by taking into account the living area of the four apartments for rent as introduced in Example 6.3. The collected pairwise judgments are numerically defined by the DM as follows:

- X_1 is two times better than X_2;
- X_1 is seven times better than X_3;
- X_1 is seven times better than X_4;
- X_2 is five times better than X_3;
- X_2 is five times better than X_4;
- X_3 is as good as X_4.

It is interesting to observe the fact that the living area of X_2 is two times larger than the constructed area of X_3 (refer to Example 6.3 for the living area of each apartment) and does not imply that X_2 is considered two times better than X_3. The quantification of the provided judgments is reflected by the following reciprocal matrix:

$$\boldsymbol{M} = \begin{bmatrix} 1 & 2 & 7 & 7 \\ 1/2 & 1 & 5 & 5 \\ 1/7 & 1/5 & 1 & 1 \\ 1/7 & 1/5 & 1 & 1 \end{bmatrix} \tag{6.5}$$

Next, it is necessary to verify the consistency of the obtained multiplicative preference relation. For this purpose, we can utilize the index of inconsistency originally proposed by Saaty (refer again to Chapter 3). Since the maximal eigenvalue of the matrix given by (6.5) is 4.0159, the index of inconsistency calculated with the use of (3.10) is

$$\nu = \frac{(4.0159 - 4)}{3} = 0.0053 \tag{6.6}$$

Therefore, as ν is much lower than the threshold 0.1, the inconsistency level of (6.5) is acceptable.

6.1.5 Fuzzy Preference Relations

As discussed in Chapter 5, a fuzzy nonstrict preference relation $\boldsymbol{R}(X_k, X_l)$ indicates the degree to which the alternative X_k is at least as good as X_l, by means of its membership function $\boldsymbol{R}(X_k, X_l) : \mathbf{X} \times \mathbf{X} \to [0, 1]$.

When fuzzy preference relations are constructed by direct assessment, a DM is supposed to indicate to what extent X_k is better than X_l by supplying a subjective value from the unit interval. Different encoding schemes can be utilized to represent the preference strength of one alternative over another. Here, we consider two schemes. One of them is reflected by a nonreciprocal fuzzy preference relation (NRFPR) (Fodor and Roubens, 1994; Ekel, 2002), which has a correspondence with the notion of fuzzy nonstrict preference relation associated with the fuzzy preference structures studied in Chapter 5. Further, it is also coherent with a rational approach for deriving fuzzy preference relations from the fuzzy estimates provided by a DM to evaluate each alternative, which will be described below. The other encoding scheme considered here is captured by an additive reciprocal fuzzy preference relation (ARFPR), which is a fuzzy preference relation satisfying the property of additive reciprocity (see expression (5.18)) (Tanino, 1984; Chiclana, Herrera, and Herrera-Viedma 1998; Chiclana, Herrera, and Herrera-Viedma, 2001). Here, the NRFPR is denoted by $\mathbf{RN}(X_k, X_l)$ and the ARFPR is denoted by $\mathbf{RR}(X_k, X_l)$.

The encoding scheme associated with the ARFPR can be summarized by the following rules:

- $\mathbf{RR}(X_k, X_l) = 0.5$ means that X_k is indifferent to X_l;
- $0.5 < \mathbf{RR}(X_k, X_l) \leq 1$ means that X_k is preferred to X_l;
- $0 \leq \mathbf{RR}(X_k, X_l) < 0.5$ means that X_l is preferred to X_k;
- the entries of the main diagonal are filled with 0.5, as each element is equal to itself and, as a result, indifferent to itself.

As a consequence of the additive reciprocity property, when a DM provides a value for $\mathbf{RR}(X_k, X_l)$, the value of $\mathbf{RR}(X_l, X_k)$ is automatically inferred as $\mathbf{RR}(X_l, X_k) = 1 - \mathbf{RR}(X_k, X_l)$. Unfortunately, as far as we know, there are no preference elicitation procedures to help a DM to directly define ARFPRs. In this way, the DM has to articulate preferences based on his/her own intuition and capabilities of quantifying coherently preference strengths with the rules presented above.

Furthermore, it is important to note that, similar to the multiplicative preference relation, ARFPR also accommodates intransitivity. However, it is desirable to collect consistent fuzzy preference relations, since, in general, the methods for the analysis of such relations also require them to be consistent, in order to guarantee high-quality outcomes. The additive transitivity property, which is given as (Tanino, 1984; Herrera-Viedma *et al.*, 2004)

$$(\mathbf{RR}(X_k, X_j) - 0.5) = (\mathbf{RR}(X_k, X_l) - 0.5) + (\mathbf{RR}(X_l, X_j) - 0.5) \quad \forall k, j, l \in \{1, 2, \ldots, n\} \tag{6.7}$$

is one of the most intuitively appealing conditions for attesting the consistency of ARFPRs in the context of individual decision-making. Chiclana, Herrera-Viedma, and Herrera (2004) demonstrated that there is a meaningful connection between the multiplicative transitivity of multiplicative preference relations and the additive transitivity of ARFPRs. This connection will be clarified in Section 6.3, when the conversion from multiplicative preference relations into ARFPRs is considered.

Similar to the case of multiplicative preference relations, in the elicitation process of ARFPRs it is also necessary to collect $n(n-1)/2$ pairwise comparisons. However, by enforcing additive transitivity, it is also possible to collect just $(n-1)$ pairwise comparisons and estimate the missing ones with the use of certain methods such as the one proposed by Herrera-Viedma *et al.* (2004). If a DM provides all pairwise comparisons and they do not satisfy (6.7), it is possible to identify which pairwise comparisons should be reviewed by the DM. Another possibility consists of using an automated method to repair (enhance) the provided judgments (without the need to ask the DM to review his/her judgments), by modifying them as slightly as possible, just to guarantee an acceptable level of consistency.

Example 6.6. In the comparison of four apartments, namely X_1, X_2, X_3, and X_4, considered for renting, by taking into account their respective location, a DM thinks that X_3 and X_2 are the best situated alternatives, as both of them are in two adjacent buildings in a very familiar neighborhood, which are very near to the DM's office. Although X_1 is on a very pleasant street, it is considered the worst located alternative, because it is too far from the DM's office. X_4 is considered an intermediary alternative, being on a calm and pleasant street, at a point that is about 14 km (or about 25 minutes of driving time) away from his office.

In this way, the DM provided the following comparisons between the pairs of alternatives:

- $\mathbf{RR}(X_1, X_2) = 0$ and $\mathbf{RR}(X_2, X_1) = 1$, as X_2 is extremely better than X_1;
- $\mathbf{RR}(X_1, X_3) = 0$ and $\mathbf{RR}(X_3, X_1) = 1$, as X_3 is extremely better than X_1;
- $\mathbf{RR}(X_1, X_4) = 0.3$ and $\mathbf{RR}(X_4, X_1) = 0.7$, as X_4 is to a moderate extent better than X_1;
- $\mathbf{RR}(X_2, X_3) = 0.5$ and $\mathbf{RR}(X_3, X_2) = 0.5$, as X_2 is as good as X_3;
- $\mathbf{RR}(X_2, X_4) = 0.8$ and $\mathbf{RR}(X_4, X_2) = 0.2$, as X_2 is strongly better than X_4;
- $\mathbf{RR}(X_3, X_4) = 0.8$ and $\mathbf{RR}(X_4, X_3) = 0.2$, as X_3 is strongly better than X_4.

These judgments are summarized by the following ARFPR, which fortunately satisfies additive transitivity (the reader is encouraged to confirm this as an exercise):

$$\mathbf{RR} = \begin{bmatrix} 0.5 & 0 & 0 & 0.3 \\ 1 & 0.5 & 0.5 & 0.8 \\ 1 & 0.5 & 0.5 & 0.8 \\ 0.7 & 0.2 & 0.2 & 0.5 \end{bmatrix} \tag{6.8}$$

Now, let us focus on the NRFPR. Its encoding scheme satisfies the following conditions:

- if $\mathbf{RN}(X_k, X_l) = 1$ and $\mathbf{RN}(X_l, X_k) = 1$, then X_k is indifferent to X_l;
- if $\mathbf{RN}(X_k, X_l) = 1$ and $\mathbf{RN}(X_l, X_k) = 0$, then X_k is strictly preferred to X_l;
- if $\mathbf{RN}(X_k, X_l) = 0$ and $\mathbf{RN}(X_l, X_k) = 1$, then X_l is strictly preferred to X_k;
- if $\mathbf{RN}(X_k, X_l) = 0$ and $\mathbf{RN}(X_l, X_k) = 0$, then X_k and X_l are not comparable;
- the entries of the main diagonal are filled with 1, due to the reflexivity of $\mathbf{RN}(X_k, X_l)$.

Intermediate judgments among the situations described above are also allowed. They can be interpreted as follows:

- if $0 \leq \mathbf{RN}(X_k, X_l) < 1$ and $\mathbf{RN}(X_l, X_k) = 1$, then X_l is weakly preferred to X_k;
- if $\mathbf{RN}(X_k, X_l) = 1$ and $0 \leq \mathbf{RN}(X_l, X_k) < 1$, then X_k is weakly preferred to X_l;
- if $0 \leq \mathbf{RN}(X_k, X_l) < 1$ and $\mathbf{RN}(X_l, X_k) = 0$, then X_k is weakly preferred to X_l, and, at the same time, X_k and X_l are to some degree considered incomparable;
- if $\mathbf{RN}(X_k, X_l) = 0$ and $0 \leq \mathbf{RN}(X_l, X_k) < 1$, then X_l is weakly preferred to X_k and, at the same time, X_k and X_l are to some degree considered incomparable.

Unfortunately, as in the case of ARFPRs, we cannot find in the current literature any preference-eliciting procedure to help a DM to define NRFPRs. In this way, the DM has to express the preferences based on his/her own abilities of quantifying preference strengths coherently with the rules presented above. Particularly, in real-world situations, it may be difficult for the DM to assign the values of $\mathbf{RN}(X_k, X_l)$ and $\mathbf{RN}(X_l, X_k)$ that are coherent with his/her own preferences. Indeed, one may find it easier to define ARFPRs rather than NRFPRs, due to the fact that $\mathbf{RR}(X_l, X_k)$ can be automatically inferred from the value of $\mathbf{RR}(X_k, X_l)$ by means of the reciprocity of ARFPRs. In the case of NRFPRs, the value of $\mathbf{RN}(X_k, X_l)$, a priori, does not say much about the value of $\mathbf{RN}(X_l, X_k)$. However, it should be mentioned that, although there is no such obvious relationship between the values of $\mathbf{RN}(X_k, X_l)$ and $\mathbf{RN}(X_l, X_k)$, there is another type of relationship between these two values. This relationship depends on how the degrees of indifference and strict preference are derived from the values assigned to $\mathbf{RN}(X_k, X_l)$ and $\mathbf{RN}(X_l, X_k)$. Next, in Example 6.7, a DM expresses his/her preferences in terms of a NRFPR. In Example 6.8, expressions (5.35) and (5.36) (which play an important role in the analysis of decision-making problems, as will be shown in Chapter 7) are utilized to extract the indifference degree and the strict preference degree from the judgments of nonstrict preference provided by the DM. As will be shown in Example 6.8, those equations determine a clear and intuitive relationship between the values of $\mathbf{RN}(X_k, X_l)$ and $\mathbf{RN}(X_l, X_k)$.

Example 6.7. By considering the criterion of quality of construction (standard of finishing), the four apartments for rent discussed in the previous example are now evaluated as follows: a DM thinks that X_1 definitely has the highest standard of finishing, exceeding expectations; both X_2 and X_3 have a high standard of finishing; and X_4 is average. In this way, the pairwise comparisons provided by the DM are reflected by the NRFPR as follows:

- $\mathbf{RN}(X_1, X_2) = 1$ and $\mathbf{RN}(X_2, X_1) = 0.5$, as X_1 is moderately better than X_2;
- $\mathbf{RN}(X_1, X_3) = 1$ and $\mathbf{RN}(X_3, X_1) = 0.5$, as X_1 is moderately better than X_3;
- $\mathbf{RN}(X_1, X_4) = 1$ and $\mathbf{RN}(X_4, X_1) = 0$, as X_1 is extremely better than X_4;
- $\mathbf{RN}(X_2, X_3) = 1$ and $\mathbf{RN}(X_3, X_2) = 1$, as X_2 is as good as X_3;
- $\mathbf{RN}(X_2, X_4) = 1$ and $\mathbf{RN}(X_4, X_2) = 0.3$, as X_2 is strongly better than X_4;
- $\mathbf{RN}(X_3, X_4) = 1$ and $\mathbf{RN}(X_4, X_3) = 0.3$, as X_3 is strongly better than X_4.

These judgments are summarized by the following fuzzy preference relation, which satisfies weak transitivity but does not satisfy the min-transitivity (the reader is invited to verify this observation):

$$\mathbf{RN} = \begin{bmatrix} 1 & 1 & 1 & 1 \\ 0.5 & 1 & 1 & 1 \\ 0.5 & 1 & 1 & 1 \\ 0 & 0.3 & 0.3 & 1 \end{bmatrix} \tag{6.9}$$

Example 6.8. By applying expression (5.35) to (6.9), we obtain the level of strict preference between the alternatives, which is reflected by the following fuzzy strict preference relation:

$$\boldsymbol{P} = \begin{bmatrix} 0 & 0.5 & 0.5 & 1 \\ 0 & 0 & 0 & 0.7 \\ 0 & 0 & 0 & 0.7 \\ 0 & 0 & 0 & 0 \end{bmatrix} \tag{6.10}$$

When we apply (5.36) to (6.9), we obtain the level of indifference between the alternatives, which is reflected by the following fuzzy indifference relation:

$$\boldsymbol{I} = \begin{bmatrix} 1 & 0.5 & 0.5 & 0 \\ 0.5 & 1 & 1 & 0.3 \\ 0.5 & 1 & 1 & 0.3 \\ 0 & 0.3 & 0.3 & 1 \end{bmatrix} \tag{6.11}$$

By analyzing (6.9)–(6.11), we can conclude that, if (5.35) and (5.36) are utilized for defining the fuzzy strict preference relation and the fuzzy indifference relation, respectively, then the difference between $\mathbf{RN}(X_k, X_l)$ and $\mathbf{RN}(X_l, X_k)$ reflects the level of strict preference between the alternatives X_k and X_l. Further, the minimum value between $\mathbf{RN}(X_k, X_l)$ and $\mathbf{RN}(X_l, X_k)$ reflects the level of indifference between the alternatives X_k and X_l. In this way, in addition to the desirable properties of (5.35) and (5.36), which are discussed in Chapter 5, these also allow for a rather intuitive assignment of values to the NRFPRs.

Regarding the preference eliciting process of NRFPRs, it is also important to indicate that, if the method utilized to analyze the fuzzy preference relations does not admit judgments of incomparability between alternatives, a DM must reconsider all judgments from which such incomparability arises. In addition to the judgment of full incomparability, $\mathbf{RN}(X_k, X_l) = 0$ and $\mathbf{RN}(X_l, X_k) = 0$, a DM should also consider the situations of partial incomparability listed above (these cases are also observable in Figure 5.6).

Finally, with reference to the consistency of NRFPRs, the min-transitivity, which is given by (5.23), and the weak transitivity, which is given by (5.59), are the most common consistency conditions, the former being a stricter condition than the latter.

6.2 Ordering of Fuzzy Quantities and the Construction of Fuzzy Preference Relations

As indicated above, a rational (viewed from the fundamental as well as psychological points of view) approach for deriving fuzzy preference relations is the use of the fuzzy estimates provided by a DM to evaluate each alternative (the third preference format considered in the previous section). In essence, the use of this approach is associated with the need to compare or rank fuzzy numbers to choose the best (largest or smallest) or worst (smallest or largest) among them.

Various works can be distinguished that were dedicated to the techniques for comparing or ranking fuzzy numbers and published before 1998 (for instance, Jain, 1976; Baas and Kwakernaak, 1977; Baldwin and Guild, 1979; Orlovsky, 1981; Yager, 1981; Dubois and Prade, 1983; Lee and Li, 1988; Tseng and Klein, 1989; Chen and Hwang, 1992; Fortemps and Roubens, 1996; Cheng, 1998; Horiuchi and Tamura, 1998).

The classification of the groups of techniques related to the ordering of fuzzy quantities can be found in Chen and Hwang (1992). In particular, the following classes of techniques can be distinguished:

- preference relations;
- use of fuzzy mean and spread characteristics;
- fuzzy scoring techniques;
- linguistic methods.

Among these classes, Horiuchi and Tamura (1998) consider the construction of fuzzy preference relations by means of pairwise comparisons as being the most practical and justified approach. Taking this into account, it is worth distinguishing the fuzzy number ranking index introduced by Orlovsky (Orlovsky, 1981), which is based on the concept of a membership function of a generalized preference relation. Given that $F(X_k)$ and $F(X_l)$ are fuzzy sets reflecting the evaluation of the objective function F or the attribute F for the alternatives X_k and X_l, respectively, the generalized preference relation between X_k and X_l is a fuzzy preference relation whose membership function is defined as

$$\eta(F(X_k), F(X_l)) = \sup_{f(X_k), f(X_l) \in F} \min(F(f(X_k)), F(f(X_l)), E(F(X_k), F(X_l))) \quad (6.12)$$

$$\eta(F(X_l), F(X_k)) = \sup_{f(X_k), f(X_l) \in F} \min(F(X_l), F(X_k), E(F(X_l), F(X_k))) \quad (6.13)$$

where $f(X_k)$ and $f(X_l)$ are real numbers reflecting the evaluation of the objective function F or of the attribute F for the alternatives X_k and X_l; $F(f(X_k))$ and $F(f(X_l))$ represent the membership functions of the fuzzy sets $F(X_k)$ and $F(X_l)$ evaluated at $f(X_k)$ and $f(X_l)$, respectively; $E(F(X_k), F(X_l))$ and $E(F(X_l), F(X_k))$ are the membership functions of the corresponding fuzzy relations which respectively reflect the essence of the preferences of X_k over X_l and of X_l over X_k (for instance, "more beautiful", "more attractive", "more flexible", etc.).

When F can be measured on a numerical scale and the essence of preference behind relation $\boldsymbol{R}$ is coherent with the natural order ($\geq$) along the axis of measured values of F, then (6.12) and (6.13), respectively, are reduced to the expressions

$$\eta(F(X_k), F(X_l)) = \sup_{\substack{f(X_k), f(X_l) \in F \\ f(X_k) \geq f(X_l)}} \min(F(f(X_k)), F(f(X_l))) \quad (6.14)$$

$$\eta(F(X_l), F(X_k)) = \sup_{\substack{f(X_k), f(X_l) \in F \\ f(X_l) \geq f(X_k)}} \min(F(f(X_k)), F(f(X_l))) \quad (6.15)$$

when F is a maximization criterion or attribute.

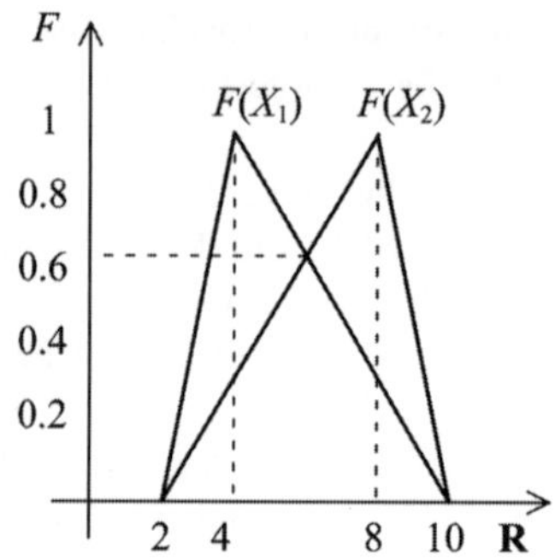

Figure 6.4 Fuzzy values of the objective function F.

If F exhibits is a minimization criterion or attribute, then relationships (6.14) and (6.15) must be modified. In this case, (6.14) stands for $f(X_k) \leq f(X_l)$ and (6.15) stands for $f(X_l) \leq f(X_k)$.

Example 6.9. Assume we are given the alternatives X_1 and X_2 along with their respective fuzzy values $F(X_1)$ and $F(X_2)$ of the objective function, as illustrated in Figure 6.4. With the use of (6.14) and (6.15), we evaluate the degrees of the preferences of X_1 over X_2 and of X_2 over X_1, in order to select the largest value between $F(X_1)$ and $F(X_2)$. For illustrative purposes, Table 6.3 shows the corresponding membership functions, evaluated just for some selected points distributed in the universe of discourse.

The formal application of (6.14) and (6.15) requires the construction of the Cartesian product of $f(X_1)$ and $f(X_2)$, as presented in Table 6.4.

The entries located on the main diagonal (marked in bold) and above it are associated with $f(X_1) \geq f(X_2)$. Taking this into account, according to (6.14), it is possible to find $\eta(F(X_1), F(X_2)) = 0.65$. On the other hand, the entries on the main diagonal and below it are associated with $f(X_2) \geq f(X_1)$. By applying (6.15), we obtain $\eta(F(X_2), F(X_1)) = 1$. Therefore, we have that X_2 is equally good (large) or better (larger) than X_1.

It is worth noting that the maximum level of the intersection of $F(X_1)$ and $F(X_2)$, calculated with the use of the min operator (which corresponds to 0.65, as can be seen in Figure 6.4), allows one to define $\eta(F(X_2), F(X_1)) = 1$ and $\eta(F(X_1), F(X_2)) = 0.65$ without the need to construct the Cartesian product of $f(X_1)$ and $f(X_2)$. Indeed, this rule of thumb for determining $\eta(F(X_1), F(X_2))$ and $\eta(F(X_2), F(X_1))$ can be used in many practical situations.

The dependencies (6.14) and (6.15) agree with the Baas–Kwakernaak (Baas and Kwakernaak, 1977), Baldwin–Guild (Baldwin and Guild, 1979), and one of the Dubois-Prade (Dubois and Prade, 1983) indices used for ranking fuzzy numbers.

Table 6.3 Membership functions of $F(X_1)$ and $F(X_2)$

R	2	3	4	5	6	7	8	9	10
$F(f(X_1))$	0	0.50	1	0.85	0.65	0.50	0.35	0.15	0
$F(f(X_2))$	0	0.15	0.35	0.50	0.65	0.85	1	0.50	0

Table 6.4 Cartesian product of $f(X_1)$ and $f(X_2)$

$F(f(X_1))/f(X_1) \rightarrow$ $F(f(X_2))/f(X_2)$ $\downarrow$	0/2	0.50/3	1/4	0.85/5	0.65/6	0.50/7	0.35/8	0.15/9	0/10
0/2	**0**	0	0	0	0	0	0	0	0
0.15/3	0	**0.15**	0.15	0.15	0.15	0.15	0.15	0.15	0
0.35/4	0	0.35	**0.35**	0.35	0.35	0.35	0.35	0.15	0
0.65/5	0	0.50	0.65	**0.65**	0.65	0.50	0.35	0.14	0
0.85/6	0	0.50	0.85	0.85	**0.65**	0.50	0.35	0.15	0
0.75/7	0	0.50	0.75	0.75	0.65	**0.50**	0.35	0.15	0
1/8	0	0.50	1	0.85	0.65	0.50	**0.35**	0.15	0
0.50/9	0	0.50	0.50	0.50	0.50	0.50	0.35	**0.15**	0
0/10	0	0	0	0	0	0	0	0	**0**

Example 6.10. We are given the alternatives X_1 and X_2 with fuzzy characterization of the objective function $F(X_1)$ and $F(X_2)$, as presented in Figure 6.5. It is necessary to evaluate the preference degrees of X_1 over X_2 and of X_2 over X_1 to select the largest value between $F(X_1)$ and $F(X_2)$. For the sake of understanding, Table 6.5 shows the corresponding membership functions, evaluated just for some points along the universe of discourse.

The Cartesian product of $f(X_1)$ and $f(X_2)$, constructed on the basis of (6.14) and (6.15), is presented in Table 6.6.

It is not difficult to verify that $\eta(F(X_1), F(X_2)) = 0.83$ and $\eta(F(X_2), F(X_1)) = 1$.

Also, it is useful to note that, by applying (5.36) and (5.35) to obtain the level of indifference and the level of strict preference between X_1 and X_2, we have that X_1 is equivalent to X_2 with the degree computed as $\min(\eta(F(X_1), F(X_2)), \eta(F(X_2), F(X_1))) = 0.65$ and that X_2 is strictly larger (better) than X_1 with the degree $\max(\eta(F(X_2), F(X_1)) - \eta(F(X_1), F(X_2)), 0) = 0.35$.

The reader should note that the generalized preference relation constructed by means of (6.14) and (6.15) is coherent with the definition of NRFPR given in Section 6.2.

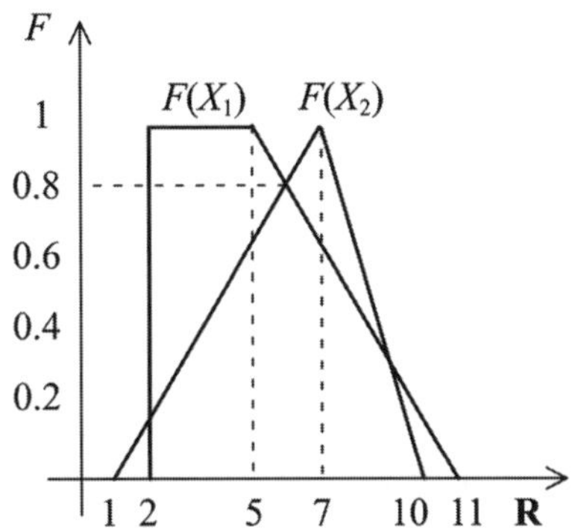

Figure 6.5 Comparison of alternatives with trapezoidal membership functions.

Table 6.5 Membership function of $F(X_1)$ and $F(X_2)$

R	1	2	3	4	5	6	7	8	9	10	11
$F(f(X_1))$	0	1	1	1	1	0.83	0.67	0.50	0.34	0.17	0
$F(f(X_2))$	0	0.17	0.34	0.50	0.67	0.83	1	0.67	0.34	0	0

Besides, by applying (5.36), we can conclude that $F(X_1)$ is equivalent or indifferent to $F(X_2)$ with the degree $\min(\eta(F(X_1), F(X_2)), \eta(F(X_2), F(X_1))) = 0.83$. At the same time, with the use of (5.35), we obtain that $F(X_2)$ is strictly larger (better) than $F(X_1)$ with the degree $\max(\eta(F(X_2), F(X_1)) - \eta(F(X_1), F(X_2)), 0) = 0.17$.

The use of this approach for comparing fuzzy quantities as well as for constructing NRFPRs is well founded. However, it is important to indicate that, in practice, there are cases where the fuzzy quantities $F(X_k)$ and $F(X_l)$ have trapezoidal membership functions that are located in such a way that it is not possible to distinguish X_k from X_l. For instance, we can observe that, for the situation shown in Figure 6.6, the alternatives are indistinguishable since

$$\eta(F(X_1), F(X_2)) = \eta(F(X_2), F(X_1)) = \alpha \tag{6.16}$$

As will be further discussed in Chapter 7, in such situations, the algorithms applied to solve the problems with fuzzy coefficients do not allow one to obtain unique solutions because they "stop" when conditions like (6.16) arise (see Examples 7.1 and 7.2 in Chapter 7). This should be considered to be a natural consequence of the existence of decision uncertainty regions, produced by a combination of the uncertainty and the relative stability of optimal solutions (Ekel and Popov, 1985; Popov and Ekel, 1987). Under these circumstances, other indices may be used as additional means for the ranking of fuzzy numbers.

Table 6.6 Cartesian product of $f(X_1)$ and $f(X_2)$

$F(f(X_1))/f(X_1)\rightarrow$ $F(f(X_2))/f(X_2)$ $\downarrow$	0/1	1/2	1/3	1/4	1/5	0.83/6	0.67/7	0.50/8	0.34/9	0.17/10	0/11
0/1	**0**	0	0	0	0	0	0	0	0	0	0
0.17/2	0	**0.17**	0.17	0.17	0.17	0.17	0.17	0.17	0.17	0.17	0
0.34/3	0	0.34	**0.34**	0.34	0.34	0.34	0.34	0.34	0.34	0.17	0
0.50/4	0	0.50	0.50	**0.50**	0.50	0.50	0.50	0.50	0.34	0.17	0
0.67/5	0	0.67	0.67	0.67	**0.67**	0.67	0.67	0.50	0.34	0.17	0
0.83/6	0	0.83	0.83	0.83	0.83	**0.83**	0.67	0.50	0.34	0.17	0
1/7	0	1	1	1	1	0.83	**0.67**	0.50	0.34	0.17	0
0.67/8	0	0.67	0.67	0.67	0.67	0.67	0.67	**0.50**	0.34	0.17	0
0.34/9	0	0.34	0.34	0.34	0.34	0.34	0.34	0.34	**0.34**	0.17	0
0/10	0	0	0	0	0	0	0	0	0	**0**	0
0/11	0	0	0	0	0	0	0	0	0	0	**0**

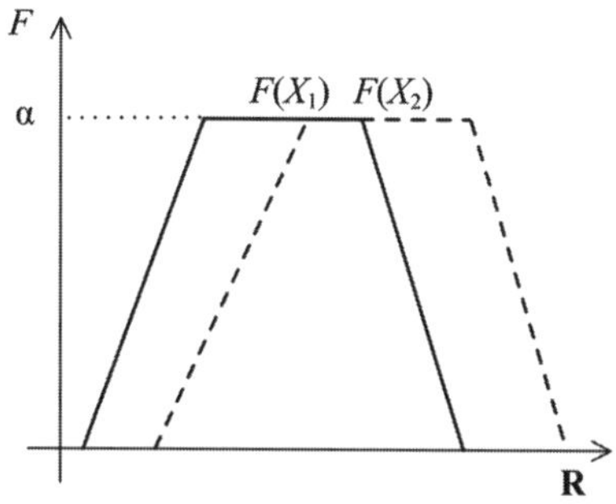

Figure 6.6 Comparison of alternatives with trapezoidal membership functions.

The reviews of techniques which have been developed for ranking fuzzy numbers can be found in Dubois and Prade (1999) and Wang and Kerre (2001). These studies cover the analysis of the fuzzy ranking indices proposed before 1998 (among more recent works in this field, it is possible to distinguish, for instance, Raj and Kumar, 1999; Modarres and Sadi-Nezhad, 2001; Facchinetti, 2002; Tran and Duckstein, 2002; Liu and Han, 2005; Abbasbandy and Asady, 2006; Wang and Lee, 2008; Abbasbandy and Hajjari, 2009; Wang and Luo, 2009).

Wang and Kerre (2001) enumerate more than 35 fuzzy number ranking indices and conclude the following: unlike the case of real numbers, fuzzy quantities have no natural order. The basic principle behind the methods for ordering fuzzy quantities consists of converting each fuzzy quantity into a real number and realizing the comparison of fuzzy quantities on the basis of the resulting real numbers. However, each approach for realizing such a conversion focuses on an intricate aspect inherent to fuzzy quantities. As a consequence, each approach suffers from some weaknesses associated with the loss of information inherent to the conversion of a fuzzy quantity to a single real number. Wang and Kerre (2001) support this point of view by citing Freeling (1980): "by reducing the whole of our analysis to a single number, we are losing much of the information we have purposely been keeping throughout calculations". Authors like Cheng (1998) and Lee-Kwang (1999) share this opinion as well. Cheng (1998) also indicates that many of the indices produce different rankings for the same problem. Other authors (Cheng, 1998; Ekel, Pedrycz, and Schinzinger, 1998; Lee-Kwang, 1999; Chen and Lu, 2002) underline that fuzzy number ranking indices occasionally result in choices which seem not to be coherent with intuition. Chen and Lu (2002) indicate that the majority of methods for the ranking of fuzzy numbers assume that membership functions of fuzzy numbers are normalized. However, this limitation is not always adequate. Tseng and Klein (1989) indicate that the ranking methods may not reflect the preferences or interests of the DMs. Further, Chen and Klein (1997) indicate that many techniques of ordering help one only to observe an order among fuzzy quantities; however, they do not permit one to measure the degree of dominance among them, requiring a significant volume of calculations. Finally, it is necessary to mention that the majority of indices for the ranking of fuzzy quantities has been proposed with the aspiration for obligatorily distinguishing the alternatives, which is often questionable because the uncertainty of information creates inherent decision uncertainty regions. Taking this into account, the possibility of identifying situations where the compared alternatives cannot be distinguished should be considered as a merit of the fuzzy number ranking index based on the underlying essence of a membership function of a generalized preference relation.

Taking all this into consideration, the fuzzy number ranking index based on the idea of a membership function of a generalized preference relation is used here for ordering fuzzy quantities and for the construction of fuzzy preference relations.

6.3 Transformation Functions and their Use for Converting Different Preference Formats into Fuzzy Preference Relations

In individual as well as in group decision-making, when different preference formats are utilized, the information must be made uniform under adequate transformation functions, before being analyzed. These transformation functions are useful in converting heterogeneous preference information, which may be qualitative or quantitative, two valued or fuzzy, ordered or nonordered, ordinal or cardinal, and even based on different types of scales (including ordinal, interval, and ratio scales), into fuzzy preference relations, which form a more general preference model, since it can be defined on all those types of scales.

In the previous section, it was shown how fuzzy preference relations can be constructed from the evaluation of the alternatives in terms of fuzzy estimates. Now, we present ways of deriving fuzzy preference relations from the evaluation of the alternatives expressed using other preference formats rather than fuzzy estimates. First, we present some transformation functions for converting the preference information from different formats, namely the ordering of the alternatives, cardinal utility values, multiplicative preference relations, and NRFPR, into the ARFPR format. The reader should be aware that it is not our intention to consider all the existing transformation functions. Here, some selected transformation functions from the current literature are presented and studied. Afterward, we derive from those selected transformation functions other transformation functions, which can be utilized for converting the preference information from those different formats (including the ARFPR) into the NRFPR format.

6.3.1 Transformation Functions for ARFPR

Before proceeding, it is worth mentioning that all transformation functions presented here for converting information from the different formats into the ARFPR preserve the weak transitivity of the original information. In this way, the ARFPRs derived from an ordered array or from a utility function, with the use of the transformation functions presented next, always satisfy weak transitivity. On the other hand, the ARFPRs derived from the NRFPR format and the multiplicative preference relation format, with the use of the transformation functions to be presented next, verify the weak transitivity as long as the collected pairwise judgments also satisfy weak transitivity.

6.3.1.1 Ordered Array → ARFPR

When the preferences are expressed by means of an ordered array $\boldsymbol{O} = [o_1 \ldots o_n]$, it is possible to convert this array into an ARFPR with the use of any function $H(x, y) : \mathbf{N} \times \mathbf{N} \to [0, 1]$ satisfying the following conditions:

- $H(o_k, o_l)$ must be a nonincreasing function of the first argument and a nondecreasing function of the second argument;
- $H(o_k, o_k) = 0.5, \forall k \in \{1, 2, \ldots, n\}$;

- $H(o_k, o_l) > 0.5$, if$o_k < o_l, \forall k, l \in \{1, 2, \ldots, n\}$;
- additive reciprocity: $H(o_k, o_l) + H(o_l, o_k) = 1, \forall k, l \in \{1, 2, \ldots, n\}$.

Here, we consider the following transformation function, which in addition to satisfying those conditions produces ARFPRs that meet the additive transitivity (Chiclana, Herrera, and Herrera-Viedma, 1998):

$$\mathbf{RR}(X_k, X_l) = H_1(o_k, o_l) = \frac{1}{2}\left(1 + \frac{o_l - o_k}{n - 1}\right) \tag{6.17}$$

Example 6.11. Given a set of five alternatives, consider that their rank ordering is given by $\boldsymbol{O} = [\,2\ 1\ 4\ 5\ 3\,]$ according to the preferences of an expert. For instance, if we apply (6.17) to the pair X_1 and X_2, we arrive at the following result:

$$\mathbf{RR}(X_1, X_2) = \frac{1}{2}\left(1 + \frac{1 - 2}{4}\right) = \frac{3}{8} \tag{6.18}$$

By applying (6.17) to the other pairs of alternatives, the following ARFPR can be constructed:

$$\mathbf{RR} = \begin{bmatrix} 0.5 & 0.375 & 0.75 & 0.875 & 0.625 \\ 0.625 & 0.5 & 0.875 & 1 & 0.75 \\ 0.25 & 0.125 & 0.5 & 0.625 & 0.375 \\ 0.125 & 0 & 0.375 & 0.5 & 0.25 \\ 0.375 & 0.25 & 0.625 & 0.75 & 0.5 \end{bmatrix} \tag{6.19}$$

One can see that (6.19) is coherent with the array provided by a DM. For instance, the second row and the second column of (6.19) reflect the fact that X_2 is the best ranked alternative as $\mathbf{RR}(X_2, X_k) > 0.5$ and $\mathbf{RR}(X_k, X_2) < 0.5$, for $k \in \{1, 3, 4, 5\}$. Further, the fourth line and the fourth column of (6.19) indicate that X_4 is the worst ranked alternative, as $\mathbf{RR}(X_4, X_k) < 0.5$ and $\mathbf{RR}(X_k, X_4) > 0.5$, for $k \in \{1, 2, 3, 5\}$.

6.3.1.2 Utility → ARFPR

Now, let us consider that the DM's opinions are expressed by means of utility values. From now on, the utility value evaluated for each alternative belonging to $\mathbf{X}$ will be represented as the array $\boldsymbol{U} = [\,u_1 \ldots u_n\,]$, in which the utility value associated with each alternative satisfies $u_k = U(X_k), \forall X_k \in \mathbf{X}$. In general, utility values normalized in the interval [0, 1] can be transformed into an ARFPR by means of any function $H(x, y) : [0, 1] \times [0, 1] \rightarrow [0, 1]$ satisfying the following conditions:

- $H(u_k, u_l)$ is a nondecreasing function of the first argument and a nonincreasing function of the second argument;
- $H(u_k, u_k) = 0.5, \forall k \in \{1, 2, \ldots, n\}$;
- $H(u_k, 0) = 0, \forall k \in \{1, 2, \ldots, n\}$, in order to reflect the fact that, if a DM judges that the utility value of an alternative is zero (which means that this alternative does not satisfy the

given criterion at all), then the other alternative should be preferred to that alternative with the maximum degree of preference (Chiclana, Herrera, and Herrera-Viedma, 1998);
- $H(u_k, u_l) > 0.5$, if $u_k > u_l, \forall k, l \in \{1, 2, \ldots, n\}$;
- additive reciprocity: $H(u_k, u_l) + H(u_l, u_k) = 1, \forall k, l \in \{1, 2, \ldots, n\}$.

Next, we present some particular transformation functions which verify those conditions for utility values assessed on different types of scales.

If the utility values are assessed on a ratio scale, the ARFPR can be derived (Tanino, 1984; Chiclana, Herrera, and Herrera-Viedma, 1998) from this array with the use of the following relationships:

$$\mathbf{RR}(X_k, X_l) = H_2(u_k, u_l) = \begin{cases} \dfrac{u_k}{u_k + u_l} & \text{if } u_k + u_l \neq 0 \\ 0.5 & \text{if } u_k = u_l = 0 \end{cases} \tag{6.20}$$

$$\mathbf{RR}(X_k, X_l) = H_3(u_k, u_l) = \begin{cases} \dfrac{u_k^2}{u_k^2 + u_l^2} & \text{if } u_k + u_l \neq 0 \\ 0.5 & \text{if } u_k = u_l = 0 \end{cases} \tag{6.21}$$

Note that both (6.20) and (6.21) define ARFPRs that do not satisfy additive transitivity, but satisfy the multiplicative one. The multiplicative transitivity, as stated in (6.4) for the multiplicative preference relation, can be written for the case of ARFPR as follows (Tanino, 1984):

$$\frac{\mathbf{RR}(X_k, X_j)}{\mathbf{RR}(X_j, X_k)} = \frac{\mathbf{RR}(X_k, X_l)}{\mathbf{RR}(X_l, X_k)} \cdot \frac{\mathbf{RR}(X_l, X_j)}{\mathbf{RR}(X_j, X_l)} \quad \forall j, k, l \in \{1, 2, \ldots, n\} \tag{6.22}$$

Substitution of (6.20) into (6.22) yields

$$\left(\frac{u_k}{u_k + u_j}\right)\left(\frac{u_k + u_j}{u_j}\right) = \left(\frac{u_k}{u_k + u_l}\right)\left(\frac{u_k + u_l}{u_l}\right)\left(\frac{u_l}{u_l + u_j}\right)\left(\frac{u_l + u_j}{u_j}\right) \tag{6.23}$$

The reader can easily confirm that this relationship is valid for every $X_k, X_j, X_l \in \mathbf{X}$. A similar development can be carried out for (6.21). The notion of multiplicative transitivity may be useful to verify the consistency of the DM's preferences, if the fuzzy preference relation is constructed in such a way that the ratio $\mathbf{RR}(X_k, X_l)/\mathbf{RR}(X_l, X_k)$ reflects how many times X_k is preferred to X_l (Tanino, 1984).

As will be shown in the following example, the main difference between (6.20) and (6.21) lies in the fact that the strength of nonstrict preference $\mathbf{RR}(X_k, X_l)$ calculated by means of (6.21) tends to be farther from the indifference judgment than the corresponding strength of nonstrict preference calculated by means of (6.20).

Example 6.12. By applying (6.20) and (6.21) to the utility array $U = [\,0.2\ 0.4\ 0.1\ 0.3\,]$, we obtain

$$\mathbf{RR} = \begin{bmatrix} 0.5 & 0.333 & 0.667 & 0.4 \\ 0.667 & 0.5 & 0.8 & 0.571 \\ 0.333 & 0.2 & 0.5 & 0.25 \\ 0.6 & 0.428 & 0.75 & 0.5 \end{bmatrix} \tag{6.24}$$

$$\mathbf{RR} = \begin{bmatrix} 0.5 & 0.2 & 0.8 & 0.308 \\ 0.8 & 0.5 & 0.941 & 0.64 \\ 0.5 & 0.059 & 0.5 & 0.1 \\ 0.692 & 0.36 & 0.9 & 0.5 \end{bmatrix} \tag{6.25}$$

As can be seen, both (6.24) and (6.25) are consonant with the utility array. However, in (6.25), more than in (6.24), the strength of nonstrict preference of each alternative over another tends to be more distant from the judgment of indifference, being much higher or much lower than 0.5.

Given a vector of cardinal utility values $\boldsymbol{U}$, defined on an interval scale normalized in [0,1], the ARFPR can be derived from this vector by means (Tanino, 1984; Chiclana, Herrera, and Herrera-Viedma, 1998) of the following expression:

$$\mathbf{RR}(X_k, X_l) = H_4(u_k, u_l) = \frac{1}{2}(1 + u_k - u_l) \tag{6.26}$$

It should be mentioned that an attractive property of (6.26) is that it constructs ARFPRs that necessarily satisfy additive transitivity.

Example 6.13. Let us recall the utility values from Example 6.2, that is, $\boldsymbol{U} = [\,1\ 0.9\ 0\ 0.2\,]$. By applying (6.26), we obtain the following fuzzy preference relation:

$$\mathbf{RR} = \begin{bmatrix} 0.5 & 0.55 & 1 & 0.9 \\ 0.45 & 0.5 & 0.95 & 0.85 \\ 0 & 0.05 & 0.5 & 0.4 \\ 0.1 & 0.15 & 0.6 & 0.5 \end{bmatrix} \tag{6.27}$$

It is worth noting that (6.27) is coherent with the vector assessed by the DM. For instance, the first line and the first column of matrix **RR** suggest that alternative X_1 is the best alternative, since $\mathbf{RR}(X_1, X_k) > 0.5$ and $\mathbf{RR}(X_k, X_1) < 0.5$ for $k = 2, \ldots, 4$. On the other hand, the third line and the third column of **RR** indicate that X_3 is the worst alternative, as $\mathbf{RR}(X_3, X_k) < 0.5$ and $\mathbf{RR}(X_k, X_3) > 0.5$ for $k \in \{1, 2, 4\}$.

6.3.1.3 Multiplicative Preference Relation → ARFPR

If the DM's preferences are defined in terms of a multiplicative relation $\boldsymbol{M}(X_k, X_l)$, $\forall (X_k, X_l) \in \mathbf{X} \times \mathbf{X}$, expressed by means of the ratio scale proposed by Saaty (see Table 3.1), the ARFPR can be obtained by means of any function $H(x)\ :\ [1/9, 9] \rightarrow [0, 1]$ satisfying the following conditions:

- $H(\boldsymbol{M}(X_k, X_l))$ should be a nondecreasing function;
- $H(1) = 0.5$;
- $H(9) = 1$;
- $H(1/9) = 0$;
- additive reciprocity: $H(x) + H(1/x) = 1,\ \forall x \in [1/9, 9]$.

Chiclana, Herrera, and Herrera-Viedma (2001) proposed the following transformation function verifying those conditions:

$$\mathbf{RR}(X_k, X_l) = H_5(\boldsymbol{M}(X_k, X_l)) = \frac{1}{2}\left(1 + \log_9 \boldsymbol{M}(X_k, X_l)\right) \tag{6.28}$$

The transformation function (6.28) can also be generalized to make it possible to use other ratio scales:

$$\mathbf{RR}(X_k, X_l) = H_6(\boldsymbol{M}(X_k, X_l)) = \frac{1}{2}\left(1 + \log_m \boldsymbol{M}(X_k, X_l)\right) \tag{6.29}$$

where m is the upper limit and $1/m$ is the lower limit of the ratio scale. Note that, if the original multiplicative preference relation satisfies multiplicative transitivity, then both (6.28) and (6.29) produce ARFPRs that verify the property of additive transitivity. A proof of this property can be found in Herrera-Viedma *et al.* (2004).

If $\boldsymbol{M}$ satisfies multiplicative transitivity, then we have that

$$\boldsymbol{M}(X_k, X_j).\boldsymbol{M}(X_j, X_l).\boldsymbol{M}(X_l, X_k) = 1 \quad \forall X_k, X_j, X_l \in \mathbf{X} \tag{6.30}$$

By taking logarithms to base m of both sides of (6.30), we obtain for all $X_k, X_j, X_l \in \mathbf{X}$

$$\log_m \boldsymbol{M}(X_k, X_j) + \log_m \boldsymbol{M}(X_j, X_l) + \log_m \boldsymbol{M}(X_l, X_k) = 0 \tag{6.31}$$

By adding 3 and dividing by 2 on both sides of (6.31), we obtain

$$\frac{1}{2}(1 + \log_m \boldsymbol{M}(X_k, X_j)) + \frac{1}{2}(1 + \log_m \boldsymbol{M}(X_j, X_l)) + \frac{1}{2}(1 + \log_m \boldsymbol{M}(X_l, X_k)) = \frac{3}{2} \tag{6.32}$$

Taking (6.29) into account, (6.32) can be rewritten as follows:

$$\mathbf{RR}(X_k, X_j) + \mathbf{RR}(X_j, X_l) + \mathbf{RR}(X_l, X_k) = \frac{3}{2} \tag{6.33}$$

The reader can easily verify that (6.33) corresponds to the additive transitivity condition given by (6.7).

Example 6.14. Let us convert the multiplicative preference relation (6.5), assessed in Example 6.5, into a fuzzy preference relation by means of (6.28):

$$\mathbf{RR} = \begin{bmatrix} 0.5 & 0.658 & 0.942 & 0.943 \\ 0.342 & 0.5 & 0.866 & 0.866 \\ 0.057 & 0.134 & 0.5 & 0.5 \\ 0.057 & 0.134 & 0.5 & 0.5 \end{bmatrix} \tag{6.34}$$

The ARFPR given by (6.34) is coherent with (6.5). For instance, the reader can see in (6.34) that X_1 is better than X_2 and is much better than X_3 and X_4, or that X_3 is as good as X_4.

6.3.1.4 NRFPR → ARFPR

When the DM's preferences are defined in terms of the NRFPR given in Section 6.2, it is possible to convert it to the ARFPR by means of any function $H(\mathbf{RN}(X_k, X_l), \mathbf{RN}(X_l, X_k)) : [0, 1] \times [0, 1] \to [0, 1]$ verifying the following conditions:

- $H(x, y)$ should be a nondecreasing function of the first argument and a nonincreasing function of the second argument;
- $H(1, 1) = 0.5$;
- $H(1, 0) = 1$;
- $H(1, x) > 0.5$, if $x < 1$, $\forall x\, [0, 1]$;
- $H(0, 1) = 0$;
- $H(x, 1) < 0.5$, if $x < 1, \forall x\, [0, 1]$;
- additive reciprocity: $H(x, y) + H(y, x) = 1, \forall x, y \in [0, 1]$.

It should be mentioned that we do not consider here the judgments of incomparability between two alternatives, since such judgments of incomparability (that is, any pair of judgments verifying $\mathbf{RN}(X_k, X_l) < 1$ and $\mathbf{RN}(X_l, X_k) < 1$) are not portrayed by the encoding scheme of ARFPRs. In general, only the judgments of total incomparability (which correspond to the extreme situation in which $\mathbf{RN}(X_k, X_l) = \mathbf{RN}(X_l, X_k) = 0$) are (in a certain way) handled by the encoding scheme of the ARFPR. A priori, this type of judgment can be represented in an ARFPR as missing values. When a DM cannot compare two alternatives, in the first moment, the corresponding entries in the matrix of the ARFPR can be left unfilled. Later, those missing numbers are estimated by using certain methods based on the assumption of additive transitivity of the ARFPR (Herrera-Viedma *et al.*, 2007) and, as a consequence, the corresponding judgments of incomparability disappear.

Here we consider the following three transformation functions verifying those conditions presented above:

$$\mathbf{RR}(X_k, X_l) = H_7(\mathbf{RN}(X_k, X_l), \mathbf{RN}(X_l, X_k)) = \frac{1}{2}(1 + \mathbf{RN}(X_k, X_l) - \mathbf{RN}(X_l, X_k)) \tag{6.35}$$

$$\begin{aligned} \mathbf{RR}(X_k, X_l) &= H_8(\mathbf{RN}(X_k, X_l), \mathbf{RN}(X_l, X_k)) \\ &= \frac{\mathbf{RN}(X_k, X_l)}{\mathbf{RN}(X_k, X_l) + \mathbf{RN}(X_l, X_k)} = \frac{1}{1 + \dfrac{\mathbf{RN}(X_l, X_k)}{\mathbf{RN}(X_k, X_l)}} \end{aligned} \tag{6.36}$$

$$\mathbf{RR}(X_k, X_l) = H_9(\mathbf{RN}(X_k, X_l), \mathbf{RN}(X_l, X_k)) \\ = \frac{\mathbf{RN}(X_k, X_l)^2}{\mathbf{RN}(X_k, X_l)^2 + \mathbf{RN}(X_l, X_k)^2} = \frac{1}{1 + \left(\dfrac{\mathbf{RN}(X_l, X_k)}{\mathbf{RN}(X_k, X_l)}\right)^2} \tag{6.37}$$

The selection of an adequate transformation function among (6.35), (6.36), and (6.37) (the construction of (6.36) was proposed by Queiroz, 2009) depends upon whether the ratio $\mathbf{RN}(X_k, X_l)/\mathbf{RN}(X_l, X_k)$ or the difference $\mathbf{RN}(X_k, X_l) - \mathbf{RN}(X_l, X_k)$ is meaningful. The transformation function given by (6.35) may be utilized when the difference $\mathbf{RN}(X_k, X_l) - \mathbf{RN}(X_l, X_k)$ is meaningful. On the other hand, both (6.36) and (6.37) may be utilized in situations in which the ratio $\mathbf{RN}(X_k, X_l)/\mathbf{RN}(X_l, X_k)$ is meaningful. For practical purposes, the main difference between (6.36) and (6.37) lies in the fact that the strength of nonstrict preference $\mathbf{RR}(X_k, X_l)$ calculated by means of (6.37) tends to be farther from the indifference judgment than the corresponding strength of nonstrict preference calculated by means of (6.36).

Another important aspect that must be stressed here is that the ARFPRs obtained with the use of (6.36) and (6.37) satisfy multiplicative transitivity (see (6.22)) as long as the corresponding NRFPR also satisfies multiplicative transitivity. Indeed, the substitution of (6.36) (or of (6.37)) into (6.22) yields

$$\left(\frac{\mathbf{RN}(X_k, X_j)}{\mathbf{RN}(X_k, X_j) + \mathbf{RN}(X_j, X_k,)}\right) \times \left(\frac{\mathbf{RN}(X_k, X_j) + \mathbf{RN}(X_j, X_k)}{\mathbf{RN}(X_j, X_k)}\right) = \left(\frac{\mathbf{RN}(X_k, X_l)}{\mathbf{RN}(X_k, X_l) + \mathbf{RN}(X_l, X_k)}\right) \\ \times \left(\frac{\mathbf{RN}(X_k, X_l) + \mathbf{RN}(X_l, X_k)}{\mathbf{RN}(X_l, X_k)}\right) \times \left(\frac{\mathbf{RN}(X_l, X_j)}{\mathbf{RN}(X_l, X_j) + \mathbf{RN}(X_j, X_l)}\right) \times \left(\frac{\mathbf{RN}(X_l, X_j) + \mathbf{RN}(X_j, X_l)}{\mathbf{RN}(X_j, X_l)}\right) \tag{6.38}$$

which can be easily simplified as the following expression of multiplicative transitivity for NRFPRs:

$$\frac{\mathbf{RN}(X_k, X_j)}{\mathbf{RN}(X_j, X_k)} = \frac{\mathbf{RN}(X_k, X_l)}{\mathbf{RN}(X_l, X_k)} \frac{\mathbf{RN}(X_l, X_j)}{\mathbf{RN}(X_j, X_l)} \tag{6.39}$$

On the other hand, it should be mentioned that the ARFPRs obtained with the use of (6.35) satisfy additive transitivity only if the corresponding NRFPR satisfies the following condition:

$$\big(\mathbf{RN}(X_k, X_j) - \mathbf{RN}(X_j, X_k)\big) + \big(\mathbf{RN}(X_j, X_l) - \mathbf{RN}(X_l, X_j)\big) \\ + (\mathbf{RN}(X_k, X_l) - \mathbf{RN}(X_l, X_k)) = 0 \tag{6.40}$$

It can be seen that by substituting (6.35) into (6.7) (which corresponds to the additive transitivity), we obtain (6.41), which can be reduced to (6.40), as the reader can easily verify:

$$\big(1 + \mathbf{RN}(X_k, X_j) - \mathbf{RN}(X_j, X_k)\big) + \big(1 + \mathbf{RN}(X_j, X_l) - \mathbf{RN}(X_l, X_j)\big) \\ + (1 + \mathbf{RN}(X_k, X_l) - \mathbf{RN}(X_l, X_k)) = 3 \tag{6.41}$$

It is worth noting that, in applications where (5.35) is utilized to obtain the level of strict preference of one alternative over the other, the difference between $\mathbf{RN}(X_k, X_l)$ and $\mathbf{RN}(X_l, X_k)$

reflects the level of strict preference of one alternative over the other, as discussed in Example 6.8. In this case, (6.41) can be interpreted as: the level of strict preference of X_k over X_j added to the level of strict preference of X_j over X_l should be equal to the level of strict preference of X_k over X_l.

Example 6.15. Let us consider that, in Example 6.7, the DM provided the NRFPR given by (6.9), in such a way that the differences $\mathbf{RN}(X_k, X_l) - \mathbf{RN}(X_l, X_k)$ have a meaning. In such a case, in order to preserve such a meaning, the conversion from NRFPR to ARFPR can be made by means of (6.35), which produces the following ARFPR:

$$\mathbf{RR} = \begin{bmatrix} 0.5 & 0.75 & 0.75 & 1 \\ 0.25 & 0.5 & 0.5 & 0.85 \\ 0.25 & 0.5 & 0.5 & 0.85 \\ 0 & 0.15 & 0.15 & 0.5 \end{bmatrix} \tag{6.42}$$

As can be seen, (6.42) is compatible with (6.9). For instance, in (6.42), it is indicated that X_1 is the best alternative, as $\mathbf{RR}(X_1, X_k) > 0.5$ and $\mathbf{RR}(X_k, X_1) < 0.5$, for $k \in \{2, 3, 4\}$. Further, the fourth row and the fourth column of matrix $\mathbf{RR}$ indicate that X_4 is the worst alternative, as $\mathbf{RR}(X_4, X_k) < 0.5$ and $\mathbf{RR}(X_k, X_4) > 0.5$, for $k \in \{1, 2, 3\}$.

Example 6.16. Now let us consider that the DM provided values for the following NRFPR in such a way that their ratios are meaningful:

$$\mathbf{RN} = \begin{bmatrix} 1 & 1 & 1 \\ 1/2 & 1 & 1 \\ 1/6 & 1/3 & 1 \end{bmatrix} \tag{6.43}$$

In this way, for instance, the ratio

$$\frac{\mathbf{RR}(X_1, X_2)}{\mathbf{RR}(X_2, X_1)} = \frac{1}{0.5} = 2$$

means that X_1 is two times better than X_2. In such a case, in order to preserve the essence of these comparisons in performing the conversion from the NRFPR to the ARFPR, this conversion may be made by means of (6.36) or (6.37). The use of (6.36) to convert the NRFPR (6.9) into an ARFPR yields

$$\mathbf{RR} = \begin{bmatrix} 0.5 & 0.667 & 0.857 \\ 0.333 & 0.5 & 0.75 \\ 0.143 & 0.250 & 0.5 \end{bmatrix} \tag{6.44}$$

The use of (6.37), rather than (6.36), produces the following ARFPR:

$$\mathbf{RR} = \begin{bmatrix} 0.5 & 0.8 & 0.972 \\ 0.2 & 0.5 & 0.9 \\ 0.027 & 0.1 & 0.5 \end{bmatrix} \tag{6.45}$$

As can be seen, both (6.44) and (6.45) are compatible with (6.43). For instance, in (6.44) and (6.45), it is indicated that X_1 is the best alternative, as $\mathbf{RR}(X_1, X_k) > 0.5$ and $\mathbf{RR}(X_k, X_1) < 0.5$, for $k \in \{2, 3\}$. Further, the third line and the third column of both (6.44) and (6.45) indicate that X_3 is the worst alternative, as $\mathbf{RR}(X_3, X_k) < 0.5$ and $\mathbf{RR}(X_k, X_3) > 0.5$, for $k \in \{1, 2\}$.

6.3.2 *Transformation Functions for NRFPRs*

It is worth indicating that the transformation functions to be presented next do not necessarily produce NRFPRs that verify the property of min-transitivity. However, similar to the transformation functions presented above, all transformation functions below preserve the weak transitivity of the original preference information. In this way, the NRFPR derived from an ordered array or from a utility function, with the use of those transformation functions, always satisfies weak transitivity. On the other hand, the NRFPR derived from an ARFPR or from a multiplicative preference relation, with the use of those transformation functions, only verifies weak transitivity if the collected pairwise judgments also satisfy weak transitivity.

6.3.2.1 ARFPR → NRFPR

Given the DM's preferences defined in terms of the ARFPR, it is possible to convert it to the NRFPR by means of the following three transformation functions:

$$\mathbf{RN}(X_k, X_l) = H_{10}(\mathbf{RR}(X_k, X_l), \mathbf{RR}(X_l, X_k)) = \begin{cases} 1 + \mathbf{RR}(X_k, X_l) - \mathbf{RR}(X_l, X_k) & \text{if } \mathbf{RR}(X_k, X_l) < 0.5 \\ 1 & \text{if } \mathbf{RR}(X_k, X_l) \geq 0.5 \end{cases} \tag{6.46}$$

$$\mathbf{RN}(X_k, X_l) = H_{11}(\mathbf{RR}(X_k, X_l), \mathbf{RR}(X_l, X_k)) = \begin{cases} \dfrac{\mathbf{RR}(X_k, X_l)}{\mathbf{RR}(X_l, X_k)} & \text{if } \mathbf{RR}(X_k, X_l) < 0.5 \\ 1 & \text{if } \mathbf{RR}(X_k, X_l) \geq 0.5 \end{cases} \tag{6.47}$$

$$\mathbf{RN}(X_k, X_l) = H_{12}(\mathbf{RR}(X_k, X_l), \mathbf{RR}(X_l, X_k)) = \begin{cases} \left(\dfrac{\mathbf{RR}(X_k, X_l)}{\mathbf{RR}(X_l, X_k)}\right)^{0.5} & \text{if } \mathbf{RR}(X_k, X_l) < 0.5 \\ 1 & \text{if } \mathbf{RR}(X_k, X_l) \geq 0.5 \end{cases} \tag{6.48}$$

Expressions (6.46)–(6.48) (the construction of (6.47) was proposed by Queiroz, 2009) represent the transformation functions that allow the reverse conversions of (6.35)–(6.37), respectively. The transformation function (6.46) may be utilized when the ARFPR is defined in such a way that the difference $\mathbf{RR}(X_k, X_l) - \mathbf{RR}(X_l, X_k)$ makes sense. The transformation functions (6.47) and (6.48) may be utilized when the ARFPR is defined in such a way that the ratio $\mathbf{RR}(X_k, X_l)/\mathbf{RR}(X_l, X_k)$, in the sense that it indicates how many times X_k is preferred to X_l.

Recalling from the previous section that the indifference judgment in those NRFPRs is reflected by the pair of judgments $\mathbf{RN}(X_k, X_l)$ and $\mathbf{RN}(X_l, X_k)$ being equal to one, the main difference between (6.47) and (6.48) is associated with the fact that each pairwise judgment $\mathbf{RN}(X_k, X_l)$ produced by (6.48) tends to be closer to the indifference judgment than each corresponding pairwise judgment produced with the use of (6.47), as can be confirmed through the following example.

Example 6.17. The use of (6.46) to convert the ARFPR (6.42) into a NRFPR yields

$$\mathbf{RN} = \begin{bmatrix} 1 & 1 & 1 & 1 \\ 0.5 & 1 & 1 & 1 \\ 0.5 & 1 & 1 & 1 \\ 0 & 0.3 & 0.3 & 1 \end{bmatrix} \tag{6.49}$$

As can be seen, (6.49) perfectly matches (6.9), which is the NRFPR from which the ARFPR given by (6.42) was converted with the use of (6.35), in Example 6.15. In this way, one can note that (6.46) performs the reverse operation of (6.35).

Example 6.18. The use of (6.47) to convert the ARFPR given by (6.44) into a NRFPR yields

$$\mathbf{RN} = \begin{bmatrix} 1 & 1 & 1 \\ 0.499 & 1 & 1 \\ 0.167 & 0.333 & 1 \end{bmatrix} \tag{6.50}$$

The use of (6.48), rather than (6.47), produces the following NRFPR:

$$\mathbf{RN} = \begin{bmatrix} 1 & 1 & 1 \\ 0.706 & 1 & 1 \\ 0.408 & 0.577 & 1 \end{bmatrix} \tag{6.51}$$

As can be seen, both (6.50) and (6.51) are compatible with (6.44), as well as with the preference relation given by (6.43), from which (6.44) was converted (refer to Example 6.16). For instance, the first line in both (6.50) and (6.51) is entirely filled with ones, which means that alternative X_1 is not worse than the remaining alternatives. Further, the third column in both (6.50) and (6.51) is also entirely filled with ones, which means that X_3 is not better than any other alternative.

Example 6.19. Let us complete the reverse operation on (6.50), by applying (6.36) to (6.50), in order to convert it back to the NRFPR from which (6.50) was converted (refer to Example 6.16). For instance, if we apply (6.36) to the pair (X_1, X_2), we obtain the following results:

$$\mathbf{RR}(X_1, X_2) = H_8(1, 0.499) = \frac{1}{1 + 0.499} = 0.667 \tag{6.52}$$

$$\mathbf{RR}(X_2, X_1) = H_8(0.499, 1) = \frac{0.499}{0.499 + 1} = 0.333 \tag{6.53}$$

By comparing (6.52)–(6.53) to their respective entries in the matrix (6.44), we can see that they match perfectly. If we proceed by applying (6.36) to the remaining pairs of alternatives, we obtain the ARFPR given by (6.44), as expected.

Next, we present some transformation functions for converting the preference information expressed in terms of the different formats directly into the NRFPR format, by substituting, into (6.46), (6.47), or (6.48), the expressions presented above for conversion from the different preference formats into an ARFPR. The choice between using transformation functions based on (6.46), (6.47), or (6.48) should consider the following aspects: (6.46) may be utilized when the difference $\mathbf{RR}(X_k, X_l) - \mathbf{RR}(X_l, X_k)$ has a meaning, which we want to preserve. Both (6.47) and (6.48) may be utilized when the ratio $\mathbf{RR}(X_k, X_l)/\mathbf{RR}(X_l, X_k)$ has a meaning that should be preserved in the conversion. Further, in the selection between (6.47) and (6.48), it should be considered that (6.48) may quantify the preference strength associated with each pairwise comparison, in such a way that it tends to be nearer to an indifference judgment. In contrast, (6.47) may quantify the preference strength associated with each pairwise comparison, in such a way that it tends to be nearer to a strict preference judgment. Therefore, in the choice between using one or the other transformation function, it is important to consider the DM's desire of accentuating, or not, his/her preference strengths for one alternative over the other. The reader can confirm this feature in Example 6.18.

6.3.2.2 Ordered Array → NRFPR

Let us begin by considering the preferences expressed in purely ordinal format. In order to convert preferences expressed in terms of an ordered array into a NRFPR, we can use the transformation function given by

$$\mathbf{RN}(X_k, X_l) = H_{13}(o_k, o_l) = \begin{cases} \dfrac{1}{2} + \dfrac{o_l - o_k}{2(n-1)} & \text{if } o_k > o_l \\ 1 & \text{if } o_k \leq o_l \end{cases} \tag{6.54}$$

which is derived by substituting (6.17) into (6.46). The advantage of using (6.17) combined with (6.46) and not with (6.47) or (6.48) is associated with the need to preserve the meaning of the differences $o_l - o_k$ among the positions of two alternatives in the conversion from the ordered array into the NRFPR.

Example 6.20. By applying (6.54) to convert the ordered array considered in Example 6.1 into a NRFPR, we obtain

$$\mathbf{RN} = \begin{bmatrix} 1 & 0.375 & 1 & 1 & 1 \\ 1 & 1 & 1 & 1 & 1 \\ 0.25 & 0.125 & 1 & 1 & 0.375 \\ 0.125 & 0 & 0.375 & 1 & 0.25 \\ 0.375 & 0.25 & 1 & 1 & 1 \end{bmatrix} \tag{6.55}$$

It is worth noting that (6.55) is coherent with the ordered array. For instance, the fact that the second row is entirely filled with ones indicates that X_2 is not worse than any other alternative; and the fact that the fourth column is entirely filled with ones indicates that X_4 is not better than any other alternative under consideration.

6.3.2.3 Utility → NRFPR

The preferences given in terms of a vector of utilities defined on a ratio scale, normalized in the interval [0, 1], can be converted into a NRFPR by means of the following expressions:

$$\mathbf{RN}(X_k, X_l) = H_{14}(u_k, u_l) = \begin{cases} \dfrac{u_k}{u_l} & \text{if } u_k < u_l \\ 1 & \text{if } u_k \geq u_l \end{cases} \tag{6.56}$$

$$\mathbf{RN}(X_k, X_l) = H_{15}(u_k, u_l) = \begin{cases} \left(\dfrac{u_k}{u_l}\right)^{0.5} & \text{if } u_k < u_l \\ 1 & \text{if } u_k \geq u_l \end{cases} \tag{6.57}$$

$$\mathbf{RN}(X_k, X_l) = H_{16}(u_k, u_l) = \begin{cases} \left(\dfrac{u_k}{u_l}\right)^{2} & \text{if } u_k < u_l \\ 1 & \text{if } u_k \geq u_l \end{cases} \tag{6.58}$$

Expressions (6.56) and (6.57), respectively, are obtained by substituting (6.20) into (6.47) and (6.48). Expression (6.58) is obtained by the substitution of (6.21) into (6.47). Note that the substitution of (6.21) into (6.48) also corresponds to (6.45). In order to preserve the meaning of the ratio u_k/u_l in the conversion from the utility values into a NRFPR, substitution of (6.20) or (6.21) into (6.47) or (6.48) is preferable to substitution of (6.20) or (6.21) into (6.46).

As can be confirmed by the following example, in the selection of a transformation function among (6.56), (6.57), or (6.58), it is important to consider that (6.57) produces pairwise judgments tending somewhat more to an indifference judgment than the other transformation functions do. In contrast, (6.58) produces pairwise judgments tending somewhat more to a strict preference than the other transformation functions do. Finally, (6.56) can be considered as an intermediate case between (6.57) and (6.58). This is a consequence of the fact that the value of $R(X_k, X_l)$, when $u_k < u_l$, can be given by one of the ratios satisfying

$$\underbrace{\left(\frac{u_k}{u_l}\right)^2}_{(6.58)} \leq \underbrace{\frac{u_k}{u_l}}_{(6.56)} \leq \underbrace{\left(\frac{u_k}{u_l}\right)^{0.5}}_{(6.57)} \tag{6.59}$$

Example 6.21. By applying (6.56), (6.57), and (6.58) to convert the utility array $\boldsymbol{U} = [\,0.2\ 0.4\ 0.1\ 0.3\,]$, which is assessed on a ratio scale, into NRFPRs, we obtain

$$\mathbf{RN} = \begin{bmatrix} 1 & 0.5 & 1 & 0.67 \\ 1 & 1 & 1 & 1 \\ 0.5 & 0.25 & 1 & 0.33 \\ 1 & 0.75 & 1 & 1 \end{bmatrix} \tag{6.60}$$

$$\mathbf{RN} = \begin{bmatrix} 1 & 0.71 & 1 & 0.82 \\ 1 & 1 & 1 & 1 \\ 0.71 & 0.5 & 1 & 0.58 \\ 1 & 0.87 & 1 & 1 \end{bmatrix} \tag{6.61}$$

$$\mathbf{RN} = \begin{bmatrix} 1 & 0.25 & 1 & 0.44 \\ 1 & 1 & 1 & 1 \\ 0.25 & 0.06 & 1 & 0.11 \\ 1 & 0.56 & 1 & 1 \end{bmatrix} \tag{6.62}$$

It is essential to note that the obtained NRFPRs are coherent with the analysis of the transformation functions (6.56), (6.57), and (6.58), presented above.

The preferences expressed in terms of a vector of utilities, defined on an interval scale, can be converted into a NRFPR by means of the following expression:

$$\mathbf{RN}(X_k, X_l) = H_{17}(u_k, u_l) = \begin{cases} 1 + u_k - u_l & \text{if } u_k < u_l \\ 1 & \text{if } u_k \geq u_l \end{cases} \tag{6.63}$$

which is obtained by means of substituting (6.26) into (6.46). Here, the benefit of considering (6.46), rather than (6.47) and (6.48), is associated with preserving the significance of the cardinal utility values defined on an interval scale when they are converted into the NRFPR.

Example 6.22. By applying (6.63) to the array of utility values defined on an interval scale $\boldsymbol{U} = [\,1\ 0.9\ 0\ 0.2\,]$, we obtain the NRFPR given by

$$\mathbf{RN} = \begin{bmatrix} 1 & 1 & 1 & 1 \\ 0.9 & 1 & 1 & 1 \\ 0 & 0.1 & 1 & 0.8 \\ 0.2 & 0.3 & 1 & 1 \end{bmatrix} \tag{6.64}$$

It is worth noting that (6.64) reflects the fact that X_1 is the best alternative (see the first row of (6.64) filled with ones); X_3 is the worst alternative (see the third column of (6.64) filled with ones); and, as $\mathbf{RN}(X_2, X_1)$ is much higher than $\mathbf{RN}(X_4, X_1)$ in (6.64), we can deduce that X_1 is somewhat better than X_2 and is much better than X_4. All these observations are coherent with the utility values under consideration.

6.3.2.4 Multiplicative Preference Relation → NRFPR

The multiplicative preference relation can be transformed into a NRFPR with the use of

$$\begin{aligned} \mathbf{RN}(X_k, X_l) &= H_{18}(\boldsymbol{M}(X_k, X_l), \boldsymbol{M}(X_l, X_k)) \\ &= \begin{cases} 1 + \dfrac{1}{2}\log_m \dfrac{\boldsymbol{M}(X_k, X_l)}{\boldsymbol{M}(X_l, X_k)} & \text{if } \log_m \boldsymbol{M}(X_k, X_l) < 0 \\ 1 & \text{if } \log_m \boldsymbol{M}(X_k, X_l) \geq 0 \end{cases} \end{aligned} \tag{6.65}$$

which is obtained by substituting (6.29) into (6.46). The advantage of considering (6.46), rather than (6.47) or (6.48), in the definition of transformation functions for converting from multiplicative preference relations into NRFPRs, lies in the fact that the meaning of the ratio $\boldsymbol{M}(X_k, X_l)/\boldsymbol{M}(X_l, X_k)$ is to some extent preserved, as can be seen in (6.65).

Example 6.23. Let us convert the multiplicative preference relation (6.5), assessed in Example 6.4, into a NRFPR. The use of (6.65), by considering that $m = 9$, yields (6.66). It is straightforward to confirm that (6.66) is coherent with (6.5). We leave this task to the reader:

$$\mathbf{RN} = \begin{bmatrix} 1 & 1 & 1 & 1 \\ 0.684 & 1 & 1 & 1 \\ 0.114 & 0.267 & 1 & 1 \\ 0.114 & 0.267 & 1 & 1 \end{bmatrix} \tag{6.66}$$

6.4 A Method for Repairing Inconsistent Judgments

The importance of collecting consistent judgments has motivated several authors to propose methods for repairing inconsistencies in the judgments expressed in terms of multiplicative preference relations or fuzzy preference relations. The aim is to obtain a consistent matrix of pairwise comparisons by modifying the original judgments as little as possible. For instance, Zeshui and Cuiping (1999) present a method for repairing multiplicative preference relations so that multiplicative transitivity is achieved, and Ma *et al.* (2006) present a method for repairing ARFPRs in order to achieve weak transitivity. An analyst can make use of such methods for revising or adjusting the inconsistent preference relations provided by a DM who is not capable of executing this task on his/her own, in accordance with the procedure represented in Figure 6.7 (Ma *et al.*, 2006).

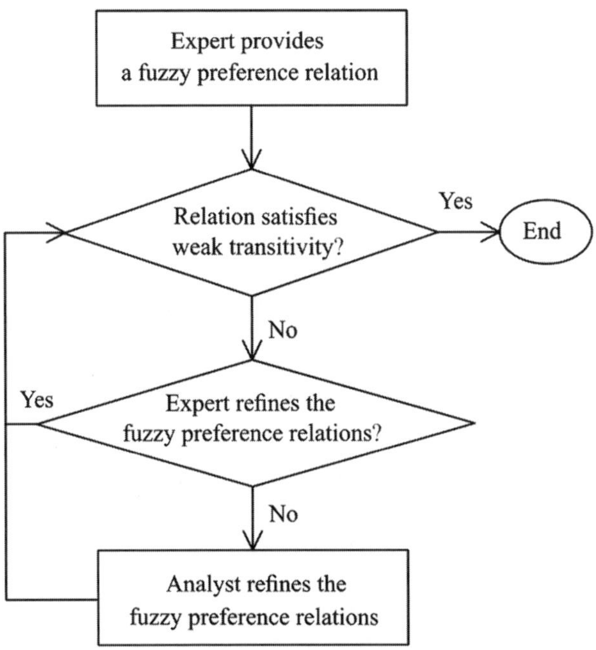

Figure 6.7 Process of consistency improvement.

As discussed in Chapter 5, at present the literature still lacks a consensus on the adequate consistency condition that needs to be satisfied by the fuzzy preference relations. Although the min-transitivity has been traditionally utilized as a consistency condition, it also has been criticized for being an excessively hard and difficult-to-meet condition for practical use (Herrera-Viedma *et al.*, 2004). In particular, the additive transitivity has been considered a reasonable condition to be applied when the fuzzy preference relation is additive reciprocal. One advantageous aspect of its use lies in the fact that the additive transitivity for reciprocal fuzzy preference relations is equivalent to Saaty's consistency for multiplicative preference relations (Herrera-Viedma *et al.*, 2004). However, it is important to mention that both types of transitivity (the additive and the multiplicative) are in conflict with their corresponding scales used to quantify the preferences. A sound explanation of the drawbacks of using the additive transitivity, as well as the multiplicative transitivity, as consistency conditions can be found in Chiclana, Herrera-Viedma, and Herrera (2004).

Finally, the weak transitivity can be seen as the minimum consistency requirement for pairwise comparisons, because it requires only that if X_k is at least as good as X_j and X_j is at least as good as X_l, then X_k should be at least as good as X_l, without taking into account the strength of these preferences. In this context, considering that a DM may not be able to make the adjustments needed to guarantee a satisfactory level of consistency within his/her judgments when they are expressed in terms of multiplicative preference relations, ARFPR, or NRFPRs, we present next the method proposed by Ma *et al.* (2006) for improving the consistency of an ARFPR until weak transitivity is achieved. Obviously, although it is supposed to be applied to ARFPRs, it can also be applied indirectly to guarantee the weak transitivity of NRFPRs or multiplicative preference relations, as long as they are previously converted into ARFPRs. The following stepwise procedure begins by considering a given ARFPR, denoted here by $\mathbf{RR}^{(0)}$.

6.4.1 Method for Repairing an Inconsistent ARFPR to Satisfy Weak Transitivity

Step 1. Set $t = 1$ and construct an ARFPR satisfying additive transitivity by means of the following expression:

$$\mathbf{RR}^{(1)}(X_k, X_l) = 0.5 + \frac{1}{n}\left(\sum_{j=1}^{n} \mathbf{RR}^{(0)}(X_k, X_j) - \mathbf{RR}^{(0)}(X_l, X_j)\right) \quad \forall X_k, X_l, X_j \in \mathbf{X} \tag{6.67}$$

where n is the number of alternatives being compared.

Step 2. If there is at least one negative entry in the constructed matrix, the entire matrix must be modified in such a way that the entries of the new matrix are given by

$$\mathbf{RR}^{(2)}(X_k, X_l) = \frac{\mathbf{RR}^{(1)}(X_k, X_l) + r}{1 + 2r} \quad \forall X_k, X_l, X_j \in \mathbf{X} \tag{6.68}$$

where r corresponds to the entry of $\mathbf{RR}^{(1)}$ which has the larger absolute value, that is, the most negative entry of $\mathbf{RR}^{(1)}$.

Step 3. Construct matrix $\mathbf{RR}^{(3)}$, called the synthesis matrix, by means of

$$\mathbf{RR}^{(3)} = (1 - t\Delta)\mathbf{RR}^{(0)} + t\Delta\mathbf{RR}^{(12)} \tag{6.69}$$

where Δ is a step size defined in the interval [0,1] and the symbol $\mathbf{RR}^{(12)}$ denotes the additive transitive matrix $\mathbf{RR}^{(1)}$ or, if $\mathbf{RR}^{(1)}$ has at least one negative entry, the additive matrix $\mathbf{RR}^{(2)}$.

Step 4. Verify whether the synthesis matrix $\mathbf{RR}^{(3)}$ satisfies weak transitivity (in Ma *et al.* (2006), two ways of conducting this test are proposed; in the next chapter, a simple test will be described). If it does, the procedure is terminated; otherwise, let $t = t + 1$ and go to Step 3.

Example 6.24. Let us apply the method described above to repair the following fuzzy preference relation:

$$\mathbf{RR}^{(0)} = \begin{bmatrix} 0.5 & 0.8 & 0.8 & 0.3 \\ 0.2 & 0.5 & 0.9 & 0.6 \\ 0.2 & 0.1 & 0.5 & 0.1 \\ 0.7 & 0.4 & 0.9 & 0.5 \end{bmatrix} \tag{6.70}$$

As can be seen, (6.70) does not satisfy weak transitivity, since we have that $\mathbf{RR}^{(0)}(X_1, X_2) > 0.5$ and $\mathbf{RR}^{(0)}(X_2, X_4) > 0.5$, although $\mathbf{RR}^{(0)}(X_4, X_1) > 0.5$.

In the execution of Step 1, the following ARFPR is obtained with the use of (6.67):

$$\mathbf{RR}^{(1)} = \begin{bmatrix} 0.5 & 0.55 & 0.875 & 0.475 \\ 0.45 & 0.5 & 0.825 & 0.425 \\ 0.125 & 0.175 & 0.5 & 0.1 \\ 0.525 & 0.575 & 0.9 & 0.5 \end{bmatrix} \tag{6.71}$$

Step 2 can be bypassed, as the relation in (6.71) does not include any negative entries. In Step 3, by applying (6.69) with $t = 1$ and $\Delta = 0.1$, we obtain the following synthesis matrix as the linear aggregation of $\mathbf{RR}^{(0)}$ and $\mathbf{RR}^{(12)}$:

$$\mathbf{RR}^{(3)} = \begin{bmatrix} 0.5 & 0.75 & 0.815 & 0.335 \\ 0.25 & 0.5 & 0.885 & 0.565 \\ 0.185 & 0.115 & 0.5 & 0.1 \\ 0.665 & 0.435 & 0.9 & 0.5 \end{bmatrix} \tag{6.72}$$

In Step 4, as (6.72) does not satisfy weak transitivity, we set $t = 2$ and go back to Step 3. In Step 3, (6.69) generates the following synthesis matrix:

$$\mathbf{RR}^{(3)} = \begin{bmatrix} 0.5 & 0.7 & 0.83 & 0.37 \\ 0.3 & 0.5 & 0.87 & 0.53 \\ 0.17 & 0.13 & 0.5 & 0.1 \\ 0.63 & 0.47 & 0.9 & 0.5 \end{bmatrix} \tag{6.73}$$

In Step 4, as (6.73) still does not satisfy weak transitivity, after setting $t = 3$, we move to Step 3. In Step 3, (6.73) is updated as

$$\mathbf{RR}^{(3)} = \begin{bmatrix} 0.5 & 0.65 & 0.845 & 0.405 \\ 0.35 & 0.5 & 0.855 & 0.495 \\ 0.155 & 0.145 & 0.5 & 0.1 \\ 0.595 & 0.505 & 0.9 & 0.5 \end{bmatrix} \tag{6.74}$$

In Step 4, as (6.74) satisfies weak transitivity, the process is terminated.

6.5 Conclusions

The input of the preference information plays a fundamental role in the decision-making process, as the recommendations are derived from the mathematical models, which are constructed in accordance with the supplied information.

In view of the fact that this information is often subjective, vague, and uncertain, it is of paramount importance to provide the experts with the means to articulate their preferences as truthfully and accurately as possible. Otherwise, if experts are forced to express their preferences using a preference format with which they do not feel comfortable, this preference information, if not correctly expressed, could negatively impact the overall multicriteria analysis.

Taking all this into account, in the current chapter we have considered five main types of preference formats and presented transformation functions that can be utilized to convert the preference information from those different formats into additive reciprocal fuzzy preference relations, as well as nonreciprocal fuzzy preference relations. The applicability of those transformation functions can be observed through several examples of multicriteria decision-making problems studied in Chapters 7, 9, and 10.

Finally, some questions of repairing inconsistencies in the judgments are also discussed in the chapter.

Exercises

Problem 6.1. Construct a NRFPR from the comparison of the alternatives X_1, X_2 and X_3, which were evaluated on a maximization criterion F as shown in Figure 6.8 (hint: use the rule of thumb given in Section 6.2).

Problem 6.2. Verify whether the fuzzy preference relation given by (6.8) satisfies additive transitivity.

Problem 6.3. Given the following definition of weak transitivity for ARFPRs:

$$\text{if } \mathbf{RR}(X_k, X_l) \geq 0.5 \text{ and } \mathbf{RR}(X_l, X_j) \geq 0.5, \text{ then } \mathbf{RR}(X_k, X_j) \geq 0.5$$

verify whether (6.8) satisfies weak transitivity.

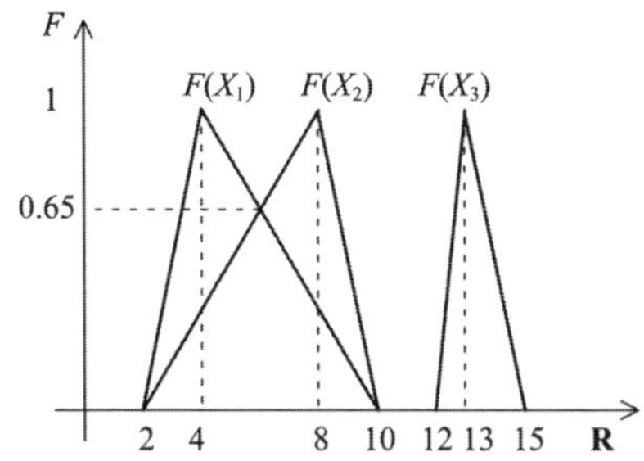

Figure 6.8 Evaluation of alternatives on criterion F.

Problem 6.4. Verify whether the fuzzy preference relation given by (6.9) satisfies (a) min-transitivity and (b) weak transitivity.

Problem 6.5. Convert the ordered array $\boldsymbol{O} = [1\ 2\ 3\ 4]$ into an ARFPR.

Problem 6.6. Use the transformation function (Chiclana, Herrera, and Herrera-Viedma, 1998)

$$\mathbf{RR}(X_k, X_l) = H(o_k, o_l) = \begin{cases} 1 & \text{if } o_l - o_k > 0 \\ 0.5 & \text{if } o_l - o_k = 0 \\ 0 & \text{if } o_l - o_k < 0 \end{cases}$$

to convert the ordered array of Problem 6.5 into an ARFPR. In analyzing the obtained ARFPR, how can one identify which is the first ranked and the worst ranked alternatives from the original ordered array? Compare this transformation function to the one given by (6.17).

Problem 6.7. Convert the array of utility values $\boldsymbol{U} = [1\ 0\ 0.2\ 0.5]$ into NRFPRs with the use of (6.56), (6.57), and (6.58). By applying (5.36) to each NRFPR, obtain the corresponding fuzzy indifference relations. Try to relate the obtained results to the main differences discussed in Section 6.3 among the transformation functions (6.56), (6.57), and (6.58).

Problem 6.8. Consider that a DM provided the following NRFPR:

$$\mathbf{RN} = \begin{bmatrix} 1 & 1 & 0.8 \\ 0.5 & 1 & 1 \\ 1 & 0.7 & 1 \end{bmatrix}$$

Verify whether it satisfies weak transitivity and, if necessary, use the method described in Section 6.4 for repairing its inconsistencies.

References

Abbasbandy, S. and Asady, B. (2006) Ranking of fuzzy numbers by sign distance. *Information Sciences*, **176** (16), 2405–2416.

Abbasbandy, S. and Hajjari, T. (2009) A new approach for ranking of trapezoidal fuzzy numbers. *Computers and Mathematics with Applications*, **57** (3), 413–419.

Baas, S.M. and Kwakernaak, H. (1977) Rating and raking of multi-aspect alternatives using fuzzy sets. *Automatica*, **13** (1), 47–58.

Baldwin, J.F. and Guild, N.C.F. (1979) Comparison of fuzzy sets on the same decision space. *Fuzzy Sets and Systems*, **2** (3), 213–231.

Belton, V. (1999) Multiple-criteria problem structuring and analysis in a value theory framework, in *Multicriteria Decision Making: Advances in MCDM Models, Algorithms, Theory* (eds T. Gal, T. Stewart, and T. Hanne), Kluwer, Dordrecht, pp. 12-2–12-29.

Chen, C. and Klein, C.M. (1997) An efficient approach to solving fuzzy MADM problems. *Fuzzy Sets and Systems*, **88** (1), 51–67.

Chen, L.H. and Lu, H.W. (2002) The preference order on fuzzy numbers. *Computers and Mathematics with Applications*, **44** (10–11), 1455–1465.

Chen, S.J. and Hwang, C.L. (1992) *Fuzzy Multiple Attribute Decision Making: Methods and Applications*, Springer-Verlag, Berlin.

Cheng, C.H. (1998) A new approach for ranking fuzzy numbers by distance methods. *Fuzzy Sets and Systems*, **95** (3), 307–313.

Chiclana, F., Herrera, F., and Herrera-Viedma, E. (1998) Integrating three representation models in fuzzy multipurpose decision making based on fuzzy preference relations. *Fuzzy Sets and Systems*, **97** (1), 33–48.

Chiclana, F., Herrera, F., and Herrera-Viedma, E. (2001) Integrating multiplicative preference relations in a multipurpose decision making model base don fuzzy preference relations. *Fuzzy Sets and Systems*, **122** (2), 277–291.

Chiclana, F., Herrera-Viedma, E., and Herrera, F. (2004) T-additive and T-multiplicative transitivity: new consistency properties for providing preference relations. Proceedings of the Tenth International Conference on Information Processing and Management of Uncertainty in Knowledge-based Systems, Perugia, pp. 403–410.

Dubois, D. and Prade, H. (1983) Ranking fuzzy numbers in the setting of possibility theory. *Information Sciences*, **30** (3), 183–224.

Dubois, D. and Prade, H. (1999) A unified view of ranking techniques for fuzzy numbers. Proceedings of the 1999 International Conference on Fuzzy Numbers, Seoul, pp. 1328–1333.

Dyer, J.S. (2005) MAUT – multiattribute utility theory, in *Multiple Criteria Decision Analysis: State of the Art Surveys* (eds J. Figueira, S. Greco, and M. Ehrgott), Springer-Verlag, New York, pp. 265–292.

Ekel, P., Pedrycz, W., and Schinzinger, R. (1998) A general approach to solving a wide class of fuzzy optimization problems. *Fuzzy Sets and Systems*, **97** (1), 49–66.

Ekel, P.Ya. (2002) Fuzzy sets and models of decision making. *Computers and Mathematics with Applications*, **44** (7), 863–875.

Ekel, P.Ya. and Popov, V.A. (1985) Consideration of the uncertainty factor in problems of modelling and optimizing electrical networks. *Power Engineering*, **23** (2), 45–52.

Facchinetti, G. (2002) Ranking functions induced by weighted average of fuzzy numbers. *Fuzzy Optimization and Decision Making*, **1** (3), 313–327.

Farquhar, P.H. and Keller, L.R. (1989) Preference intensity measurement. *Annals of Operations Research*, **19** (1–4), 205–217.

Fodor, J.C. and Roubens, M. (1994) *Fuzzy Preference Modelling and Multicriteria Decision Support*, Kluwer, Boston, MA.

Fortemps, P. and Roubens, M. (1996) Ranking and defuzzification methods based on area compensation. *Fuzzy Sets and Systems*, **82** (3), 319–330.

Freeling, S. (1980) Fuzzy sets and decision analysis. *IEEE Transactions on Systems, Man, and Cybernetics*, **10** (3), 341–354.

Harker, P.T. and Vargas, L.G. (1987) The theory of ratio scale estimation Saaty's Analytic Hierarchy Process. *Management Science*, **33** (11), 1383–1402.

Herrera-Viedma, E., Alonso, S., Chiclana, F., and Herrera, F. (2007) A consensus model for group decision making with incomplete fuzzy preference relations. *IEEE Transactions on Fuzzy Systems*, **15** (5), 863–877.

Herrera-Viedma, E., Herrera, F., Chiclan, F., and Luque, M. (2004) Some issues on consistency of fuzzy preference relations. *European Journal of Operational Research*, **154** (1), 98–109.

Horiuchi, K. and Tamura, N. (1998) VSOP fuzzy numbers and their fuzzy ordering. *Fuzzy Sets and Systems*, **93** (2), 197–210.

Jain, R. (1976) Decision making in the presence of fuzzy variables. *IEEE Transactions on Systems, Man, and Cybernetics*, **6** (10), 698–703.

Keeney, R.L. and Raiffa, H. (1976) *Decisions with Multiple Objectives: Preferences and Values Tradeoffs*, John Wiley & Sons, Inc., New York.

Lee, E.S. and Li, R.L. (1988) Comparison of fuzzy numbers based on the probability measure of fuzzy events. *Computers and Mathematics with Applications*, **15** (3), 887–896.

Lee-Kwang, H. (1999) A method for ranking fuzzy numbers and its application to decision making. *IEEE Transactions on Fuzzy Systems*, **7** (1), 677–685.

Liu, X.W. and Han, S.L. (2005) Ranking fuzzy numbers with preference weighting function expectations. *Computers and Mathematics with Applications*, **49** (11–12), 1731–1753.

Ma, J., Fan, Z.P., Jiang, Y.P., *et al.* (2006) A method for repairing the inconsistency of fuzzy preference relations. *Fuzzy Sets and Systems*, **157** (1), 20–33.

Modarres, M. and Sadi-Nezhad, S. (2001) Ranking fuzzy numbers by preference ratio. *Fuzzy Sets and Systems*, **118** (3), 429–436.

Orlovsky, S.A. (1981) *Problems of Decision Making with Fuzzy Information*, Nauka, Moscow (in Russian).

Popov, V.A. and Ekel, P.Ya. (1987) Fuzzy set theory and problems of controlling the design and operation of electric power systems. *Soviet Journal of Computer and System Sciences*, **25** (4), 92–99.

Queiroz, J.C.B. (2009) Models and methods of decision making support for strategic management. PhD thesis (in Portuguese). Federal University of Minas Gerais, Belo Horizonte.

Raj, P.A. and Kumar, D.N. (1999) Ranking alternatives with fuzzy weights using maximizing set and minimizing set. *Fuzzy Sets and Systems*, **105** (3), 365–375.

Saaty, T. (1980) *The Analytic Hierarchy Process*, McGraw-Hill, New York.

Salo, A.A. and Hämäläinen, R.P. (1997) On the measurement of preferences in the Analytic Hierarchy Process. *Journal of Multi-criteria Decision Analysis*, **6** (6), 309–319.

Tanino, T. (1984) Fuzzy preference orderings in group decision making. *Fuzzy Sets and Systems*, **12** (1), 117–131.

Tran, L. and Duckstein, L. (2002) Comparison of fuzzy numbers using a fuzzy distance measure. *Fuzzy Sets and Systems*, **130** (3), 331–341.

Tseng, T.Y. and Klein, C.M. (1989) New algorithm for the ranking procedure in fuzzy decision making. *IEEE Transactions on Systems, Man, and Cybernetics*, **19** (5), 1289–1296.

Von Winterfeldt, D. and Edwards, W. (1986) *Decision Analysis and Behavioral Research*, Cambridge University Press, Cambridge.

Wang, X. and Kerre, E.E. (2001) Reasonable properties for the ordering of fuzzy quantities (I) and (II). *Fuzzy Sets and Systems*, **118** (3), 387–405.

Wang, Y.J. and Lee, H.S. (2008) The revised method of ranking fuzzy numbers with an area between the centroid and original points. *Computers and Mathematics with Applications*, **55** (9), 2033–2042.

Wang, Y.M. and Luo, Y. (2009) Area ranking of fuzzy numbers based on positive and negative ideal points. *Computers and Mathematics with Applications*, **58** (9), 1767–1779.

Yager, R.R. (1981) A procedure for ordering fuzzy sets of the unit interval. *Information Sciences*, **24** (2), 143–161.

Zeshui, X. and Cuiping, W. (1999) A consistency improving method in the analytic hierarchy process. *European Journal of Operational Research*, **116** (2), 443–449.

Zhang, Q., Chen, J.C.H., and Chong, P.P. (2004) Decision consolidation: criteria weight determination using multiple preference formats. *Decision Support Systems*, **38** (2), 247–258.

Zhang, Q., Wang, Y., and Yang, Y. (2007) Fuzzy multiple attribute decision making with eight types of preference information on alternatives. Proceedings of the 2007 IEEE Symposium on Computational Intelligence in Multicriteria Decision Making, Honolulu, pp. 288–293.

7

Discrete Models of Multicriteria Decision-Making and their Analysis

In this chapter, we discuss the essence and present methods used to analyze problems of multicriteria evaluation, comparison, choice, prioritization, and/or ordering of alternatives. There exist two classes of situations which give rise to these problems. The first one is associated with the direct statement of multiattribute decision-making problems when the consequences associated with solutions to problems cannot be estimated with the use of a single criterion. The second class is related to problems that may be solved on the basis of a single criterion; however, if the uncertainty of information does not permit a unique solution to be obtained, it is possible to include additional criteria and thereby convert these problems into tasks of multiattribute decision-making. In the chapter, we describe diverse techniques of multiattribute analysis of alternatives in a fuzzy environment (techniques aimed at the analysis of $\langle \mathbf{X}, \boldsymbol{R} \rangle$ models), developed on the basis of fuzzy preference modeling. Although these techniques are directly related to individual decision-making, they can be and are applied to procedures of group decision-making. It is shown that the discussed techniques can lead to different solutions. However, this diversity of solutions is to be considered natural and the most appropriate technique has to be selected by taking into account the essence of the problem, the possible sources of information, and its associated uncertainty.

7.1 Optimization Problems with Fuzzy Coefficients and their Analysis

In Chapter 1, we discussed the general issues related to the necessity of setting up and solving multicriteria problems. In particular, one of the classes of situations which call for the application of a multicriteria approach is associated with problems that may be solved on the basis of a single criterion. However, if uncertainty of available information does not allow us to derive unique solutions, it is possible to transform these problems to multicriteria decision-making by applying some additional criteria, including those of a qualitative character, in order to reduce the decision uncertainty regions. Taking this into account, let us consider a model which includes fuzzy coefficients present in an objective function and constraints. There exist numerous problems related to system design, planning, operation, and control, which can be formalized within the framework of this type of model. Further, although there are diverse formulations of optimization problems with fuzziness (Dubois and Prade, 1980; Orlovsky, 1981; Delgado *et al.*, 1994; Zimmermann, 1996; Zimmermann, 2008), in the opinion of Orlovsky (1981) and Pedrycz and Gomide (1998) the problems with fuzzy coefficients in objective functions and constraints are to be considered as general problems of fuzzy mathematical programming. They can be formulated as follows:

$$\text{maximize } F(x_1, x_2, \ldots, x_n) \tag{7.1}$$

subject to constraints

$$G_j(x_1, x_2, \ldots, x_n) \subseteq B_j, \quad j = 1, 2, \ldots, m \tag{7.2}$$

where the objective function (7.1) and constraints (7.2) include fuzzy coefficients.

The fundamental question which arises when solving optimization problems under uncertainty is how to account for constraints of a different nature and, primarily, the functional constraints. For simplicity of our considerations, we start with just a single constraint of the following form:

$$\sum_{i=q}^{n} G_i x_i \subseteq B \tag{7.3}$$

where G_i, $i = 1, 2, \ldots, n$, and B are fuzzy numbers with their membership functions $G_i(g_i)$, $i = 1, 2, \ldots, n$, and $B(b)$, respectively.

An approach to handling constraints of the form (7.3) was proposed a long time ago (Negoita and Ralescu, 1975). In particular, if certain conditions are satisfied (specifically, with regard to the convexity of the fuzzy coefficients G_i, $i = 1, 2, \ldots, n$, and B), and we assume the possibility of introducing order

$$0 \leq \alpha_1 < \alpha_2 < \cdots < \alpha_k < \cdots < \alpha_l \leq \min\left(\min_{1 \leq i \leq n} \sup G_i(g_i), B(b)\right) \tag{7.4}$$

then the constraint (7.3) can be modified to obtain the following system of numeric inclusions:

$$\sum_{i=1}^{n} G_{i,\alpha_k} x_i \subseteq B_{\alpha_k}, \quad k = 1, 2, \ldots, l \tag{7.5}$$

where G_{i,α_k} and B_{α_k}, $k = 1, 2, \ldots, l$, are sets of the α_k-level (α-cuts) of G_i, $i = 1, 2, \ldots, n$, and B, respectively.

Considering the definition of sets of an α_k-level, see Chapter 2, from (7.5) we obtain

$$\sum_{i=1}^{n} [g_{i_1,\alpha_k}, g_{i_2,\alpha_k}] x_i \subseteq [b_{1,\alpha_k}, b_{2,\alpha_k}], \quad k = 1, 2, \ldots, l \tag{7.6}$$

which means that

$$\sum_{i=1}^{n} g_{i_2,\alpha_k} x_i \leq b_{2,\alpha_k}, \quad k = 1, 2, \ldots, l \tag{7.7}$$

and

$$\sum_{i=1}^{n} g_{i_1,\alpha_k} x_i \geq b_{1,\alpha_k}, \quad k = 1, 2, \ldots, l \tag{7.8}$$

Using the principle of explicit domination, see (7.12), (7.13) below, we can reduce the dimensionality of the sets of inequalities (7.7) and (7.8). As a result of normalization (Ekel, Pedrycz, and Schinzinger, 1998), carried out in accordance with the expression

$$h_{i,\alpha_k} = g_{i,\alpha_k} \frac{b}{b_{\alpha_k}}, \quad k = 1, 2, \ldots, l, \quad i = 1, 2, \ldots, n \tag{7.9}$$

we can consider, instead of (7.7) and (7.8), the sets of constraints

$$\sum_{i=1}^{n} h_{i_2,\alpha_k} x_i \leq b, \quad k = 1, 2, \ldots, l \tag{7.10}$$

and

$$\sum_{i=1}^{n} h_{i_1,\alpha_k} x_i \geq b, \quad k = 1, 2, \ldots, l \tag{7.11}$$

respectively. In (7.9)–(7.11), $b > 0$ is a normalization factor.

If, as a result of analyzing the set of constraints (7.10) with $h_{i_2,\alpha_k} \geq 0$, it turns out that

$$h_{i_2,\alpha_q} \leq h_{i_2,\alpha_p}, \quad q \neq p, \quad i = 1, 2, \ldots, n \tag{7.12}$$

then the pth constraint, for a purposeful increase in the variables $x_i, i = 1, 2, \ldots, n$, is disturbed earlier than the qth constraint. For this reason, the qth constraint can be eliminated from further consideration.

In a similar way, the condition of eliminating the qth constraint from consideration in the case of analyzing the set of constraints (7.11) becomes

$$h_{i_1,\alpha_q} \geq h_{i_1,\alpha_p}, \quad q \neq p, \quad i = 1, 2, \ldots, n \tag{7.13}$$

According to the essence of the optimization problem, one may replace constraints (7.2) with constraints

$$g_j(x_1, x_2 \ldots, x_n) \leq b_j, \quad j = 1, 2, \ldots, m' \geq m \tag{7.14}$$

or constraints

$$g_j(x_1,x_2,\ldots,x_n) \geq b_j, \quad j = 1, 2, \ldots, m'' \geq m \tag{7.15}$$

Hence, as regards the problem with constraints containing fuzzy coefficients, one can obtain an equivalent nonfuzzy analog of the problem whose dimension is reduced by using the principle of explicit domination (7.12) or (7.13).

The solution to problems containing fuzzy coefficients present in objective functions alone is possible by a modification of traditional mathematical programming methods (Ekel, Pedrycz, and Schinzinger, 1998; Ekel, 2002).

When using optimization methods for fuzzy problems, one needs to compare solutions at the levels of the objective function (in essence, we compare or rank corresponding fuzzy numbers to choose the largest or smallest one). If we consider a problem of linear programming and the corresponding modification of the simplex method for its solution, it is necessary to compare the coefficients of nonbasic variables to zero at any cycle of the optimization process.

Taking into account the discussion covered in Chapter 6, we can apply the fuzzy number ranking index introduced by Orlovsky (1981) to complete this comparison. However, let us keep in mind that if the membership functions of the solutions (fuzzy numbers) F_1 and F_2 being compared are trapezoidal or flat fuzzy numbers, then these solutions can be indistinguishable considering the condition (6.16). In such situations, algorithms based on the modification of traditional optimization methods do not allow unique solutions to be obtained because they "stop" when conditions such as (6.16) arise (Ekel, Pedrycz, and Schinzinger, 1998; Galperin and Ekel, 2005). This is natural because a combination of the uncertainty and the relative stability of optimal solutions can produce decision uncertainty regions. This is illustrated by the following simple example (Galperin and Ekel, 2005) where we apply an appropriate modification of the simplex method of linear programming.

Example 7.1. Consider the following problem:

$$\text{maximize } F(x_1, x_2) = C_1x_1 + C_2x_2 \tag{7.16}$$

subject to

$$G_{11}x_1 + G_{12}x_2 \subseteq B_1 \tag{7.17}$$

$$G_{21}x_1 + G_{22}x_2 \subseteq B_2 \tag{7.18}$$

$$x_1 \geq 0, \ x_2 \geq 0 \text{ (numeric)} \tag{7.19}$$

where all coefficients in (7.16)–(7.18) are trapezoidal fuzzy numbers defined as

$$C_1 = \{1.2, 1.3, 1.6, 1.7\}, \ C_2 = \{2.1, 2.2, 2.7, 2.8\}, \ G_{11} = \{9, 10, 11, 12\},$$
$$G_{12} = \{5, 6, 8, 9\}, \ B_1 = \{24, 29, 49, 53\}, \ G_{21} = \{6, 7, 9, 10\},$$
$$G_{22} = \{6, 7, 9, 10\}, \text{ and } B_2 = \{25, 29, 48, 52\}.$$

Taking into account, that G_{11}, G_{12}, B_1, G_{21}, G_{22}, and B_2 are trapezoidal fuzzy numbers, it is sufficient to consider constraints (7.17) and (7.18) for $\alpha_1 = 0$ and $\alpha_2 = \alpha_l = 1$. For this reason, using (7.10) and (7.11), we can rewrite (7.17) as follows:

$$12x_1 + 9x_2 \leq 53 \tag{7.20}$$

$$11x_1 + 8x_2 \leq 49 \tag{7.21}$$

and

$$9x_1 + 5x_2 \geq 24 \tag{7.22}$$

$$10x_1 + 6x_2 \geq 29 \tag{7.23}$$

Similarly, we can replace the constraint (7.18) by the following inequalities:

$$10x_1 + 10x_2 \leq 52 \tag{7.24}$$

$$9x_1 + 9x_2 \leq 48 \tag{7.25}$$

and

$$6x_1 + 6x_2 \geq 25 \tag{7.26}$$

$$7x_1 + 7x_2 \geq 29 \tag{7.27}$$

Taking into account that we have to maximize the objective function with positive coefficients, it is possible to ignore (7.22), (7.23) and (7.26), (7.27). The principle of explicit domination (7.12) applied to (7.20) and (7.21) allows us to eliminate (7.21) from further consideration. The application of the principle of explicit domination (7.12) to (7.24), and (7.25) results in the elimination of (7.25).

Finally, introducing the slack variables $x_3 \geq 0$ and $x_4 \geq 0$, we transform (7.20) and (7.24) to the form

$$12x_1 + 9x_2 + x_3 \leq 53 \tag{7.28}$$

$$10x_1 + 10x_2 + x_4 \leq 52 \tag{7.29}$$

respectively.

To apply the modification of the version of the simplex method given in Rao (1996), we have to consider the minimization problem instead of (7.16). Thus we have the problem

$$\text{minimize } [-F(x_1, x_2)] = -C_1x_1 - C_2x_2 \tag{7.30}$$

subject to (7.28), (7.29), and

$$x_i \geq 0, \quad i = 1, \ldots, 4 \tag{7.31}$$

Applying the above-mentioned modification of the simplex method with the realization of necessary operations, discussed in Chapter 3 for the fuzzy coefficients of the objective function, in the first cycle we obtain: $x_1 = 4.42$, $x_4 = 7.85$ (basic variables) and $x_2 = 0$, $x_3 = 0$

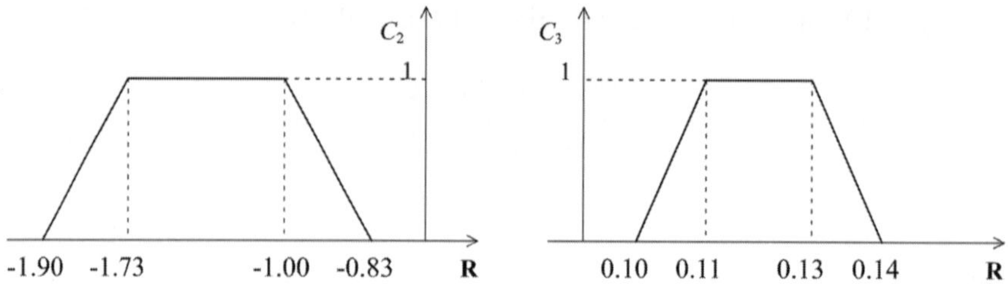

Figure 7.1 Coefficients of nonbasic variables (the first cycle).

(nonbasic variables) with $C_2 = \{-1.90, -1.73, -1.00, -0.83\}$, $C_3 = \{0.10, 0.11, 0.13, 0.14\}$ (see Figure 7.1). Since $C_2 < 0$, we continue and realize the second cycle: $x_1 = 2.07, x_2 = 3.13$ (basic variables) and $x_3 = 0$, $x_4 = 0$ (nonbasic variables) with $C_3 = \{-0.53, -0.47, -0.20, -0.13\}$, $C_4 = \{0.33, 0.40, 0.69, 0.76\}$ (see Figure 7.2). Since $C_3 < 0$, we are able to continue, obtaining in the third cycle: $x_2 = 5.20, x_3 = 6.22$ (basic variables) and $x_1 = 0, x_4 = 0$ (nonbasic variables) with $C_1 = \{0.40, 0.60, 1.40, 1.60\}$, $C_4 = \{-0.15, -0.02, 0.51, 0.64\}$ (see Figure 7.3). Taking into account that, when comparing C_4 to zero, the situation (6.16) takes place, the simplex method "stops"; that is, it "cannot identify" if the optimal solution has been obtained or not.

In an attempt to overcome this type of situation or, at least, to contract the decision uncertainty regions to the highest extent, the approach based on formulating and solving one and the same problem within the framework of mutually related models has been proposed (Ekel, Pedrycz, and Schinzinger, 1998; Ekel, 2002). In particular, problem (7.1) with constraints (7.2) approximated by (7.14), and the problem

$$\text{minimize } F(x_1, x_2, \ldots, x_n) \tag{7.32}$$

subject to the same constraints (7.2), approximated by (7.15), can serve as a mutually related model.

This approach is applicable for solving continuous as well as discrete optimization problems. To understand its essence, let us proceed with the analysis of a certain discrete problem.

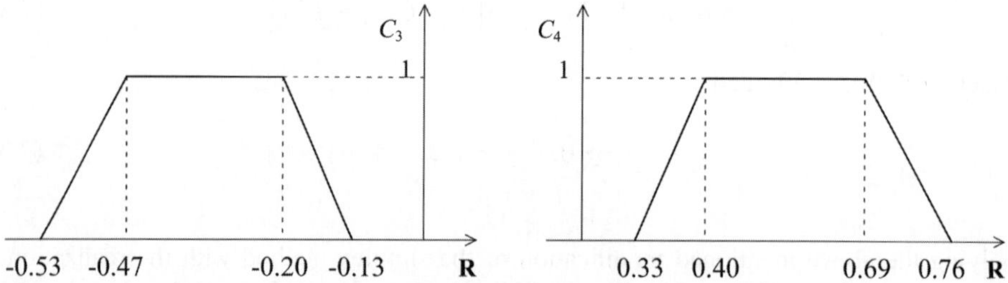

Figure 7.2 Coefficients of nonbasic variables (the second cycle).

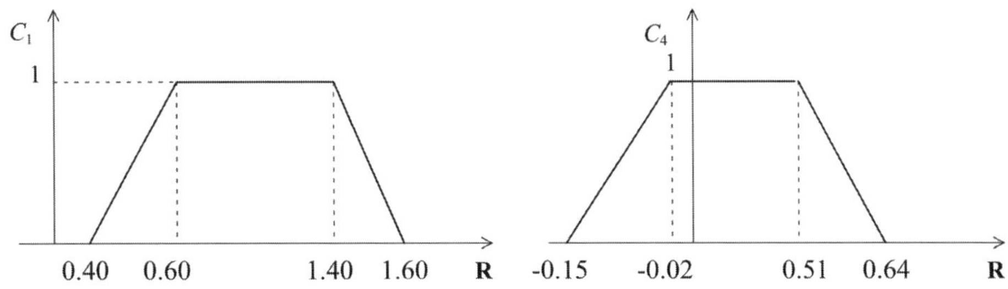

Figure 7.3 Coefficients of nonbasic variables (the third cycle).

The desirability of allowing for constraints on the discrete nature of variables in the form of discrete sequences

$$x_{s_i}, \alpha_{s_i}, \beta_{s_i}, \ldots, s_i = 1, 2, \ldots, r_i \tag{7.33}$$

has been validated by Zorin and Ekel (1980); here $\alpha_{s_i}, \beta_{s_i}, \ldots$ are technical and economic characteristics required for the formation of objective functions, constraints, and their increments that correspond to the sth standard value of the variable x_i.

It is expedient to use discrete sequences of the type (7.33) because the characteristics $\alpha_{s_i}, \beta_{s_i}, \ldots$ cannot always be fitted closely to the analytical relationships in terms of x_{s_i}, but in discrete sequences of the type (7.33) these characteristics may be treated as exact. Furthermore, a flexible formalization of combinatorial types of problems is possible on the basis of the discrete sequences because they can be different for different variables. Examples of this flexible usage of the discrete sequences are presented in Zorin and Ekel (1980) and Ekel and Schuffner Neto (2006).

Taking the above into consideration with respect to the expediency of using discrete sequences, and by analogy with the problem (7.1), (7.2), the maximization problem can be formulated as follows.

Assume we are given discrete sequences of the type (7.33) (which, depending on the formulation of the problem, could be either increasing or decreasing). From these sequences of discrete values it is necessary to choose elements such that the objective

$$\text{maximize} F(x_{s_1}, \alpha_{s_1}, \beta_{s_1}, \ldots, x_{s_2}, \alpha_{s_2}, \beta_{s_2}, \ldots, x_{s_n}, \alpha_{s_n}, \beta_{s_n}, \ldots) \tag{7.34}$$

is met while satisfying the constraints

$$g_j(x_{s_1}, \alpha_{s_1}, \beta_{s_1}, \ldots, x_{s_2}, \alpha_{s_2}, \beta_{s_2}, \ldots, x_{s_n}, \alpha_{s_n}, \beta_{s_n}, \ldots) \subseteq B_j, \quad j = 1, 2, \ldots, m \tag{7.35}$$

Given a maximization problem of the type (7.33)–(7.35) considered above, and by analogy with (7.32), we can formulate a mutually related problem with the objective

$$\text{minimize} F(x_{s_1}, \alpha_{s_1}, \beta_{s_1}, \ldots, x_{s_2}, \alpha_{s_2}, \beta s2, \ldots, x_{s_n}, \alpha_{s_n}, \beta_{s_n}, \ldots) \tag{7.36}$$

while satisfying the constraints (7.35).

Taking the above into account, the constraints (7.35) may be reduced to the set of nonfuzzy (numeric) constraints

$$g_j(x_{s_1}, \alpha_{s_1}, \beta_{s_1}, \ldots, x_{s_2}, \alpha_{s_2}, \beta_{s_2}, \ldots, x_{s_n}, \alpha_{s_n}, \beta_{s_n}, \ldots) \leq b_j, \quad j = 1, 2, \ldots, m' \geq m \tag{7.37}$$

and

$$g_j(x_{s_1}, \alpha_{s_1}, \beta_{s_1}, \ldots, x_{s_2}, \alpha_{s_2}, \beta_{s_2}, \ldots, x_{s_n}, \alpha_{s_n}, \beta_{s_n}, \ldots) \geq b_j, \quad j = 1, 2, \ldots, m'' \geq m \tag{7.38}$$

Let us consider an example from Ekel, Pedrycz, and Schinzinger (1998) to demonstrate the analysis of the mutually related models (7.34), (7.37) (with the increasing (decreasing) discrete sequences (7.33)) and (7.36), (7.38) (with the decreasing (increasing) discrete sequences (7.33)) and the results based on its application.

Example 7.2. Assume that we are given the discrete sequence

$$\begin{array}{llll} & x_{s_i} & \alpha_{s_i} & \beta_{s_i} \\ s_i = 1: & 0, & 0, & 26 \\ s_i = 2: & 1, & 3, & 25 \\ s_i = 3: & 2, & 6, & 23 \\ s_i = 4: & 3, & 9, & 19 \\ s_i = 5: & 4, & 12, & 14 \\ s_i = 6: & 5, & 15, & 9 \end{array} \tag{7.39}$$

From this sequence it is necessary to choose elements which maximize the objective function

$$F(x_1, x_2) = [(C_1\alpha_{s_1} + \beta_{s_1}) + (C_2\alpha_{s_2} + \beta_{s_2})] \tag{7.40}$$

subject to the following set of constraints:

$$G_{11}x_{s_1} + G_{12}x_{s_2} \subseteq B_1 \tag{7.41}$$

$$G_{21}x_{s_1} + G_{22}x_{s_2} \subseteq B_2 \tag{7.42}$$

where all coefficients in (7.40)–(7.42) are trapezoidal fuzzy numbers defined as

$$C_1 = \{1.2, 1.4, 1, 5, 1.7\},\ C_2 = \{2.1, 2.4, 2.5, 2.8\},\ G_{11} = \{5, 6, 8, 9\},$$
$$G_{12} = \{9, 10, 11, 12\},\ B_1 = \{24, 29, 49, 53\},\ G_{21} = \{6, 7, 9, 10\},$$
$$G_{22} = \{6, 7, 9, 10\},\ \text{and } B_2 = \{25, 29, 48, 52\}.$$

As presented in Example 7.1, it is sufficient to consider the constraints (7.41) and (7.42) for $\alpha_1 = 0$ and $\alpha_2 = \alpha_l = 1$. In this regard, we can write for (7.41)

$$12x_{s_1} + 9x_{s_2} \leq 53 \tag{7.43}$$

$$11x_{s_1} + 8x_{s_2} \leq 49 \tag{7.44}$$

and

$$9x_{s_1} + 5x_{s_2} \geq 24 \tag{7.45}$$
$$10x_{s_1} + 6x_{s_2} \geq 29 \tag{7.46}$$

Similarly, we may go from the constraints (7.42) to the following inequalities:

$$10x_{s_1} + 10x_{s_2} \leq 52 \tag{7.47}$$
$$9x_{s_1} + 9x_{s_2} \leq 48 \tag{7.48}$$

and

$$6x_{s_1} + 6x_{s_2} \geq 25 \tag{7.49}$$
$$7x_{s_1} + 7x_{s_2} \geq 29 \tag{7.50}$$

The principle of explicit domination (7.12) applied to (7.43) and (7.44) results in the elimination of (7.44) from further consideration. The application of the principle of explicit domination (7.12) to (7.47) and (7.48) eliminates (7.48).

Similarly, applying the principle of explicit domination (7.13) to (7.45) and (7.46), we eliminate (7.45) from further consideration. The application of the principle of explicit domination (7.13) to (7.49) and (7.50) eliminates (7.50).

Thus the problem becomes reduced to maximizing (7.40) subject to the constraints (7.43) and (7.47). At the same time, the mutually related problem consists of minimizing

$$F(x_1, x_2) = [-(C_1\alpha_{s_1} + \beta_{s_1}) - (C_2\alpha_{s_2} + \beta_{s_2})] \tag{7.51}$$

subject to the constraints (7.46) and (7.49) with the use of the discrete sequence that is decreasing on s_i:

$$\begin{array}{lccc} & x_{s_i} & \alpha_{s_i} & \beta_{s_i} \\ s_i = 1: & 5, & 15, & 26 \\ s_i = 2: & 4, & 12, & 14 \\ s_i = 3: & 3, & 9, & 19 \\ s_i = 4: & 2, & 6, & 23 \\ s_i = 5: & 1, & 3, & 25 \\ s_i = 6: & 0, & 0, & 26 \end{array} \tag{7.52}$$

The process of solving the problem (7.39), (7.40), (7.43), and (7.47) on the basis of modifying the generalized algorithms of discrete optimization (Ekel, Pedrycz, and Schinzinger, 1998; Ekel and Schuffner Neto, 2006) is presented in Ekel, Pedrycz, and Schinzinger (1998). In particular, the process "stops" when we meet a situation where it is impossible to distinguish two solutions $X_1 = \{x_1 = 2, x_2 = 3\}$ and $X_2 = \{x_1 = 1, x_2 = 4\}$. At the same time, the solution of the mutually related problem (7.52), (7.51), (7.46), and (7.49) leads to the solution $X_3 = \{x_1 = 0, x_2 = 5\}$. As is shown in Ekel, Pedrycz, and Schinzinger (1998), there are no more

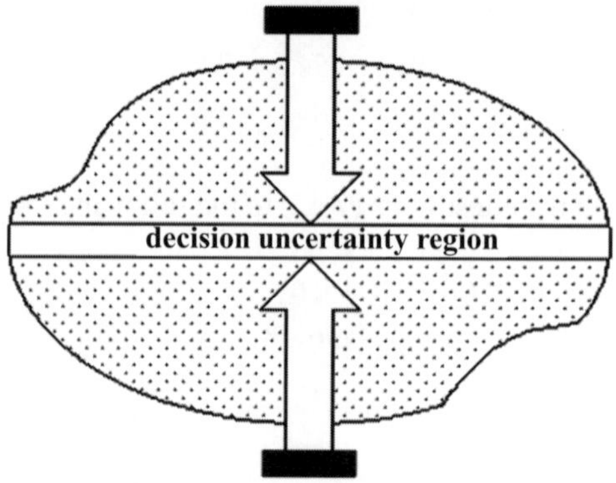

Figure 7.4 Cutting dominated alternatives.

solutions which are competitive. Thus, the decision uncertainty region $X = \{X_1, X_2, X_3\}$ is a formal solution to the problem (7.39)–(7.42).

Schematically, the demonstrated approach can be reflected as in Figure 7.4: the solutions dominated by the initial objective function are cut off from below as well as from above to the highest degree.

Thus, it was demonstrated that the uncertainty of information, particularly reflected by fuzzy coefficients in objective functions and constraints of monocriteria problems, generates the decision uncertainty regions. Their contraction, as indicated above, is possible on the basis of reducing the problem to multicriteria decision-making by applying additional criteria, including those of a qualitative character. It is natural that the problem of the evaluation, comparison, choice, prioritization, and/or ordering of alternatives can be initially stated as a multicriteria problem.

7.2 Discrete Models ($\langle \mathbf{X}, \boldsymbol{R} \rangle$ Models) of Multiattribute Decision-Making

The problem of multiattribute analysis of alternatives in a fuzzy environment can be formulated as follows. Assume we are given a set $\mathbf{X}$ of alternatives coming from the decision uncertainty region and/or predetermined alternatives, which are to be examined by q criteria of a quantitative and/or qualitative nature. The problem of decision-making, as elaborated in Chapter 1, may be presented as a pair $\langle \mathbf{X}, \boldsymbol{R} \rangle$ where $\boldsymbol{R} = [\, \boldsymbol{R}_1 \; \boldsymbol{R}_2 \dots \boldsymbol{R}_p \dots \boldsymbol{R}_q]$ is a vector of fuzzy preference relations (Orlovsky, 1981; Fodor and Roubens, 1994), which can be presented as follows:

$$\boldsymbol{R}_p(X_k, X_l) : \mathbf{X} \times \mathbf{X} \rightarrow [0, 1], \quad k,l = 1, 2, \dots, n, \quad p = 1, 2, \dots, q \tag{7.53}$$

where $\boldsymbol{R}_p(X_k, X_l) : \mathbf{X} \times \mathbf{X} \to [0, 1]$ is a membership function of the pth fuzzy preference relation.

In Chapter 6, we analyzed the use of different preference formats for the presentation of initial information for decision-making and the rationality of utilizing fuzzy preference relations for a uniform preference representation (along with the questions of converting different preference forms into fuzzy preference relations). Taking this into account, henceforth in this chapter we present approaches for carrying out the evaluation, comparison, choice, prioritization, and/or ordering of alternatives on the basis of information reflected by (7.53).

Below, we discuss six different techniques of multiattribute analysis of alternatives in a fuzzy environment (techniques of analysis of $\langle \mathbf{X}, \boldsymbol{R} \rangle$ models). The *first*, *second*, and the *third* techniques are directly based on the notion of the Orlovsky choice function (Orlovsky, 1978; Orlovsky, 1981). The *fourth* technique is also based on applying the notion of the Orlovsky choice function. However, it allows a DM to present information related to the importance of criteria considered in a fuzzy form, particularly in the NRFPR (Nonreciprocal Fuzzy Preference Relation) form. The *fifth* technique should be considered as a generalized version of applying the Orlovsky choice function related to the use of the ordered weighted average (OWA) operator (Yager, 1988; Yager, 1995; Chiclana *et al.*, 1996, Grabisch, Orlovski, and Yager, 1998). Finally, the *sixth* technique is based on the construction and exploitation of a specific type of fuzzy preference relation named the outranking relation (Roy, 1968; Brans and Vincke, 1985; Roy, 1991; Bouyssou, 1997).

In this chapter, to be consistent with the notation utilized in Chapter 6, we use $\boldsymbol{R}$ to denote fuzzy nonstrict preference relations. We anticipate that, as we only make use of nonreciprocal fuzzy preference relations within this chapter, we do not use any particular notation to indicate whether the fuzzy preference relation under consideration is an additive reciprocal fuzzy preference relation or a nonreciprocal fuzzy preference relation.

7.3 Basic Techniques of Analysis of $\langle \mathbf{X}, \boldsymbol{R} \rangle$ Models

In this section, we present three techniques of the analysis of $\langle \mathbf{X}, \boldsymbol{R} \rangle$ models, which are based on the application of the notion of the Orlovsky choice function. This notion was introduced by Orlovsky (Orlovsky, 1978; Orlovsky, 1981) and afterward studied by many researchers. For instance, it was shown in Barrett, Patanalk, and Salles (1990) that the Orlovsky choice function possesses many interesting and desirable properties. Its axiomatic characterization is given, for example, in Banerjee (1993), Bouyssou (1997), and Sengupta (1998).

At first, let us consider the situation of setting up a single fuzzy nonstrict preference relation $\boldsymbol{R}$. It can be processed to construct a fuzzy strict preference relation $\boldsymbol{P}$. In particular, $(X_k, X_l) \in \boldsymbol{P}$ means that X_k is strictly better than X_l (or X_k dominates X_l, that is, $X_k \succ X_l$).

As discussed in Chapter 5, with the use of the operations on fuzzy sets, it is possible to define the fuzzy strict preference relation $\boldsymbol{P}$ exclusively in terms of the fuzzy nonstrict preference relation $\boldsymbol{R}$ (a conceptual definition of $\boldsymbol{P}$ in terms of $\boldsymbol{R}$ is given by expression (5.33)). For practical purposes, one possible manner of deriving the fuzzy strict preference relation $\boldsymbol{P}$ from a fuzzy nonstrict preference relation is by means of expression (5.35). As will be shown next, (5.35) plays an important role in this chapter, as it allows one to carry out the choice or ranking of the alternatives. In particular, one can note that $\boldsymbol{P}(X_l, X_k)$, $\forall X_k \in \mathbf{X}$, is the membership function of the fuzzy set of all X_k which are strictly dominated by X_l. Naturally, the complementary relation $\boldsymbol{P}^c(X_l, X_k) = 1 - \boldsymbol{P}(X_l, X_k), \forall X_k \in \mathbf{X}$ (refer to (5.3) for the definition of complementary relation), gives the fuzzy set of alternatives which are not dominated by X_l. Therefore, in order to meet the set of alternatives from $\mathbf{X}$ that are not dominated by other alternatives, it is sufficient to find the fuzzy preference relation that corresponds to the intersection of all $\boldsymbol{P}^c(X_l, X_k)$, $X_k \in \mathbf{X}$, on all $X_l \in \mathbf{X}$ (Orlovsky, 1981). This intersection, which corresponds to the fuzzy set of nondominated alternatives, can be implemented as follows:

$$\mathbf{ND}(X_k) = \min_{X_l \in \mathbf{X}}(1 - \boldsymbol{P}(X_l, X_k)) = 1 - \max_{X_l \in \mathbf{X}} \boldsymbol{P}(X_l, X_k) \tag{7.54}$$

In this way, (7.54) allows one to evaluate the level of nondominance of each alternative X_k. Considering that it is natural to choose alternatives providing the highest level of nondominance, one can choose alternatives $\mathbf{X}^{ND}$ in accordance with the following expression:

$$\mathbf{X}^{ND} = \left\{ X_k^{ND} | X_k^{ND} \in \mathbf{X}, \mathbf{ND}(X_k^{ND}) = \max_{X_k \in \mathbf{X}} \mathbf{ND}(X_k) \right\} \tag{7.55}$$

Example 7.3. Consider the fuzzy nonstrict preference relation

$$\boldsymbol{R} = \begin{bmatrix} 1 & 0.6 & 0.5 & 0.4 \\ 0 & 1 & 0.4 & 0.8 \\ 0.1 & 0.6 & 1 & 0 \\ 0.6 & 0.2 & 0.8 & 1 \end{bmatrix} \tag{7.56}$$

defined on a set of alternatives $\mathbf{X} = \{X_1, X_2, X_3, X_4\}$, which are to be ordered and the best one is to be selected.

Applying (5.35) to (7.56), we can obtain the membership function of the fuzzy strict preference relation

$$\boldsymbol{P} = \begin{bmatrix} 0 & 0.6 & 0.4 & 0 \\ 0 & 0 & 0 & 0.6 \\ 0 & 0.2 & 0 & 0 \\ 0.2 & 0 & 0.8 & 0 \end{bmatrix} \tag{7.57}$$

Then on the basis of (7.54), we obtain the membership function of the fuzzy set of nondominated alternatives

$$\mathbf{ND} = [0.8 \quad 0.4 \quad 0.2 \quad 0.4] \tag{7.58}$$

This permits us to determine $X_1 \succ X_2 \sim X_4 \succ X_3$. Finally, $\mathbf{X}^{ND} = \{X_1\}$.

Following Orlovsky (1981), it is possible to introduce the notion of a set of nonfuzzy, nondominated alternatives. In particular, if $\max_{X_k \in \mathbf{X}} \mathbf{ND}(X_k) = 1$, then the set of alternatives

$$\mathbf{X}^{NFND} = \left\{ X_k^{NFND} | X_k^{NFND} \in \mathbf{X},\ \mathbf{ND}(X_k^{NFND}) = 1 \right\} \tag{7.59}$$

is nonfuzzy and nondominated and can be considered as a nonfuzzy solution to the problem expressed in terms of fuzzy sets.

If the fuzzy preference relation $\boldsymbol{R}$ satisfies weak transitivity, then we have $\mathbf{X}^{NFND} \neq \varnothing$. Taking this into consideration, it should be noted that when the preferences of a DM are expressed by means of the ordering of alternatives, utility values, or fuzzy estimates and are subsequently transformed into fuzzy preference relations with the use of adequate transformation functions, selected among the ones presented in Chapter 6, then $\mathbf{X}^{NFND}$ is nonempty, since those transformation functions guarantee weak transitivity of the resulting fuzzy preference relation. However, it is possible to have $\mathbf{X}^{NFND} = \varnothing$ if a DM provides his/her preferences as a multiplicative preference relation or a fuzzy preference relation that does not satisfy weak transitivity. In such cases, we should recall from Chapter 6 that the transformation functions for converting multiplicative preference relations or ARFPRs into NRFPRs transmit the existent inconsistencies to the fuzzy preference relation. Hence, by considering the cardinality of $\mathbf{X}^{NFND}$, it is possible to detect contradictions in the expert's estimates. In real-world applications, weak transitivity is a requisite to be satisfied by the preferences of each DM, since it is a necessary consistency condition to guarantee the rationality of the decisions based on applying the Orlovsky choice function (Sengupta, 1998). Particularly, an unreasonable effect of the lack of weak transitivity in a fuzzy nonstrict preference relation is associated with the fact that it may cause the Orlovsky choice function to violate an axiom referred to in the literature as *independence of rejected alternatives*. According to this axiom, if the Orlovsky choice function indicates a subset $\mathbf{X}^1 \subset \mathbf{X}$ as the best solution for the problem, the exclusion from the set $\mathbf{X}$ of any alternative not belonging to $\mathbf{X}^1$ should not affect the formation of the choice set obtained with the use of the Orlovsky choice function. In the following example a violation of the independence of rejected alternatives is demonstrated.

Example 7.4. A DM must rank a set $\mathbf{X} = \{X_1, X_2, X_3\}$ of alternatives and select the best one. The DM's preferences are articulated as the fuzzy nonstrict preference relation

$$\boldsymbol{R} = \begin{bmatrix} 1 & 1 & 0.8 \\ 0.98 & 1 & 1 \\ 1 & 0.8 & 1 \end{bmatrix} \tag{7.60}$$

which does not satisfy the property of weak transitivity. We should have $\boldsymbol{R}(X_2, X_1) \geq \boldsymbol{R}(X_1, X_2)$ in (7.60), since we have $\boldsymbol{R}(X_2, X_3) > \boldsymbol{R}(X_3, X_2)$ and $\boldsymbol{R}(X_3, X_1) > \boldsymbol{R}(X_1, X_3)$. But, instead, we have $\boldsymbol{R}(X_2, X_1) = 0.98$, which is lower than $\boldsymbol{R}(X_1, X_2) = 1$.

Applying (5.35) to (7.60), we obtain the membership function of the fuzzy strict preference relation

$$\boldsymbol{P} = \begin{bmatrix} 0 & 0.02 & 0 \\ 0 & 0 & 0.2 \\ 0.2 & 0 & 0 \end{bmatrix} \tag{7.61}$$

Then, with the use of (7.54), we obtain the membership function of the fuzzy set of nondominated alternatives

$$\mathbf{ND} = [\,0.8 \quad 0.98 \quad 0.8\,] \tag{7.62}$$

Although (7.62) allows us to order $X_2 \succ X_1 \sim X_3$ and to select $\mathbf{X}^{ND} = \{X_2\}$, it yields $\mathbf{X}^{NFND} = \varnothing$. Further, according to the axiom of independence of rejected alternatives, the set $\mathbf{X}^{ND}$ should remain the same if X_1 or X_3 is excluded from $\mathbf{X}$. However, when X_3 is excluded from $\mathbf{X}$, the choice set becomes $\mathbf{X}^{ND} = \{X_1\}$. The reader should note that, with the exclusion of X_3 from $\mathbf{X}$, the third line and the third column of (7.60) are eliminated as follows:

$$\boldsymbol{R} = \begin{bmatrix} 1 & 1 \\ 0.98 & 1 \end{bmatrix} \tag{7.63}$$

Similarly, the third line and the third column of (7.61) are eliminated as follows:

$$\boldsymbol{P} = \begin{bmatrix} 0 & 0.02 \\ 0 & 0 \end{bmatrix} \tag{7.64}$$

Consequently, the membership function of the fuzzy set of nondominated alternatives defined in accordance with (7.54) is updated to

$$\mathbf{ND} = \begin{bmatrix} 1 & 0.98 \end{bmatrix} \tag{7.65}$$

which allows us to order $X_1 \succ X_2$ and to select $\mathbf{X}^{ND} = \{X_1\}$.

Now, let us consider that the DM has repaired (7.60) to satisfy weak transitivity as follows:

$$\boldsymbol{R} = \begin{bmatrix} 1 & 0.98 & 0.8 \\ 1 & 1 & 1 \\ 1 & 0.8 & 1 \end{bmatrix} \tag{7.66}$$

Then the membership function of the fuzzy strict preference relation becomes

$$\boldsymbol{P} = \begin{bmatrix} 0 & 0 & 0 \\ 0.02 & 0 & 0.2 \\ 0.2 & 0 & 0 \end{bmatrix} \tag{7.67}$$

and, with the use of (7.54), we obtain the membership function of the fuzzy set of nondominated alternatives

$$\mathbf{ND} = [\,0.8 \quad 1 \quad 0.8\,] \tag{7.68}$$

which permits us to determine the order $X_2 \succ X_1 \sim X_3$. Finally, $\mathbf{X}^{ND} = \{X_2\} = \mathbf{X}^{NFND}$, because it is a nonfuzzy solution. The reader is invited to confirm that, when we consider (7.66) rather than (7.60), the exclusion of the rejected alternatives X_1 or X_3 does not cause any change to the choice set.

The expressions (5.35), (7.54), and (7.55) may be utilized to solve choice problems, as well as other problems, related to the evaluation, comparison, choice, prioritization, and/or ordering of alternatives with a single criterion. These expressions may also be applied when $\boldsymbol{R}$ is a vector of fuzzy preference relations.

Let us consider the *first technique* for dealing with a vector of fuzzy preference relations $\boldsymbol{R}$ (Orlovsky, 1981). The expressions (5.35), (7.54), and (7.55) are applicable if we take the intersection $\boldsymbol{R} = \bigcap_{p=1}^{q} \boldsymbol{R}_p$ with the membership function

$$\boldsymbol{R}(X_k, X_l) = \min_{1 \leq p \leq q} \boldsymbol{R}_p(X_k, X_l), \quad X_k, X_l \in \mathbf{X} \tag{7.69}$$

The reader should note that the use of an intersection operator to aggregate the fuzzy preference relations in (7.69) reflects the need to satisfy all criteria simultaneously in a strict sense; that is, we can interpret (7.69) as the need to satisfy F_1 and F_2 and ... F_p and ... and F_q. When using (7.69), the set $\mathbf{X}^{ND}$ fulfils the role of a Pareto set (Orlovsky, 1981). Its contraction is possible on the basis of differentiating the importance of $\boldsymbol{R}_p$, $p = 1, \ldots, q$, with the use of the following convolution:

$$\boldsymbol{T}(X_k, X_l) = \sum_{p=1}^{q} \lambda_p \boldsymbol{R}_p(X_k, X_l), \quad X_k, X_l \in \mathbf{X} \tag{7.70}$$

where λ_p, $p = 1, \ldots, q$, are importance factors of the corresponding criteria, defined as (4.15) and (4.16).

The construction of $\boldsymbol{T}(X_k, X_l)$, $X_k, X_l \in \mathbf{X}$, allows us to obtain the membership function $\mathbf{NDT}(X_k)$ of the fuzzy set of nondominated alternatives according to an expression similar to (7.54). The intersection

$$Q(X_k) = \min(\mathbf{ND}(X_k), \mathbf{NDT}(X_k)), \quad X_k \in \mathbf{X} \tag{7.71}$$

provides us with a set of alternatives with the highest level of nondominance

$$\mathbf{X}^{ND} = \left\{ X_k^{ND} | X_k^{ND} \in X, \, Q(X_k^{ND}) = \sup_{X_k \in X} Q(X_k) \right\} \tag{7.72}$$

Expressions (7.54) and (7.55) can serve as the basis for building the *second technique*, which is of a lexicographic character. It is based on the step-by-step application of criteria for comparing alternatives. The technique permits one (Ekel, Pedrycz, and Schinzinger, 1998; Ekel, 2001; Ekel, 2002) to construct a sequence $\mathbf{X}^1, \mathbf{X}^2, \ldots, \mathbf{X}^q$ so that $\mathbf{X} \supseteq \mathbf{X}^1 \supseteq \mathbf{X}^2 \supseteq \cdots \supseteq \mathbf{X}^q$. This is accomplished by using the following expressions:

$$\mathbf{ND}^p(X_k) = \min_{X_l \in \mathbf{X}^{p-1}} (1 - \boldsymbol{P}_p(X_l, X_k)) = 1 - \max_{X_l \in \mathbf{X}^{p-1}} \boldsymbol{P}_p(X_l, X_k), \quad p = 1, 2, \ldots, q \tag{7.73}$$

$$\mathbf{X}^p = \left\{ \mathbf{X}_k^{ND,p} | X_k^{ND,p} \in \mathbf{X}^{p-1}, \, \mathbf{ND}^p(X_k^{ND,p}) = \max_{X_l \in \mathbf{X}^{p-1}} \mathbf{ND}^p(X_k) \right\} \tag{7.74}$$

Finally, it is possible to present the *third technique* (Ekel and Schuffner Neto, 2006; Ekel, Martini, and Palhares, 2008). In particular, the utilization of (7.54) in the form

$$\mathbf{ND}(X_k) = 1 - \max_{X_l \in \mathbf{X}} \boldsymbol{P}_p(X_l, X_k), \quad p = 1, 2, \ldots, q \tag{7.75}$$

allows us to construct the membership functions of the fuzzy set of nondominated alternatives for each fuzzy preference relation. The fuzzy sets $\mathbf{ND}_p(X_k)$, $p = 1, 2, \ldots, q$, can be aggregated with the use of an intersection operator in order to reflect the need to satisfy F_1 and F_2 and $\ldots$ F_p and $\ldots$ and F_q (the reader should note that the membership functions $\mathbf{ND}_p(X_k)$ play a role identical to membership functions replacing objective functions $F_p(x)$, $p = 1, \ldots, q$, when analyzing $\langle \mathbf{X}, \mathbf{M} \rangle$ models). Therefore, one constructs

$$\mathbf{ND}(X_k) = \min_{1 \le p \le q} \mathbf{ND}_p(X_k) \tag{7.76}$$

to obtain $\mathbf{X}^{ND}$. If it is necessary to differentiate the importance of different preference relations, it is possible to transform (7.76) as follows:

$$\mathbf{ND}(X_k) = \min_{1 \le p \le q} (\mathbf{ND}_p(X_k))^{\lambda_p} \tag{7.77}$$

The use of (7.77) does not require normalization of λ_p, $p = 1, \ldots, q$, in the way similar to (4.16).

Example 7.5. This example is an immediate continuation of Example 7.2, where the solution of the discrete optimization problem with fuzzy coefficients in the objective function and constraints has generated the decision uncertainty region $\mathbf{X} = \{X_1, X_2, X_3\}$. The indistinguishable alternatives are to be compared with the application of the three criteria. The first criterion ($p = 1$) demands the minimization of $F_1(X_k)$. The second criterion ($p = 2$) and the third criterion ($p = 3$) demand the maximization of $F_2(X_k)$ and $F_3(X_k)$, respectively.

Without discussing the question of constructing fuzzy nonstrict preference relations which correspond to these criteria, assume that they are as follows:

$$\boldsymbol{R}_1 = \begin{bmatrix} 1 & 1 & 1 \\ 1 & 1 & 1 \\ 0.94 & 0.94 & 1 \end{bmatrix} \tag{7.78}$$

$$\boldsymbol{R}_2 = \begin{bmatrix} 1 & 0.94 & 0.14 \\ 1 & 1 & 0.94 \\ 1 & 1 & 1 \end{bmatrix} \tag{7.79}$$

$$\boldsymbol{R}_3 = \begin{bmatrix} 1 & 1 & 1 \\ 0.94 & 1 & 1 \\ 0.94 & 1 & 1 \end{bmatrix} \tag{7.80}$$

Let us begin by considering the solution of the problem on the basis of the *first technique*. The intersection of the fuzzy nonstrict preference relations (7.78)–(7.80) constructed on the basis of (7.69) is as follows:

$$R = \begin{bmatrix} 1 & 0.94 & 0.14 \\ 0.94 & 1 & 0.94 \\ 0.94 & 0.94 & 1 \end{bmatrix} \tag{7.81}$$

Applying (5.35), we can construct the fuzzy strict preference relation

$$P = \begin{bmatrix} 0 & 0 & 0 \\ 0 & 0 & 0 \\ 0.80 & 0 & 0 \end{bmatrix} \tag{7.82}$$

which, according to (7.54), generates

$$\mathbf{ND} = [\,0.2 \quad 1 \quad 1\,] \tag{7.83}$$

and $\mathbf{X}^{ND} = \{X_2, X_3\}$.

Let us consider the application of the *second technique*, where the criteria are arranged, for example, in the following order of importance: $p = 1$, $p = 2$, and $p = 3$.

By subsequently applying (5.35), (7.73), and (7.74), on the basis of the fuzzy nonstrict preference relation (7.78), we obtain

$$P_1 = \begin{bmatrix} 0 & 0 & 0.06 \\ 0 & 0 & 0.06 \\ 0 & 0 & 0 \end{bmatrix} \tag{7.84}$$

$$\mathbf{ND}^1 = [\,1 \quad 1 \quad 0.94\,] \tag{7.85}$$

and $\mathbf{X}^1 = \{X_1, X_2\}$. Thus, the alternatives X_1 and X_2 are to be considered for a subsequent analysis.

For the second step, we can construct the fuzzy nonstrict preference relation, the fuzzy strict preference relation, and the fuzzy set of nondominated alternatives only for alternatives X_1 and X_2, as follows:

$$R^2 = \begin{bmatrix} 1 & 0.94 \\ 1 & 1 \end{bmatrix} \tag{7.86}$$

$$P_2 = \begin{bmatrix} 0 & 0 \\ 0.06 & 0 \end{bmatrix} \tag{7.87}$$

$$\mathbf{ND}^2 = [\,0.94 \quad 1\,] \tag{7.88}$$

and $\mathbf{X}^2 = \{X_2\}$.

Finally, let us consider the application of the *third technique*. The membership function of the set of nondominated alternatives for the first fuzzy preference relation R_1 is (7.85).

The fuzzy nonstrict preference relation (7.79) generates the following membership function of nondominated alternatives:

$$\mathbf{ND}_2 = [\,0.14 \quad 0.94 \quad 1\,] \tag{7.89}$$

and the fuzzy nonstrict preference relation (7.80) leads to

$$\mathbf{ND}_3 = [\,1 \quad 0.94 \quad 0.94\,] \tag{7.90}$$

The intersection of (7.85), (7.89), and (7.90) according to (7.76) allows us to construct

$$\mathbf{ND} = [0.\,14 \quad 0.94 \quad 0.94\,] \tag{7.91}$$

to obtain $\mathbf{X}^{ND} = \{X_2, X_3\}$.

In characterizing the described techniques of decision-making in a fuzzy environment, it should be noted that the application of the *second technique* may lead to solutions different from the results obtained on the basis of the *first technique*. However, solutions based on the *first technique* and the *third technique*, which share a single generic basis, may also be different. At the same time, the *third technique* is preferential from the substantial point of view. In particular, the use of the *first technique* can lead to choosing alternatives with the degree of nondominance equal to one, though these alternatives are not the best ones from the point of view of all preference relations. The *third technique* can generate this result only for alternatives that are the best solutions from the point of view of all fuzzy preference relations. It should be stressed that the possibility of obtaining different solutions on the basis of different approaches (as demonstrated by the example above) is to be considered natural, and the choice of the approach is a prerogative of a DM.

The described techniques are of a universal nature and have already been used to solve problems in power engineering (Canha *et al.*, 2007), naval engineering (Botter and Ekel, 2005), and management (Berredo *et al.*, 2005). These techniques have been implemented within the framework of an interactive system for multicriteria decision-making (that is, the MDMS), which will be described in the next section.

7.4 Interactive Decision-Making System for Multicriteria Analysis of Alternatives in a Fuzzy Environment

The MDMS has been developed in the C++ programming language and is executed in the graphical environment of the Microsoft Windows operating system. Below, we show several typical windows that appear in the process of initial data preparation (we concentrate here on the use of two preference formats for the input of preferences: fuzzy estimates and nonreciprocal fuzzy preference relations) and also several typical windows that appear in the process of multicriteria decision-making.

An initial window (see Figure 7.5) permits the decision-making process to be started by indicating the Technique to be used and by defining the number of Alternatives and the number of Criteria. The screenshot in Figure 7.5 reflects the input information for Example 7.5.

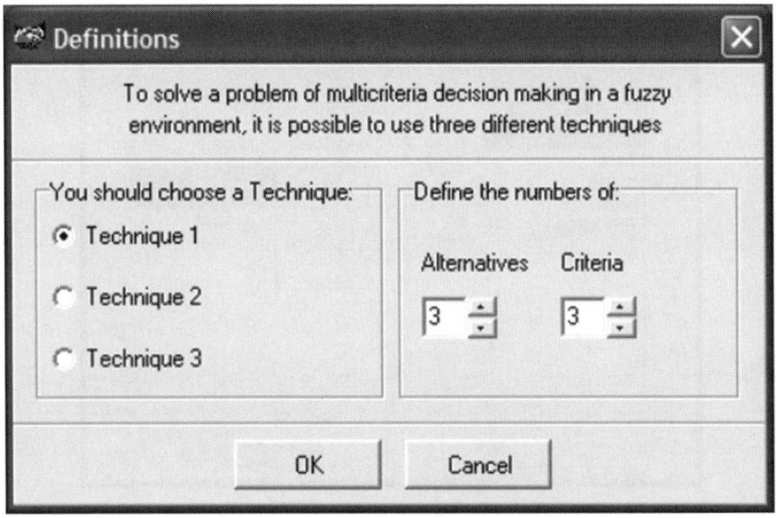

Figure 7.5 Initial window.

The next two windows (see Figures 7.6 and 7.7) are used to specify the character of the problem (Maximization or Minimization) for the corresponding criterion and to define fuzzy estimates for evaluating the alternatives with respect to this criterion (the fuzzy estimates given in Figure 7.7 are associated with the first criterion considered in the decision-making problem studied in Example 7.5).

The estimate "Very Small" for evaluating the first alternative with respect to the first criterion is shown in Figure 7.8. The analytical description of the estimate and its parameters is given at "Estimate description and its parameters". The type of the estimate shape (the

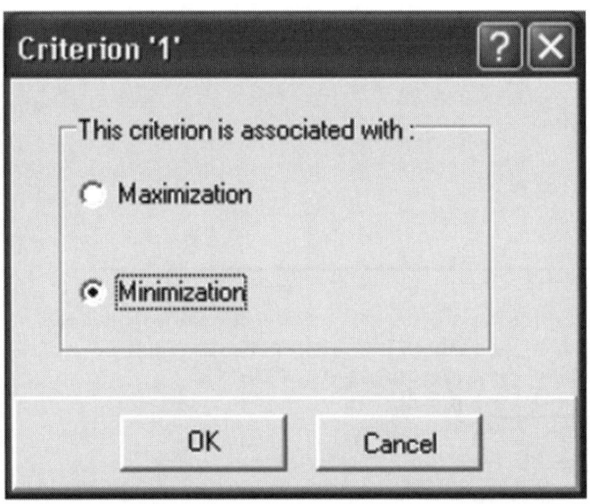

Figure 7.6 Defining the nature of the problem (maximization or minimization).

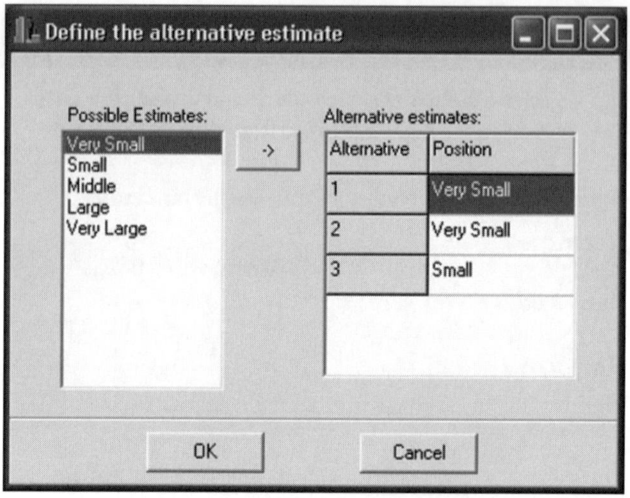

Figure 7.7 Forming the estimates of alternatives.

basic estimate shape) can be changed by the corresponding choice at "Choose the estimate shape" (see Figure 7.8), as shown in Figure 7.9. The estimate shape can be modified by changing parameters at "Estimate description and its parameters" (see Figure 7.8) as shown in Figure 7.10. The estimate shape can also be modified on an experimental basis by clicking on one of six buttons at the upper left corner of the screen (see Figure 7.8). Finally, the MDMS

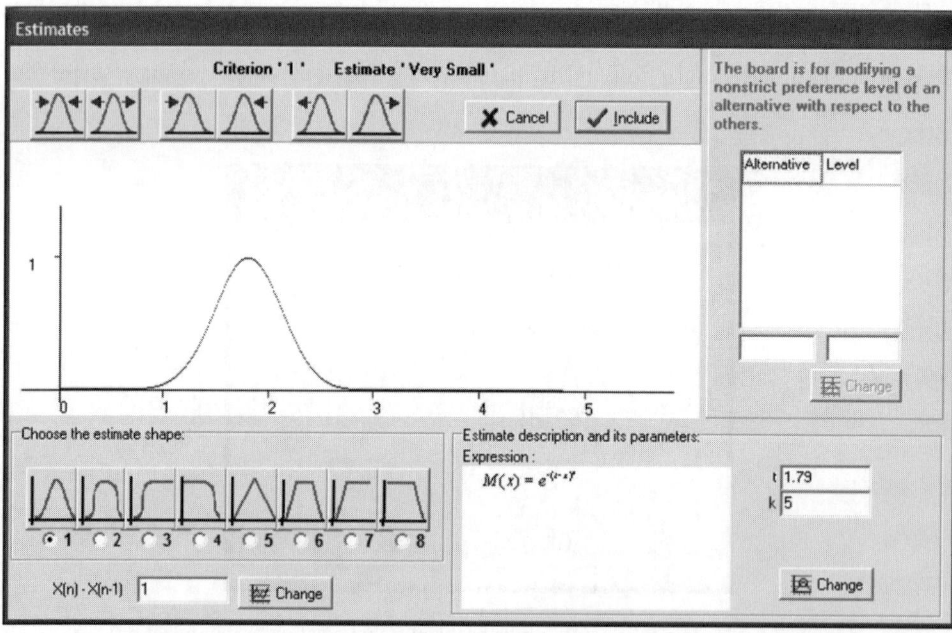

Figure 7.8 Estimate "Very Small".

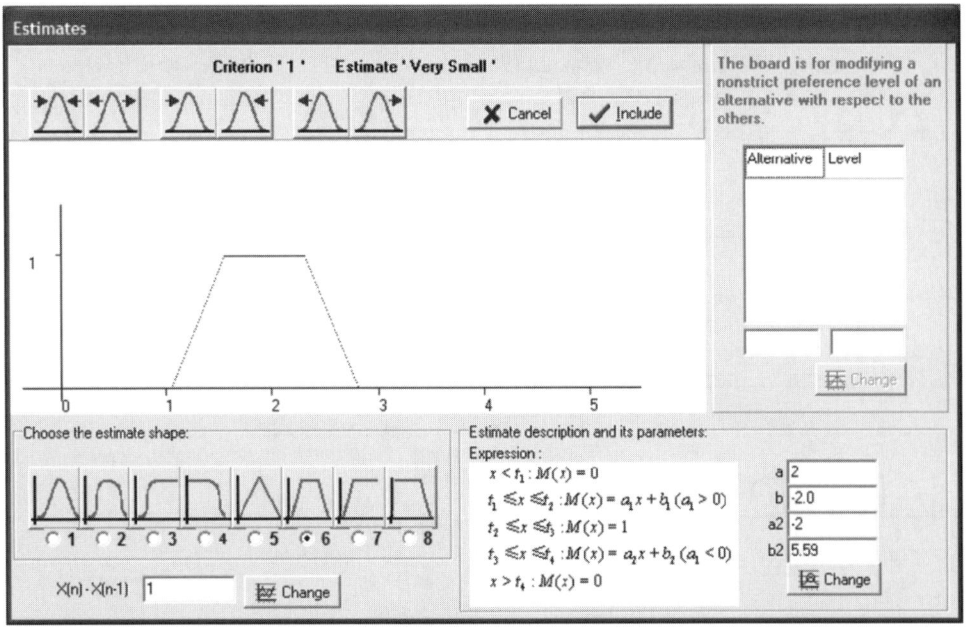

Figure 7.9 Changed type of estimate "Very Small".

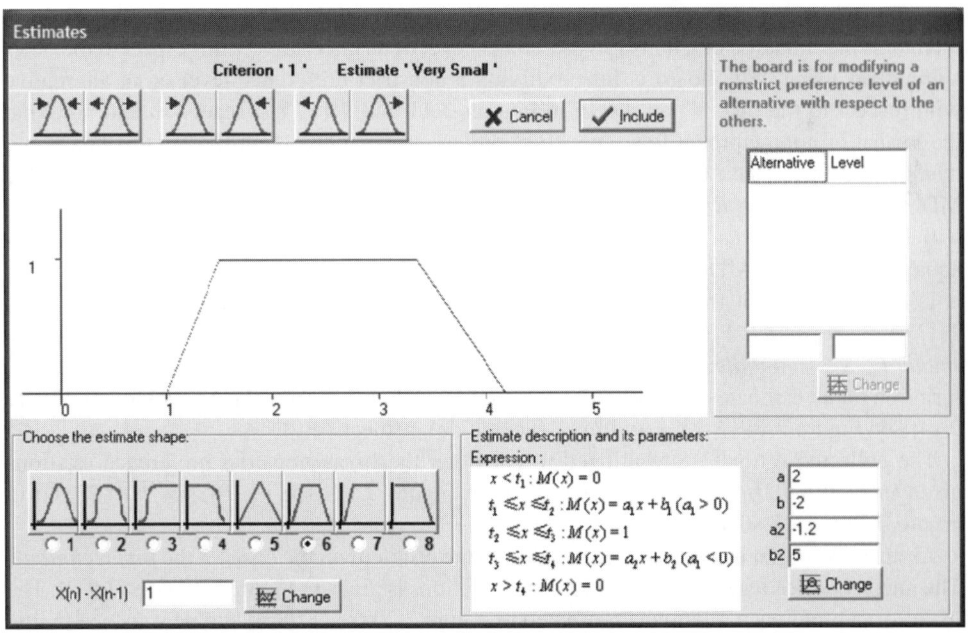

Figure 7.10 Modified estimate of the expression "Very Small".

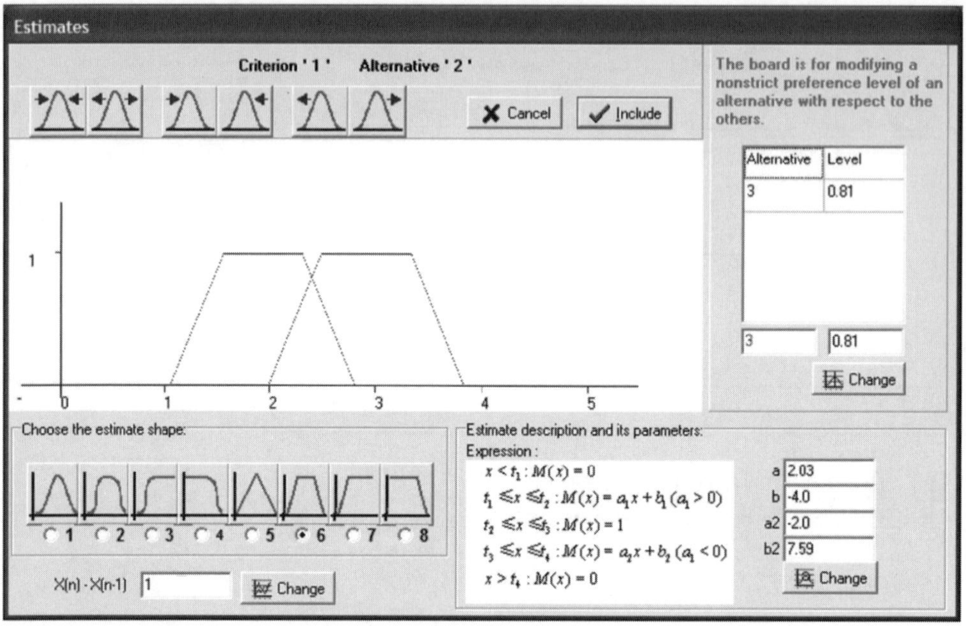

Figure 7.11 Initial preference level.

facilitates modification of a nonstrict preference level of an alternative with respect to the other alternatives using "The board is for modifying a nonstrict preference level of an alternative with respect to the others", as shown in Figures 7.11 and 7.12. The reader should note that the format of nonreciprocal fuzzy nonstrict preference relations is utilized in this board and that a DM does not have to specify both $\boldsymbol{R}(X_k, X_l)$ and $\boldsymbol{R}(X_l, X_k)$. If X_k is better than X_l, the MDMS assigns one to $\boldsymbol{R}(X_k, X_l)$ and lets the DM adjust the value of $\boldsymbol{R}(X_l, X_k)$ in conformity with his/her preferences. Similarly, if X_l is better than X_k, then $\boldsymbol{R}(X_l, X_k)$ is automatically set equal to one and the MDMS lets the DM adjust the value of $\boldsymbol{R}(X_k, X_l)$.

Let us assume that the alternatives from Example 7.5 have been evaluated as follows: $F_1(X_1) =$ *very small*, $F_1(X_2) =$ *very small*, $F_1(X_3) =$ *small*, $F_2(X_1) =$ *very small*, $F_2(X_2) =$ *small*, $F_2(X_3) =$ *middle*, $F_3(X_1) =$ *large*, $F_3(X_2) =$ *middle*, and $F_3(X_3) =$ *middle*. The description for all estimates is trapezoidal (see Figure 7.13 ($F_1(X_1)$ and $F_1(X_2)$ coincide), Figure 7.14, and Figure 7.15 ($F_3(X_2)$ and $F_3(X_3)$ coincide), respectively).

The collected estimates are utilized to construct the fuzzy nonstrict preference relations given in Figure 7.16 (see (7.78)) for $p = 1$, in Figure 7.17 (see (7.79)) for $p = 2$, and in Figure 7.18 (see (7.80)) for $p = 3$.

As an example, let us consider the solution to the problem on the basis of the *first technique*. The intersection of fuzzy nonstrict preference relations is given in Figure 7.19 (see (7.81)). The fuzzy strict preference relation is presented in Figure 7.20 (see (7.82)) and permits one to find the membership function of the fuzzy set of nondominated alternatives given in Figure 7.21 (see (7.83)). Thus, the solution, obtained on the basis of the *first technique*, is $\mathbf{X}^{ND} = \{X_2, X_3\}$.

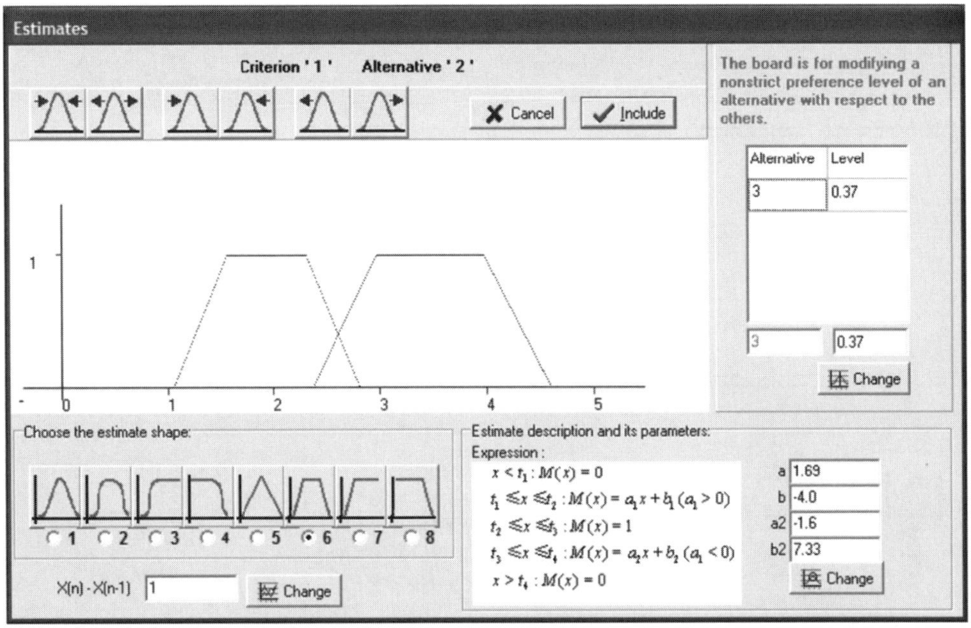

Figure 7.12 Modified preference level.

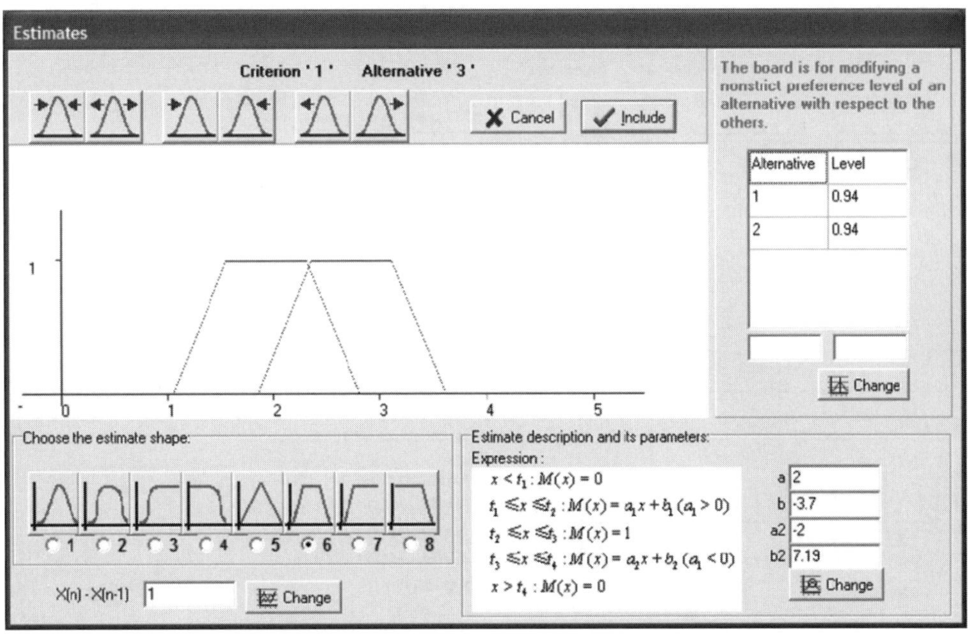

Figure 7.13 Estimates for the first criterion.

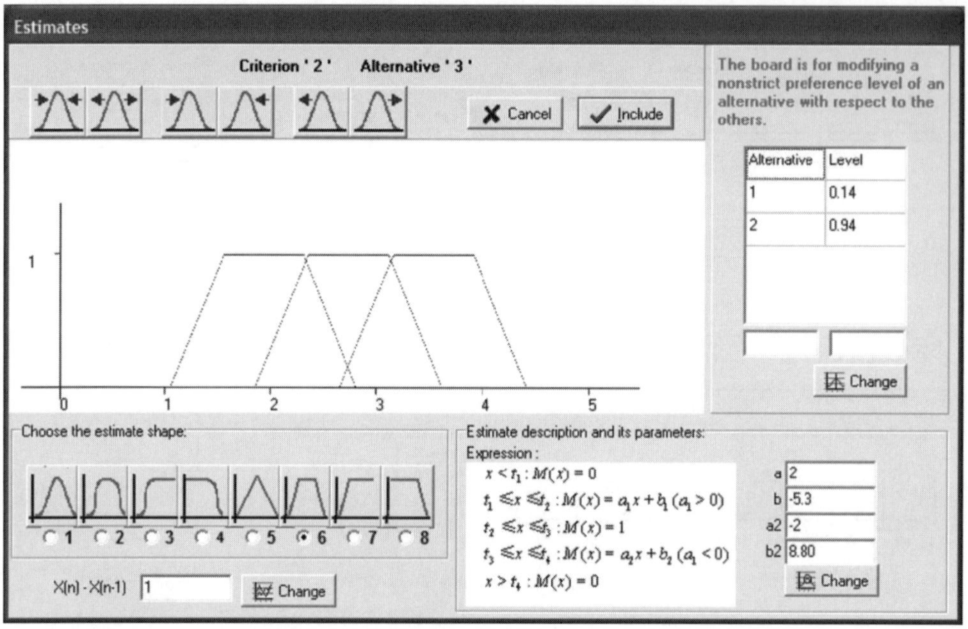

Figure 7.14 Estimates for the second criterion.

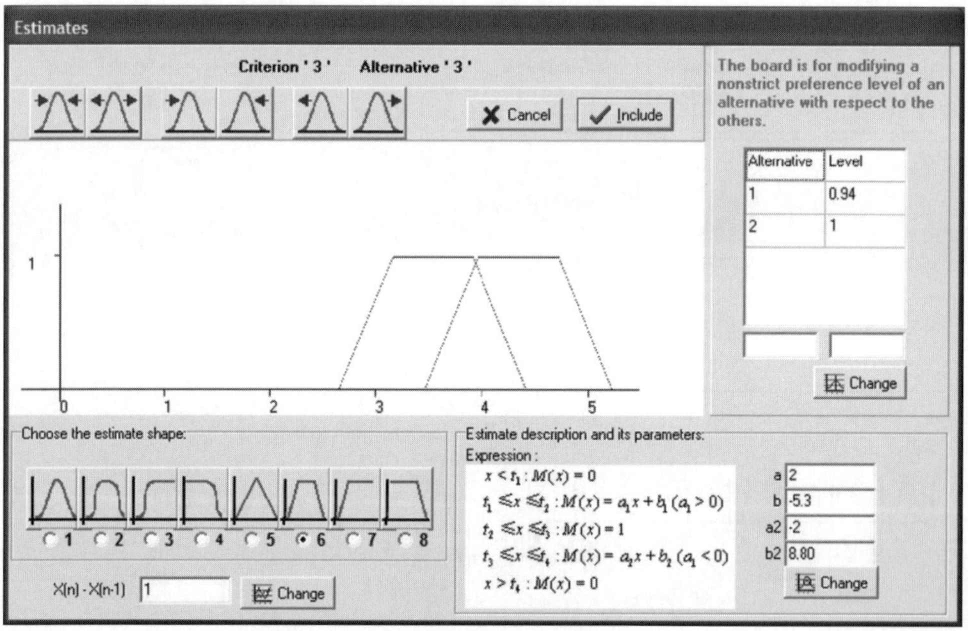

Figure 7.15 Estimates for the third criterion.

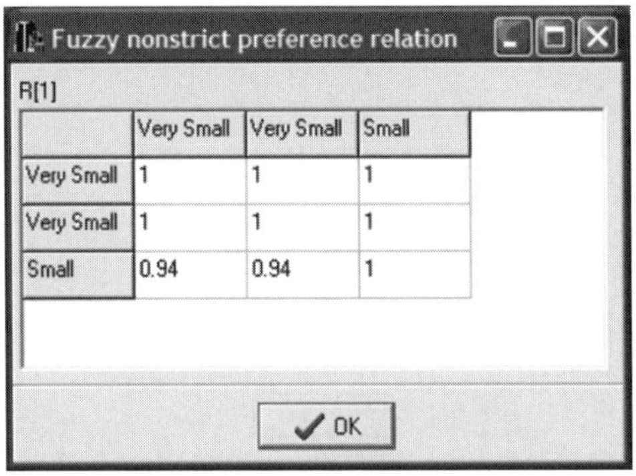

	Very Small	Very Small	Small
Very Small	1	1	1
Very Small	1	1	1
Small	0.94	0.94	1

Figure 7.16 Fuzzy nonstrict preference relation for the first criterion.

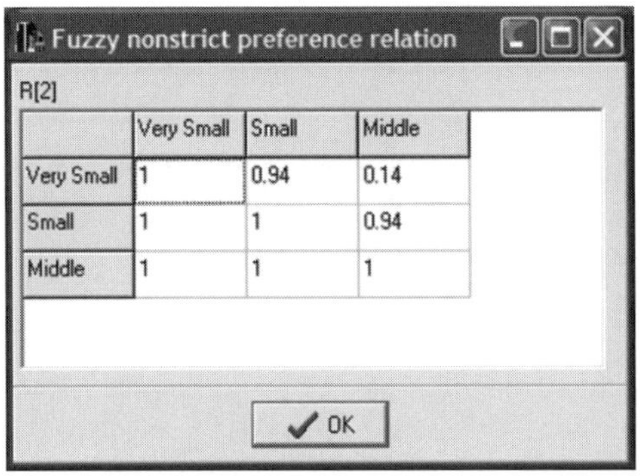

	Very Small	Small	Middle
Very Small	1	0.94	0.14
Small	1	1	0.94
Middle	1	1	1

Figure 7.17 Fuzzy nonstrict preference relation for the second criterion.

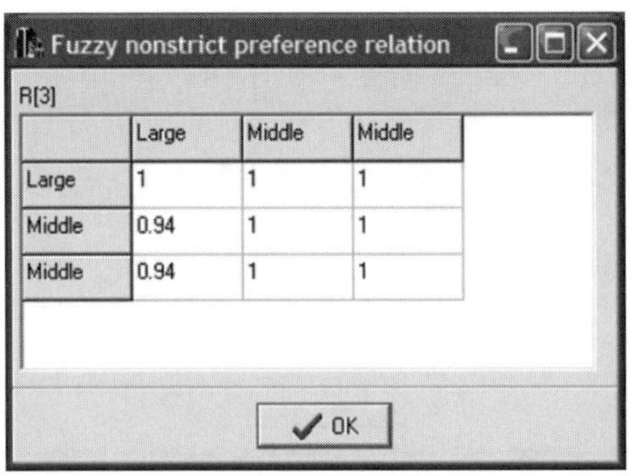

	Large	Middle	Middle
Large	1	1	1
Middle	0.94	1	1
Middle	0.94	1	1

Figure 7.18 Fuzzy nonstrict preference relation for the third criterion.

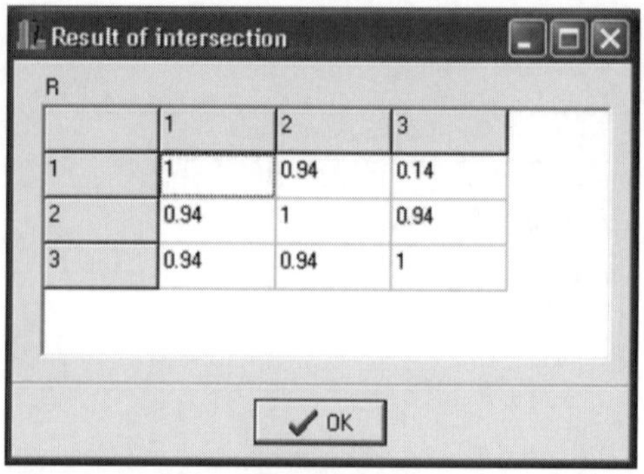

Figure 7.19 Intersection of fuzzy nonstrict preference relations.

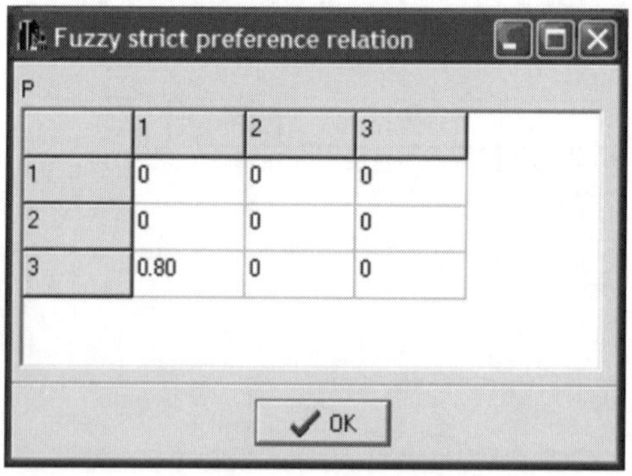

Figure 7.20 Fuzzy strict preference relation.

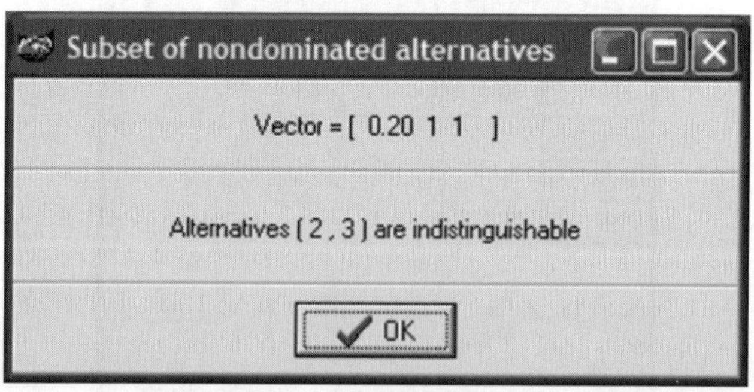

Figure 7.21 Membership function of the fuzzy set of nondominated alternatives.

7.5 Multicriteria Analysis of Alternatives with Fuzzy Ordering of Criteria

Actually, all the techniques for analyzing $\langle \mathbf{X}, \boldsymbol{R} \rangle$ models, which were described in Section 7.3, require the explicit direct or indirect ordering of the criteria. Consequently, it is necessary to distinguish the results of Orlovsky (1981), which allow a DM to present information related to the importance of the criteria in the form of a NRFPR:

$$\Lambda(\lambda_p, \lambda_t) : \Lambda \times \Lambda \rightarrow [0, 1], \quad p, t = 1, 2, \ldots, q \tag{7.92}$$

With the membership functions of the fuzzy sets of nondominated alternatives for all preference relations (7.75), it is possible to construct the following fuzzy preference relation induced by the preference relations (7.75) and (7.92):

$$\boldsymbol{R}_{\Lambda}(X_k, X_l) = \max_{\lambda_p, \lambda_t \in \Lambda} \min_{X_k, X_l \in \mathbf{X}} (\mathbf{ND}_p(X_k), \mathbf{ND}_t(X_l), \Lambda(\lambda_p, \lambda_t)), \quad p, t = 1, 2, \ldots, q \tag{7.93}$$

The fuzzy preference relation (7.93) can be considered as a result of aggregating the family of $\boldsymbol{R}_p$, $p = 1, 2, \ldots, q$, with the use of information reflecting the relative importance of criteria given in the form of (7.92). Applying (5.35) and (7.54) to (7.93), it is possible to construct the fuzzy set of nondominated alternatives $\overline{\mathbf{ND}}_{\Lambda}(X_k)$. As shown in Orlovsky (1981), the set $\overline{\mathbf{ND}}_{\Lambda}(X_k)$ is to be modified in accordance with the following relationship:

$$\mathbf{ND}_{\Lambda}(X_k) = \min(\overline{\mathbf{ND}}_{\Lambda}(X_k), \boldsymbol{R}_{\Lambda}(X_k, X_k)) \tag{7.94}$$

Example 7.6. We are given a set of alternatives $\mathbf{X} = \{X_1, X_2, X_3\}$ which are to be compared by applying three criteria. The corresponding fuzzy nonstrict preference relations are as follows:

$$\boldsymbol{R}_1 = \begin{bmatrix} 1 & 1 & 1 \\ 0.8 & 1 & 0.6 \\ 0.4 & 0.6 & 1 \end{bmatrix} \tag{7.95}$$

$$\boldsymbol{R}_2 = \begin{bmatrix} 1 & 0.4 & 0.3 \\ 1 & 1 & 0.7 \\ 0.9 & 0.6 & 1 \end{bmatrix} \tag{7.96}$$

$$\boldsymbol{R}_3 = \begin{bmatrix} 1 & 0.8 & 0.7 \\ 0.4 & 1 & 0.8 \\ 0.3 & 0.6 & 1 \end{bmatrix} \tag{7.97}$$

The information related to the importance of criteria is presented in the following form:

$$\Lambda = \begin{bmatrix} 1 & 0.8 & 0.7 \\ 1 & 1 & 0.5 \\ 0.9 & 0.7 & 1 \end{bmatrix} \tag{7.98}$$

Applying (5.35) to (7.95)–(7.97), we can construct the fuzzy strict preference relations

$$\boldsymbol{P}_1 = \begin{bmatrix} 0 & 0.2 & 0.6 \\ 0 & 0 & 0 \\ 0 & 0 & 0 \end{bmatrix} \tag{7.99}$$

$$\boldsymbol{P}_2 = \begin{bmatrix} 0 & 0 & 0 \\ 0.6 & 0 & 0.1 \\ 0.6 & 0 & 0 \end{bmatrix} \tag{7.100}$$

and

$$\boldsymbol{P}_3 = \begin{bmatrix} 0 & 0.4 & 0.4 \\ 0 & 0 & 0.2 \\ 0 & 0 & 0 \end{bmatrix} \tag{7.101}$$

for the first, second, and third criteria, respectively. Applying (7.54) to (7.99)–(7.101), we obtain the following membership functions of the fuzzy sets of nondominated alternatives:

$$\mathbf{ND}_1 = [1 \quad 0.8 \quad 0.4] \tag{7.102}$$

$$\mathbf{ND}_2 = [0.4 \quad 1 \quad 0.9] \tag{7.103}$$

and

$$\mathbf{ND}_3 = [1 \quad 0.6 \quad 0.6] \tag{7.104}$$

for the first, second, and third criterion, respectively.

Applying (7.93) to process (7.102)–(7.104) together with (7.98), we obtain

$$\boldsymbol{R}_\wedge = \begin{bmatrix} 1 & 0.8 & 0.8 \\ 1 & 1 & 0.9 \\ 0.9 & 0.9 & 0.4 \end{bmatrix} \tag{7.105}$$

By making use of (5.35), we construct corresponding fuzzy strict preference relation

$$\boldsymbol{P}_\wedge = \begin{bmatrix} 0 & 0 & 0 \\ 0.2 & 0 & 0 \\ 0.1 & 0 & 0 \end{bmatrix} \tag{7.106}$$

which, according to (7.54), leads to

$$\overline{\mathbf{ND}}_\wedge = [0.8 \quad 1 \quad 1] \tag{7.107}$$

Finally, the application of (7.94) and taking into account that

$$\boldsymbol{R}_\wedge(X_k, X_k) = [1 \quad 1 \quad 0.4] \tag{7.108}$$

give rise to

$$\overline{\mathbf{ND}}_{\wedge} = [\,0.8 \quad 1 \quad 0.4\,] \tag{7.109}$$

7.6 Multicriteria Analysis of Alternatives with the Concept of Fuzzy Majority

The *fifth technique* for the analysis of $\langle \mathbf{X}, \boldsymbol{R} \rangle$ models presented here utilizes the concept of fuzzy majority as the aggregation rule for considering the criteria.

In real-world applications, there are cases where it is not reasonable to assume that a good alternative must simultaneously satisfy all criteria. A DM may consider such a requirement as being too hard (restrictive) and may prefer a softer one, such as: a good alternative must satisfy "most" criteria or "at least half" the criteria or a "few" criteria, for instance. It is important to underline an essential difference between both requirements: in the softer one, it suffices to satisfy some criteria to a high level, without the need to identify which of them should be satisfied at this high level. In situations like this, a DM is not obliged to distinguish the priority or the importance of any criterion.

The use of an AND operator to aggregate all criteria is suitable when it is a necessary condition that a good alternative X_k must simultaneously satisfy F_1 and F_2 and ... and F_q. Such aggregation is noncompensatory, in the sense that the high level of satisfaction of some criteria does not relieve the remaining ones from the requirement of being satisfied. On the other hand, the use of an OR operator is interesting when each alternative is supposed to satisfy at least one criterion – in other words, when it is sufficient to require that X_k satisfies F_1 or F_2 or ... or F_q. Such aggregation is extremely compensatory in the sense that the high level of satisfaction of any criterion is sufficient (independently of which criterion is satisfied to a high level).

However, as already mentioned, there are cases where the relationship among the criteria does not correspond to a "pure" AND or to a "pure" OR operation. The ordered weighted average (OWA) operator, originally proposed by Yager (1988), allows us to deal with such intermediate situations, as it unifies the OR and AND operators in one parameterized operator.

An OWA operator of dimension n corresponds to a mapping function $[0, 1]^n \to [0, 1]$, which aggregates a set of n normalized values $a_1, a_2, \ldots, a_n$, in such a way that

$$\text{OWA}(a_1, a_2, \ldots, a_n) = \sum_{i=1}^{n} w_i b_i \tag{7.110}$$

where b_i is the ith largest value among $a_1, a_2, \ldots, a_n$ and the set of weights $w_1, w_2, \ldots, w_n$ satisfies the conditions $w_i \in [0, 1]$ and $\sum_{i=1}^{n} w_i = 1$ (Yager, 1988).

Example 7.7. The aggregation of three normalized values, $a_1 = 0.12$, $a_2 = 0.56$, and $a_3 = 0.42$, using OWA with the weights $w_1 = 0.5$, $w_2 = 0.2$, and $w_3 = 0.3$, results in

$$\text{OWA}(0.12, 0.56, 0.42) = 0.56(0.5) + 0.42(0.2) + 0.12(0.3) = 0.4 \tag{7.111}$$

Table 7.1 Equivalence between OWA and some aggregation operators

Aggregation operator	Weights of OWA
Min	$w_1 = 1,\, w_i = 0,\, i = 2, 3, \ldots, n$
Max	$w_n = 1,\, w_i = 0,\, i = 1, 2, \ldots, n-1$
Median	if n is odd, $w_{(n+1)/2} = 1,\, w_i = 0$ otherwise
	if n is odd, $w_{n/2} = 1,\, w_i = 0$ otherwise
kth-order statistics	$w_k = 1,\, w_i = 0,\, i = 1, 2, \ldots, n \wedge i \neq k$
Arithmetic mean	$w_i = 1/n,\, i = 1, 2, \ldots, n$

As can be seen in Example 7.7, w_1 is associated with the first largest element to be aggregated, which is 0.56; w_2 is associated with the second largest element to be aggregated, which corresponds to 0.42; and so forth. Hence, it must be clear that each weight w_i is associated with the ith ordered position rather than to a particular element. Because of this particularity, a relevant issue concerning the use of OWA corresponds to the specification of adequate values for these weights (Chiclana, Herrera, and Herrera-Viedma, 1998). It is interesting to note that, by properly adjusting them, it is possible to set the degree of AND and OR inherent to the parameterized aggregation operator. For instance, when the weights are adjusted as $w_n = 1$, $w_i = 0$, for $i \neq q$, $i = 1, 2, \ldots, n-1$, the OWA operator is equivalent to the max operator, which corresponds to a "pure" OR aggregation. When the weights are adjusted as $w_1 = 1$, $w_i = 0$, $i = 2, 3, \ldots, n$, the OWA operator is equivalent to the min operator, which corresponds to a "pure" AND aggregation. As Table 7.1 shows, in addition to min and max, some other common operators can be implemented as specific cases of OWA (Yager, 1995). However, the major attraction of using OWA in the context of multicriteria analysis possibly corresponds to the fact that it allows a DM to indirectly specify the weights by using linguistic quantifiers.

A fuzzy quantifier corresponds to a fuzzy set $Q(r)$, which reflects the level at which the $r \in [0, 1]$ portion of objects satisfies the concept reflected by Q. Figure 7.22 shows some examples of linguistic quantifiers. In Yager (1995), fuzzy quantifiers of regular increasing monotone (RIM) type are used to specify the weights of OWA operators. A RIM fuzzy

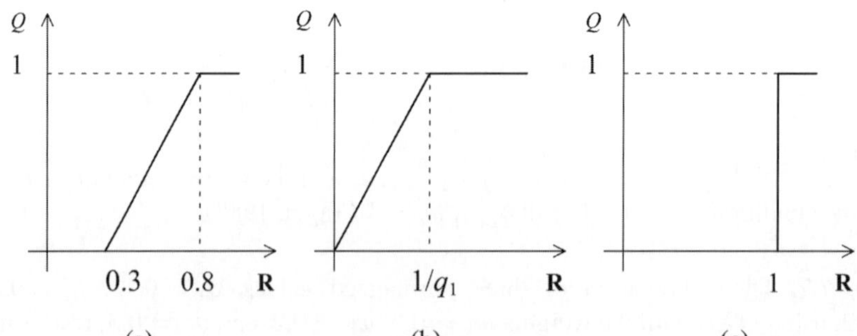

Figure 7.22 Fuzzy quantifiers: (a) most; (b) at least q_1; (c) all.

quantifier satisfies some mathematical conditions necessary to guarantee that, as more criteria are satisfied, the overall satisfaction of a DM cannot decrease:

- $Q(0) = 0$
- $Q(1) = 1$
- if $r_1 > r_2$, then $Q(r_1) > Q(r_2)$.

Once a suitable quantifier has been chosen, the value of each weight can be determined as follows:

$$w_i = Q\left(\frac{i}{n}\right) - Q\left(\frac{i-1}{n}\right), \quad i = 1, 2, \ldots, n \tag{7.112}$$

In the analysis of $\langle \mathbf{X}, \boldsymbol{R} \rangle$ models, having at hand the nonstrict preference relations for each criterion, it is possible to obtain a global nonstrict preference relation on the basis of the OWA operator as shown next (Grabisch, Orlovski, and Yager, 1998):

$$\text{OWA}(\boldsymbol{R}_1(X_k, X_l), \boldsymbol{R}_2(X_k, X_l), \ldots, \boldsymbol{R}_q(X_k, X_l)) = \sum_{p=1}^{q} w_p b_p \tag{7.113}$$

where b_p is the pth largest element from $\boldsymbol{R}_1(X_k, X_l), \boldsymbol{R}_2(X_k, X_l), \ldots, \boldsymbol{R}_q(X_k, X_l)$ and, as mentioned previously, the weights w_p, $p = 1, 2, \ldots, q$, must be nonnegative and satisfy $\sum_{p=1}^{q} w_p = 1$.

Example 7.8. Considering the three fuzzy nonstrict preference relations from Example 7.6, given by (7.95)–(7.97), and the fuzzy quantifier "most", the global nonstrict preference matrix can be obtained on the basis of the OWA operator. But, first, it is necessary to calculate the weights with the use of (7.112) (here, we consider the fuzzy set $Q(r)$ for the quantifier "most" as shown in Figure 7.22), which gives

$$w_1 = Q(1/3) - Q(0) = 0.066 - 0 = 0.066 \tag{7.114}$$

$$w_2 = Q(2/3) - Q(1/3) = 0.733 - 0.066 = 0.667 \tag{7.115}$$

$$w_3 = Q(1) - Q(2/3) = 1 - 0.733 = 0.266 \tag{7.116}$$

Then, on the basis of (7.113), we obtain

$$\boldsymbol{R} = \begin{bmatrix} 1 & 0.706 & 0.613 \\ 0.706 & 1 & 0.679 \\ 0.406 & 0.6 & 1 \end{bmatrix} \tag{7.117}$$

In order to clarify how the weights of OWA affect the construction of the global nonstrict preference, let us observe, for instance, how the values of positions $\boldsymbol{R}(X_2, X_1)$ and $\boldsymbol{R}(X_2, X_3)$

from (7.117) were obtained

$$\boldsymbol{R}(X_2, X_1) = 1 \cdot 0.066 + 0.8 \cdot 0.667 + 0.4 \cdot 0.266 = 0.706 \tag{7.118}$$

$$\boldsymbol{R}(X_2, X_3) = 0.8 \cdot 0.066 + 0.7 \cdot 0.667 + 0.6 \cdot 0.266 = 0.679 \tag{7.119}$$

It is worth noting in Example 7.8 that, when $\boldsymbol{R}(X_2, X_1)$ was considered, the weights w_1, w_2, and w_3 were associated to the respective criteria F_2, F_1, and F_3, as a consequence of the ranking $\boldsymbol{R}_2(X_2, X_1) > \boldsymbol{R}_1(X_2, X_1) > \boldsymbol{R}_3(X_2, X_1)$. Conversely, when $\boldsymbol{R}(X_2, X_3)$ was considered, the weights w_1, w_2, and w_3 were respectively associated to criteria F_3, F_2, and F_1, as a consequence of the ranking $\boldsymbol{R}_3(X_2, X_1) > \boldsymbol{R}_2(X_2, X_1) > \boldsymbol{R}_1(X_2, X_1)$. Thus, as can be seen, when OWA is utilized to aggregate across all criteria, given a pair of alternatives X_k and X_l, a high level of global preference of one alternative over another is achieved if, for some criteria (it does not matter which of them), a high level of nonstrict preference $\boldsymbol{R}_p(X_k, X_l)$ is observed.

Having at hand the global nonstrict preference matrix, it is possible to exploit the global relation and obtain a global ranking of the alternatives with the use of the quantifier guided dominance degree (QGDD)

$$\text{QGDD}(X_k) = \text{OWA}(\boldsymbol{R}(X_k, X_l)), \quad l = 1, 2, \ldots, n, l \neq k \tag{7.120}$$

and of the quantifier guided nondominance degree (QGNDD)

$$\text{QGNDD}(X_k) = \text{OWA}(1 - \boldsymbol{P}(X_l, X_k)), \quad l = 1, 2, \ldots, n, l \neq k \tag{7.121}$$

both of them introduced by Chiclana *et al.* (1996). Whereas QGDD(X_k) reflects the dominance level of X_k over the other alternatives, in a fuzzy majority sense (or other fuzzy quantifier) QGNDD(X_k) reflects the level at which X_k is not dominated by a fuzzy majority (or other fuzzy quantifier) of the remaining alternatives. It is interesting to observe that if we choose the fuzzy quantifier "all", which will make OWA work as the min operator, then QGNDD corresponds to the Orlovsky nondominated degree. QGNDD can be used for choosing the best alternative and, in case of indifference between two or more alternatives, QGDD can be used to distinguish them.

Example 7.9. As a continuation of the previous example, it is possible to exploit the global nonstrict preference relation given by (7.117), with the use of QGNDD(X_k), calculated by taking into account the fuzzy quantifier "most". It is worth noting that the calculus of QGNDD or QGDD for each alternative requires the use of $n-1$ weights $w_1, w_2, \ldots, w_{n-1}$, as the diagonal of the global nonstrict preference and strict preference matrices are not considered either in (7.120) or in (7.121), respectively. In this way, we consider

$$w_1 = Q(1/2) - Q(0) = 0.4 - 0 = 0.4 \tag{7.122}$$

$$w_2 = Q(1) - Q(1/2) = 1 - 0.4 = 0.6 \tag{7.123}$$

which produces

$$\text{QGNDD}(X_k) = [\,1 \quad 1 \quad 0.844\,] \tag{7.124}$$

A priori, we cannot distinguish X_1 and X_2, but the choice degree reflected by QGDD allows us to choose X_2 as the best alternative:

$$\text{QGDD}(X_k) = [\,0.650 \quad 0.689 \quad 0.483\,] \tag{7.125}$$

Example 7.10. In order to illustrate the importance of choosing a fuzzy quantifier that adequately reflects the requirement desired by a DM, consider three fuzzy nonstrict preference relations from Example 7.5, given by (7.78)–(7.80). The global nonstrict preference matrix can be obtained on the basis of OWA and the fuzzy quantifier "all". First, it is necessary to calculate the weights with the use of (7.112) (here, we consider the fuzzy set $Q(r)$ for the quantifier "all" as shown in Figure 7.22). We obtain

$$w_1 = Q(1/3) - Q(0) = 0 \tag{7.126}$$

$$w_2 = Q(2/3) - Q(1/3) = 0 \tag{7.127}$$

$$w_3 = Q(1) - Q(2/3) = 1 \tag{7.128}$$

Then, making use of (7.113), we have

$$\boldsymbol{R} = \begin{bmatrix} 1 & 0.94 & 0.14 \\ 0.94 & 1 & 0.94 \\ 0.94 & 0.94 & 1 \end{bmatrix} \tag{7.129}$$

which coincides with (7.81), as can be seen.

The computations of weights w_1 and w_2, realized on the basis of the quantifier "all", to be utilized with both QGNDD and QGDD return the results

$$w_1 = Q(1/2) - Q(0) = 0 \tag{7.130}$$

$$w_2 = Q(1) - Q(1/2) = 1 \tag{7.131}$$

Finally, on the basis of (7.112) and (7.113), we can obtain the QGNDD and QGDD, respectively:

$$\text{QGNDD}(X_k) = [\,0.2 \quad 1 \quad 1\,] \tag{7.132}$$

$$\text{QGDD}(X_k) = [\,0.14 \quad 0.94 \quad 0.94\,] \tag{7.133}$$

As we can see, a priori, neither QGNDD nor QGDD allow us to distinguish X_2 and X_3. However, with the use of a more relaxed quantifier such as "most", instead of "all", we can obtain the global nonstrict preference matrix

$$\boldsymbol{R} = \begin{bmatrix} 1 & 0.984 & 0.771 \\ 0.984 & 1 & 0.984 \\ 0.944 & 0.984 & 1 \end{bmatrix} \tag{7.134}$$

as well as the QGNDD and QGDD, respectively, given by

$$\text{QGNDD}(X_k) = [\,0.896 \quad 1 \quad 1\,] \tag{7.135}$$

and

$$\text{QGDD}(X_k) = [\,0.856 \quad 0.984 \quad 0.960\,] \tag{7.136}$$

Now the alternatives are formally ranked as follows:$X_2 \succ X_3 \succ X_1$. It is interesting to note that, in this problem, by using the fuzzy quantifier "most" instead of "all", it became possible to rank all alternatives, without the need to differentiate the importance of the criteria.

7.7 Multicriteria Analysis of Alternatives Based on an Outranking Approach (Fuzzy Promethee)

This section focuses on an approach for multicriteria decision-making based on the construction and exploitation of a binary fuzzy relation named the outranking relation. Formally, the outranking relation corresponds to a binary fuzzy relation, such as the fuzzy preference relation that we have been studying. However, it should be stressed that the notion of outranking relation can be defined, constructed, and exploited independently of the theory of fuzzy preference structures or of the notion of the fuzzy strict preference relation. Indeed, the outranking approach differs from the other approaches considered in this chapter in two main aspects: the outranking approach allows a DM to define the shape of the outranking relation in a very direct way, just by adjusting a few input parameters; and the exploitation of the outranking relations constructed by a DM is based on their representation as weighted graphs rather than on the concept of the fuzzy strict preference relation or fuzzy nondominance set.

The concept of the outranking relation and the first methods for constructing and exploiting outranking relations were developed by French researchers. Consequently, the outranking approach for decision-making is usually referred to as the European or French School. Here, we selected a fuzzy version (Geldermann, Spengler, and Rentz, 2000) of a popular method from the French School, called Promethee II (Brans and Vincke, 1985), to show how the outranking approach can be utilized to solve multicriteria decision-making problems in a fuzzy environment. For other instances of fuzzy versions of the Promethee II method, the reader can refer to the relevant literature (Goumas and Lygerou, 2000; Halouani, Chabchoub, and Martel, 2009; Wei-xiang and Bang-yi, 2010).

In Promethee II, the DM preferences, being restricted to a single criterion F_p, are modeled through a nondecreasing function $S_p(X_k, X_l)$ named the preference function, which reflects the preference level of X_k over X_l, according to the following rules:

- if $S_p(X_k, X_l) = 0$, both alternatives are considered indifferent to each other;
- if $S_p(X_k, X_l) = 1$, X_k is strictly preferred to X_l.

These preference functions are defined in terms of the difference

$$d = F_p(X_k) - F_p(X_l) \tag{7.137}$$

in such a way that they transform the difference in the evaluations of two alternatives into a preference intensity between 0 and 1. In the fuzzy version of Promethee II considered here, the evaluation of an alternative X_k over $F_p(X_k)$ is a fuzzy set. Therefore, the difference $D = F_p(X_k) - F_p(X_l)$ also corresponds to a fuzzy set. For the arithmetic operations on fuzzy numbers (sets), the reader should refer to Chapter 3.

The original (nonfuzzy) version of Promethee II admits six different generalized models for the preference function, which cover most of the scenarios encountered in real-world applications (Brans and Vincke, 1985). Next, those six generalized models are defined for the difference d given by (7.137). We want to draw attention to the fact that, by applying the extension principle, these functions can be extended to deal with a fuzzy set D (rather than a real number d):

- Usual criterion

$$S_p(X_k, X_l) = S_p(d) = \begin{cases} 0 & d \le 0 \\ 1 & d > 0 \end{cases} \tag{7.138}$$

- Quasi-criterion

$$S_p(X_k, X_l) = S_p(d) = \begin{cases} 0 & d \le a_p \\ 1 & d > a_p \end{cases} \tag{7.139}$$

- Pseudo-criterion or level criterion

$$S_p(X_k, X_l) = S_p(d) = \begin{cases} 0 & d \le a_p \\ 0.5 & a_p < d \le b_p \\ 1 & d > b_p \end{cases} \tag{7.140}$$

- Linear criterion

$$S_p(X_k, X_l) = S_p(d) = \begin{cases} \dfrac{d}{b_p} & d \le b_p \\ 1 & d > b_p \end{cases} \tag{7.141}$$

- Criterion with linear preference and an indifference region

$$S_p(X_k, X_l) = S_p(d) = \begin{cases} 0 & d \le a_p \\ \dfrac{d - a_p}{b_p - a_p} & a_p < d \le b_p \\ 1 & d > b_p \end{cases} \tag{7.142}$$

- Gaussian criterion:

$$S_p(X_k, X_l) = S_p(d) = \begin{cases} 0, & d \le 0 \\ 1 - \exp\left(\dfrac{-d}{2\sigma_p^2}\right), & d > 0 \end{cases} \tag{7.143}$$

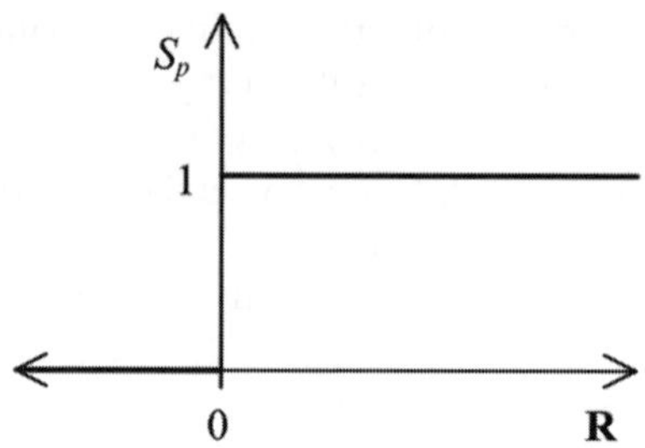

Figure 7.23 Usual criterion.

Graphical representations of the preference functions given by (7.138), (7.139), and (7.140) are shown in Figures 7.23, 7.24, and 7.25, respectively. These three preference functions are particularly easy to define. Although the quasi-criterion and the pseudo-criterion require the DM to fix some thresholds according to his/her preferences, in each case, the required parameters have an intuitive significance. Graphical representations of (7.141), (7.142), and (7.143) are shown in Figures 7.26, 7.27, and 7.28, respectively. As can be seen, these three preference functions present a smooth transition between indifference and strict preference, which permits the DM to make judgments at different levels of preference. In the visualization of the linear criterion, the slope of the preference function will depend on the value of the preference threshold b_p. In the case of the Gaussian criterion, σ_p is the distance between the origin and the point of inflection of the curve $S_p(X_k, X_l)$ (see Figure 7.28). Finally, in Figure 7.29, one can see the extension principle being applied to obtain an image of the fuzzy set $D = (d_1, d_2, d_3)$ with a triangular membership function, under a preference function $S_p(d)$, defined as given by (7.142).

The global preferences of a DM are reflected by the weighted mean of the preference functions restricted to each criterion, as follows:

$$\Pi(X_k, X_l) = \sum_{p=1}^{q} \lambda_p S_p(X_k, X_l) \tag{7.144}$$

where λ_p, $p = 1, 2, \ldots, q$, are the importance factors associated with each criterion. Usually they are numbers satisfying conditions (4.15), (4.16). Promethee II does not provide specific guidelines for determining these importance factors, but assumes that a DM is able to weight the criteria appropriately. Again, the reader should be aware of the need to implement the operations of multiplication and addition between fuzzy sets in (7.144).

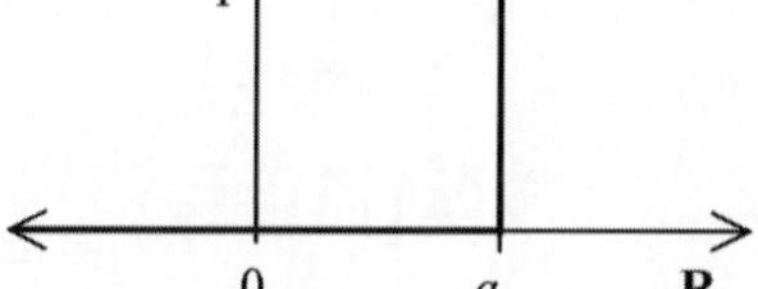

Figure 7.24 Quasi-criterion.

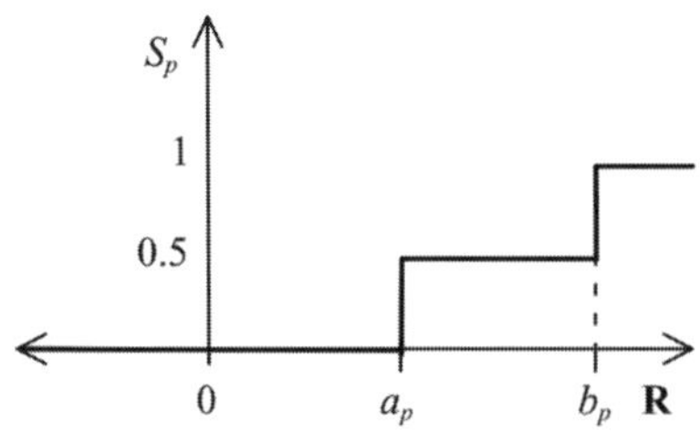

Figure 7.25 Pseudo-criterion.

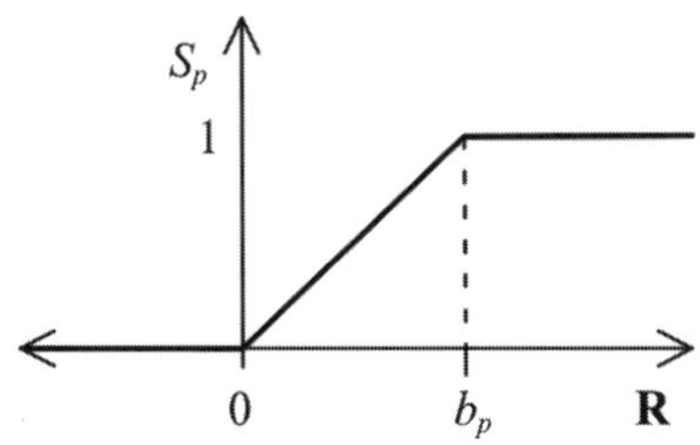

Figure 7.26 Criterion with linear preference.

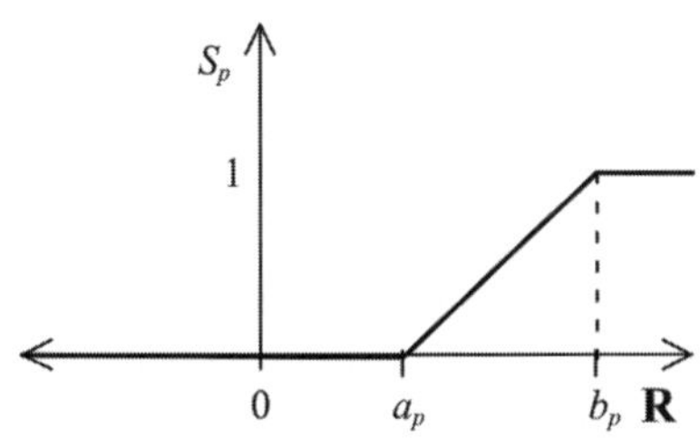

Figure 7.27 Criterion with linear preference and an indifference region.

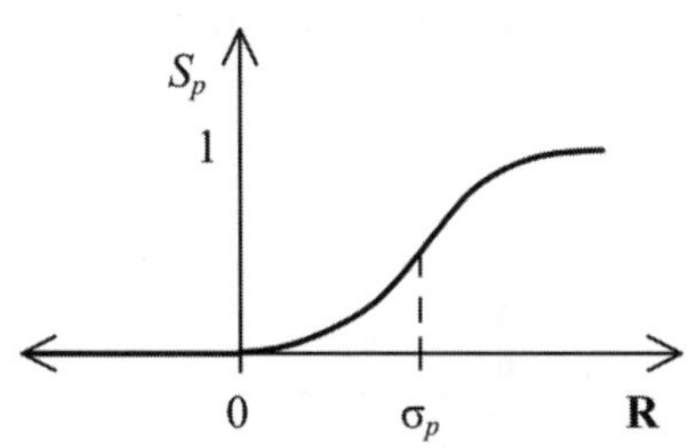

Figure 7.28 Criterion with Gaussian preference function.

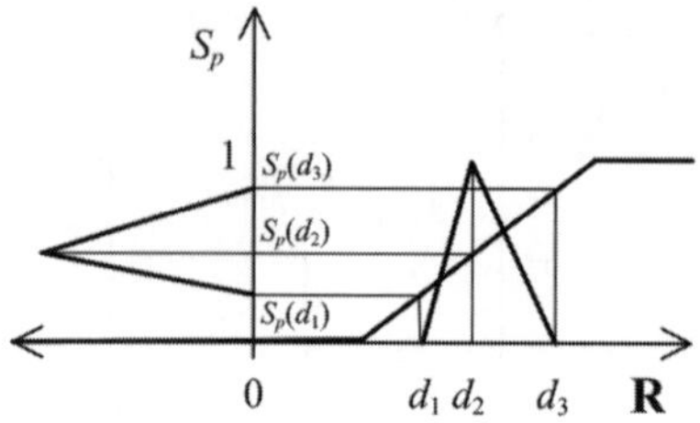

Figure 7.29 The use of the extension principle to a preference function given by (7.142).

It is worth noting that $\Pi(X_k, X_l)$ reflects the global or aggregated preference level of X_k over X_l. From the matrix representation of the global preferences for all pairs of alternatives, a graph can be drawn in such a way that each alternative is represented by a node and each global preference relation is represented by an arc. Hence, between each pair of nodes (X_k, X_l), there are always two arcs, one associated with $\Pi(X_k, X_l)$ and the other with $\Pi(X_l, X_k)$, as shown in Figure 7.30. Leaving ϕ_{out} and entering flows ϕ_{in} at each node, which are given by

$$\phi_{out}(X_k) = \sum_{\forall X_l \in \mathbf{X}} \Pi(X_k, X_l) \tag{7.145}$$

and

$$\phi_{in}(X_k) = \sum_{\forall X_l \in \mathbf{X}} \Pi(X_l, X_k) \tag{7.146}$$

play an important role in Promethee II. These two reflect a measure of the outranking quality and of the outranked quality of X_k and the difference between the leaving and the entering flows, that is, the net flow of each node $X_k \in \mathbf{X}$

$$\phi(X_k) = \phi_{out}(X_k) - \phi_{in}(X_k) \tag{7.147}$$

is utilized to derive a complete ranking of the alternatives. Here, as in Geldermann, Spengler, and Rentz (2000), the ranking of the alternatives is derived from the "defuzzified" net flow.

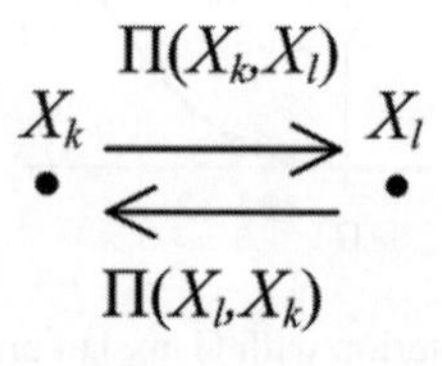

Figure 7.30 Arcs symbolizing the global preference relations between X_k and X_l.

For converting fuzzy sets into real values (that is, for the operation of "defuzzification"), the center of area

$$\phi_{defuzz} = \frac{\int x.\phi(x)\,\mathrm{d}x}{\int \phi(x)\,\mathrm{d}x} \tag{7.148}$$

is utilized.

Finally, a complete ranking can be induced on the basis of the defuzzified net flow as follows:

- if $\phi_{defuzz}(X_k) > \phi_{defuzz}(X_l)$, then X_k is preferred to X_l;
- if $\phi_{defuzz}(X_k) = \phi_{defuzz}(X_l)$, then X_k is indifferent to X_l.

Example 7.11. The multicriteria decision-making problem related to selecting a site for the construction of a new hospital is studied in Vahidnia, Alesheikh, and Alimohammadi (2009). Here, we consider a simplified version of this problem, in which six sites are to be ranked, by taking into account the following four criteria:

1. Distance from arterial streets (minimization criterion)
2. Cost of land (minimization criterion)
3. Population density (maximization criterion)
4. Average travel time to arrive at the nearest existing hospital (maximization criterion).

The fifth criterion which could be considered is associated with the pollution level present at each site. However, as the sites considered here do not significantly differ with respect to this criterion, we decided to eliminate it from the subsequent multicriteria analysis. It is worth noting that both F_1 and F_2 are to be minimized whereas both F_3 and F_4 are to be maximized. Figure 7.31 presents the fuzzy scores utilized for evaluating the third criterion, F_3, the population density. Table 7.2 gives the evaluation matrix of the alternatives.

Next, this problem is studied by means of the fuzzy version of Promethee II. The input parameters provided by a DM for the execution of this method are listed in Table 7.3. As in

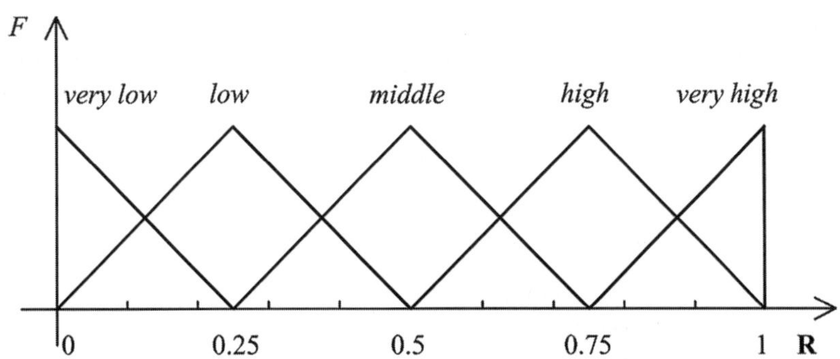

Figure 7.31 Valuation of criterion F_3 (population density).

Table 7.2 Evaluation matrix of the alternatives

	$F_1(X_k)$ (m)	$F_2(X_k)$ (\$/m^2)	$F_3(X_k)$	$F_4(X_k)$ (minutes)
X_1	0	28	*middle*	22
X_2	350	24	*high*	17
X_3	150	18	*high*	12
X_4	500	15	*middle*	10
X_5	50	8	*low*	7
X_6	300	10	*very high*	5

Vahidnia, Alesheikh, and Alimohammadi (2009), a DM considered the average travel time to be the most important criterion and the population density to be the least important criterion for the decision.

Table 7.4 shows the leaving ϕ_{out}, entering ϕ_{in}, and net ϕ flows, which are determined for each alternative by means of (7.145), (7.146), and (7.147), respectively. Figure 7.32 shows the fuzzy sets corresponding to the net flow of each alternative. After converting the fuzzy net flows into real-valued net flows with the use of (7.148) (refer to Table 7.5), the final ranking of the alternatives is as follows: $X_4 \succ X_2 \succ X_6 \succ X_3 \succ X_5 \succ X_1$.

7.8 Application Examples

The first example given below demonstrates the application of three basic techniques for analyzing $\langle \mathbf{X}, \boldsymbol{R} \rangle$ models.

Example 7.12. The problem of substation planning in a power system taking into account the uncertainty of information has been considered in Fontoura Filho, Ales, and Tortelly (1994). Its practical application is associated with a group of 138/13.8 kV substations of a power utility. In particular, a careful analysis has been carried out to select a solution from three alternatives on the basis of their total costs, where the uncertainty of interest rates is modeled as trapezoidal membership functions. Details about the membership functions of alternative costs are given in Table 7.6 and are also illustrated in Figure 7.33.

It is evident that the selection of the most preferable alternative is hampered: the difference between the alternatives X_1 and X_2 is equal to 0.38% for the left bounds of the corresponding membership functions for a certainty of 70% that is accepted in Fontoura Filho, Ales, and

Table 7.3 Input parameters of Promethee provided by a DM

	Preference function specification	Importance factor
F_1	Linear criterion, $b_1 = 50\,\text{m}$	0.25
F_2	Linear criterion, $b_2 = 5\,\$/\text{m}^2$	0.25
F_3	Usual criterion	0.2
F_4	Quasi-criterion, $a_4 = 3\,\text{min}$	0.3

Table 7.4 Leaving ϕ_{out}, entering ϕ_{in} and net ϕ flows

	ϕ_{out}	ϕ_{in}	ϕ
X_1	(2.75, 2.95, 3.55)	(1.2, 1.8, 2.2)	(0.55, 1.15, 2.35)
X_2	(1.65, 2.25, 2.65)	(2.3, 2.5, 3.1)	(−1.45, −0.25, 0.35)
X_3	(1.85, 2.45, 2.85)	(1.75, 1.95, 2.55)	(−0.7, 0.5, 1.1)
X_4	(0.95, 1.15, 1.75)	(2.35, 2.95, 3.35)	(−2.4, −1.8, −0.6)
X_5	(2.1, 2.1, 2.5)	(1.35, 2.15, 2.15)	(−0.05, −0.05, −1.15)
X_6	(1.7, 2.5, 2.5)	(2.05, 2.05, 2.45)	(−0.75, 0.45, 0.45)

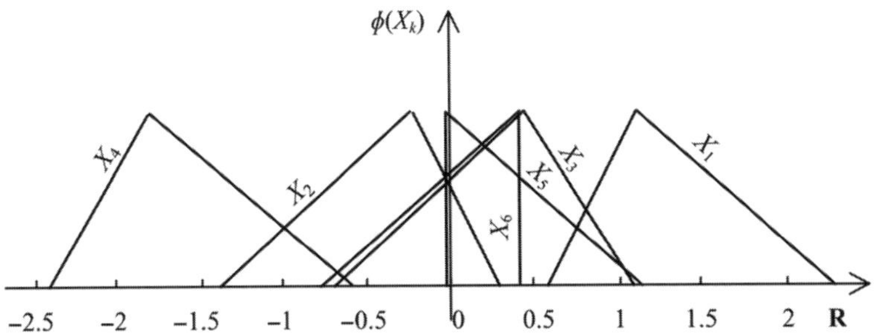

Figure 7.32 Fuzzy net flow.

Table 7.5 Defuzzified net ϕ_{defuzz} flows

	X_1	X_2	X_3	X_4	X_5	X_6
ϕ_{defuzz}	1.35	−0.45	0.3	−1.6	0.35	0.05

Table 7.6 Total costs (US$ thousand) of alternatives

Alternative	1	2	3	4
1	20 291	22 007	22 769	27 054
2	21 058	21 831	22 378	25 865
3	21 977	22 749	23 098	24 276

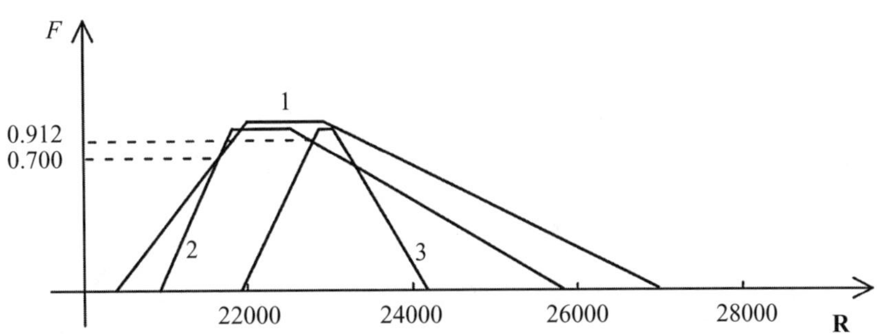

Figure 7.33 Total costs of alternatives.

Tortelly (1994), but does not give grounds for proceeding with a convincing decision. This may also be illustrated by analyzing a fuzzy nonstrict preference relation

$$\boldsymbol{R}_1 = \begin{bmatrix} 1 & 1 & 1 \\ 1 & 1 & 1 \\ 1 & 0.912 & 1 \end{bmatrix} \tag{7.149}$$

constructed on the basis of Figure 7.33. Applying (5.35) to (7.149), we obtain the following fuzzy strict preference relation:

$$\boldsymbol{P} = \begin{bmatrix} 0 & 0 & 0 \\ 0 & 0 & 0.088 \\ 0 & 0 & 0 \end{bmatrix} \tag{7.150}$$

Using (7.54), we can obtain the membership function of the fuzzy set of nondominated alternatives

$$\mathbf{ND} = [\,1 \quad 1 \quad 0.912\,] \tag{7.151}$$

which indicates that the alternatives X_1 and X_2 are indistinguishable.

Taking this into account, we can consider the indices "*Flexibility of development*" and "*Damage to agriculture*" as additional criteria denoted by $F_2(X_k)$ and $F_3(X_k)$. The membership functions corresponding to the normalized fuzzy values $S(F) =$ {*very small, small, middle, large, very good*} of the linguistic variables *Flexibility of development* and *Damage to agriculture*, which can be used to estimate $F_2(X_k)$ and $F_3(X_k)$, are illustrated in Figure 7.34.

Assume that the alternatives have received the following estimates: $F_2(X_1) =$*large*, $F_2(X_2) =$*large*, $F_2(X_3) =$*very large* for the second criterion and $F_3(X_1) =$*small*, $F_3(X_2) =$*middle*, $F_3(X_3) =$*large* for the third criterion. Taking this into account as well as the need to maximize $F_2(X_k)$ and to minimize $F_3(X_k)$, it is possible to construct the matrices of

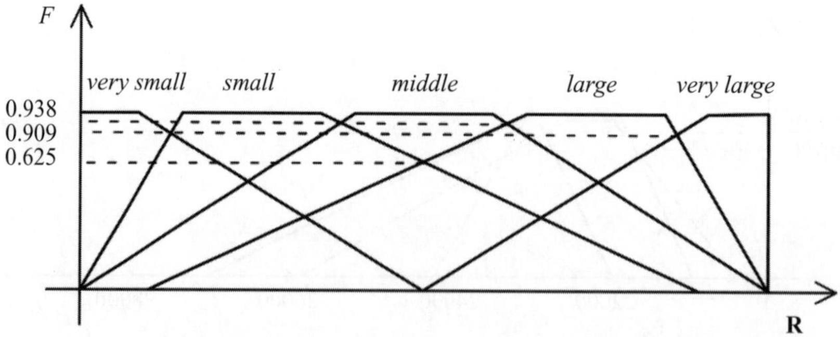

Figure 7.34 Membership functions for normalized fuzzy values.

the fuzzy nonstrict preference relations

$$R_2 = \begin{bmatrix} 1 & 1 & 0.909 \\ 1 & 1 & 0.909 \\ 1 & 1 & 1 \end{bmatrix} \tag{7.152}$$

and

$$R_3 = \begin{bmatrix} 1 & 1 & 1 \\ 0.938 & 1 & 1 \\ 0.625 & 0.938 & 1 \end{bmatrix} \tag{7.153}$$

for the second and third criterion, respectively.

Applying the *first technique*, we obtain the intersection of (7.149), (7.152), and (7.153) as follows:

$$R = \begin{bmatrix} 1 & 1 & 0.909 \\ 0.938 & 1 & 0.909 \\ 0.625 & 0.912 & 1 \end{bmatrix} \tag{7.154}$$

Applying (5.35) to (7.154), we construct the fuzzy strict preference relation

$$P = \begin{bmatrix} 0 & 0.062 & 0.274 \\ 0 & 0 & 0 \\ 0 & 0.03 & 0 \end{bmatrix} \tag{7.155}$$

which permits us, using (7.54), to obtain the membership function of the fuzzy set of nondominated alternatives

$$\mathbf{ND} = [\,1 \quad 0.938 \quad 0.716\,] \tag{7.156}$$

The alternative 1 has the maximum degree of nondominance and it is natural to consider it as the solution, that is, $\mathbf{X}^{ND} = \{X_1\}$. Thus, we have obtained the solution without applying the convolution (7.70).

Let us consider the *second technique* with the criteria being arranged in the following order of importance: $p = 1, p = 2$, and $p = 3$.

Following (5.35), (7.54), and (7.55), we obtain, on the basis of (7.149), the result coinciding with (7.151). This is obvious, and $\mathbf{X}^1 = \{X_1, X_2\}$. Thus, the alternatives X_1 and X_2 may be considered for a sequent analysis, and from (7.152) we can proceed with the second step

$$R^2 = \begin{bmatrix} 1 & 1 \\ 1 & 1 \end{bmatrix} \tag{7.157}$$

which leads to

$$\mathbf{ND}^2 = [1 \quad 1] \tag{7.158}$$

and $\mathbf{X}^2 = \{X_1, X_2\}$. The second step does not allow us to narrow down the decision uncertainty region.

From (7.153), we can proceed with the third step

$$\boldsymbol{R} = \begin{bmatrix} 1 & 1 \\ 0.938 & 1 \end{bmatrix} \tag{7.159}$$

Then

$$\mathbf{ND}^3 = [1 \quad 0.938] \tag{7.160}$$

and $\mathbf{X}^3 = \{X_1\}$.

If the criteria were arranged in another order, one can anticipate that it would be possible to obtain another solution. For instance, it is not difficult to verify that if $p = 2$, $p = 3$, and $p = 1$, then $\mathbf{X}^3 = \{X_3\}$.

Finally, let us consider the application of the *third technique*. The membership function of the set of nondominated alternatives for the first fuzzy preference relation $\mathbf{ND}_1 = \{X_k\}$ is as given by (7.151). Using (7.152), we construct

$$\mathbf{ND_2} = [0.909 \quad 0.909 \quad 1] \tag{7.161}$$

In an analogous way, the use of (7.153) leads to

$$\mathbf{ND}_3 = [1 \quad 0.938 \quad 0.625] \tag{7.162}$$

The intersection of (7.151), (7.161), and (7.162) gives rise to

$$\mathbf{ND} = [0.909 \quad 0.909 \quad 0.625] \tag{7.163}$$

and this implies that $\mathbf{X}^{ND} = \{X_1, X_2\}$.

In this manner, the *third technique* does not permit one to choose a unique alternative. It allows only the exclusion of alternative X_3 from further consideration: information given by (7.151), (7.161), and (7.162) is not sufficient to choose a unique alternative.

Thus, the *first technique* allows one to choose the alternative X_1. The *second technique* indicates the alternative X_1 (for the order of importance: $p = 1, p = 2$, and $p = 3$) as well. The *third technique* only permits one to eliminate the alternative X_3.

The second example below demonstrates the applicability of the results presented in Section 7.3 to analyze problems with nonfuzzy preference relations as well.

Example 7.13. The direct development of an integrated project by an enterprise presents considerable difficulties. There exist several ways around these difficulties:

1. Training proper professionals.
2. Inviting new professionals capable of developing a project.
3. Contracting another enterprise with the necessary profile.

The decision made by a manager is to be based on applying the following criteria:

1. Financial expenditures
2. Project development quality
3. Project development duration.

Thus, we have three alternatives X_1, X_2, and X_3, which are to be analyzed from the point of view of these three criteria. Let us apply the *first technique*, which is given in Section 7.3, by taking into account, if necessary, the following importance factors: $\lambda_1 = 0.6$, $\lambda_2 = 0.2$, and $\lambda_3 = 0.2$.

Assume that the manager thinks that X_1 is as good as X_2 and that X_1 is extremely better than X_3 from the point of view of the first criterion. These judgments permit the following nonstrict preference relation to be constructed:

$$\boldsymbol{R}_1 = \begin{bmatrix} 1 & 1 & 1 \\ 1 & 1 & 0 \\ 0 & 0 & 1 \end{bmatrix} \tag{7.164}$$

The preferences expressed from the point of view of the second criterion are the following: X_2 is extremely better than X_1 and X_3 is extremely better than X_1. This allows the construction of the second nonstrict preference relation

$$\boldsymbol{R}_2 = \begin{bmatrix} 1 & 0 & 0 \\ 1 & 1 & 0 \\ 1 & 0 & 1 \end{bmatrix} \tag{7.165}$$

Finally, the preferences from the point of view of the third criterion are presented as follows: X_1 is as good as X_2 and X_3 is extremely better than X_1. This allows construction of the third nonstrict preference relation

$$\boldsymbol{R}_3 = \begin{bmatrix} 1 & 1 & 0 \\ 1 & 1 & 0 \\ 1 & 0 & 1 \end{bmatrix} \tag{7.166}$$

The intersection of (7.164)–(7.166) leads to the formation of the following nonstrict preference relation:

$$\boldsymbol{R} = \begin{bmatrix} 1 & 0 & 0 \\ 1 & 1 & 0 \\ 0 & 0 & 1 \end{bmatrix} \tag{7.167}$$

With the use of (5.35), the following strict preference relation is derived from (7.167):

$$\boldsymbol{P} = \begin{bmatrix} 0 & 0 & 0 \\ 1 & 0 & 0 \\ 0 & 0 & 0 \end{bmatrix} \tag{7.168}$$

It also provides

$$\mathbf{ND} = [0 \quad 1 \quad 1] \tag{7.169}$$

on the basis of (7.54). In such a way, we can bring the convolution (7.70) into consideration by applying the importance factors given above as follows:

$$\boldsymbol{T} = 0.6\begin{bmatrix} 1 & 1 & 1 \\ 1 & 1 & 0 \\ 0 & 0 & 1 \end{bmatrix} + 0.2\begin{bmatrix} 1 & 0 & 0 \\ 1 & 1 & 0 \\ 1 & 0 & 1 \end{bmatrix} + 0.2\begin{bmatrix} 1 & 1 & 0 \\ 1 & 1 & 0 \\ 1 & 0 & 1 \end{bmatrix} = \begin{bmatrix} 1 & 0.8 & 0.6 \\ 1 & 1 & 0 \\ 0.4 & 0 & 1 \end{bmatrix} \tag{7.170}$$

It is not difficult to see that (7.170) generates the following strict preference relation:

$$\boldsymbol{P} = \begin{bmatrix} 0 & 0 & 0.2 \\ 0.2 & 0 & 0 \\ 0 & 0 & 0 \end{bmatrix} \tag{7.171}$$

The membership function of the set of nondominated alternatives, which corresponds to (7.171), is as follows:

$$\mathbf{NDT} = [0.8 \quad 1 \quad 0.8] \tag{7.172}$$

Finally, the intersection of (7.169) and (7.172), realized in accordance with (7.71), leads to

$$\boldsymbol{Q} = [0 \quad 1 \quad 0.8] \tag{7.173}$$

Given this, we define $\mathbf{X}^{ND} = \{X_2\}$, meaning that a suitable alternative is to recommend inviting new professionals capable of developing the project.

The third example demonstrates the analysis of $\langle \mathbf{X}, \boldsymbol{R} \rangle$ models with fuzzy ordering of criteria.

Example 7.14. The problem of choosing a local reactive power source at a power system bus with reactive power shortage is considered in Orudjev (1983). The following alternatives are considered:

1. Controlled thyristor reactor with constantly connected capacitor banks.
2. Controlled thyristor reactor with capacitor banks connected through the reactor.
3. Synchronous compensator.
4. Capacitor banks with smooth thyristor control.

The decision is to be made on the basis of applying the following criteria:

1. Reliability
2. Investment
3. Speed of control.

The fuzzy nonstrict preference relations corresponding respectively to the first, second, and third criteria are as follows:

$$\boldsymbol{R}_1 = \begin{bmatrix} 1 & 0.7 & 0.4 & 0.8 \\ 0 & 1 & 0.2 & 1 \\ 0.5 & 0.3 & 1 & 0.1 \\ 0.8 & 0.4 & 0.2 & 1 \end{bmatrix} \tag{7.174}$$

$$\boldsymbol{R}_2 = \begin{bmatrix} 1 & 0.1 & 0.5 & 0.8 \\ 0 & 1 & 0.8 & 0.6 \\ 0.7 & 0.4 & 1 & 0.7 \\ 0.4 & 0.8 & 0.2 & 1 \end{bmatrix} \tag{7.175}$$

$$\boldsymbol{R}_3 = \begin{bmatrix} 1 & 0.9 & 0.12 & 0.3 \\ 0.3 & 1 & 0.8 & 0.5 \\ 0.3 & 0.15 & 1 & 0.7 \\ 0.9 & 0.6 & 0.2 & 1 \end{bmatrix} \tag{7.176}$$

The information related to the importance of criteria is presented in the following form:

$$\Lambda = \begin{bmatrix} 1 & 0.8 & 0.6 \\ 0.5 & 1 & 0.7 \\ 0.2 & 0.1 & 1 \end{bmatrix} \tag{7.177}$$

The application of (5.35) to (7.174)–(7.176) results in the construction of the corresponding fuzzy strict preference relations and, then, the use of (7.54) generates the following membership functions of the fuzzy sets of nondominated alternatives:

$$\mathbf{ND}_1 = [0.9 \quad 0.3 \quad 0.9 \quad 0.4] \tag{7.178}$$

$$\mathbf{ND}_2 = [0.8 \quad 0.8 \quad 0.6 \quad 0.5] \tag{7.179}$$

and

$$\mathbf{ND}_3 = [0.4 \quad 0.4 \quad 0.35 \quad 0.5] \tag{7.180}$$

for the first, second, and third criterion, respectively.

The application of (7.93) to process (7.178)–(7.180) together with (7.177) leads to the relation

$$\boldsymbol{R}_\Lambda = \begin{bmatrix} 1 & 0.8 & 0.9 & 0.5 \\ 0.8 & 0.9 & 0.5 & 0.5 \\ 0.9 & 0.8 & 0.9 & 0.8 \\ 0.5 & 0.5 & 0.5 & 0.5 \end{bmatrix} \tag{7.181}$$

The application of (5.35) to (7.181) to construct the corresponding fuzzy strict preference relations and, then, the use of (7.54) form the following membership function of the fuzzy set of nondominated alternatives:

$$\bar{R}_{\Lambda}^{ND}(X_k) = [\,1 \quad 0.7 \quad 1 \quad 0.7\,] \tag{7.182}$$

Finally, the application of (7.94), considering that

$$\boldsymbol{R}_{\Lambda}(X_k, X_k) = [\,1 \quad 0.9 \quad 0.9 \quad 0.5\,] \tag{7.183}$$

leads to

$$\mathbf{ND}_{\Lambda} = [\,1 \quad 0.7 \quad 0.9 \quad 0.5\,] \tag{7.184}$$

Thus, a controlled thyristor reactor with constantly connected capacitor banks should be selected.

The last example below demonstrates the analysis of $\langle \mathbf{X}, \boldsymbol{R} \rangle$ models by applying the concept of majority.

Example 7.15. The multicriteria decision-making problem related to energy planning is considered in Beccali, Cellura, and Ardente (1998). It consists of the selection of the most appropriate technology in a renewable energy diffusion plan for Sardinia. Here, we will consider the following six criteria:

1. Targets of primary energy saving on a regional scale.
2. Sustainability according to greenhouse pollutant emissions.
3. Consistency of installation and maintenance requirements with local technical conditions.
4. Continuity and predictability of performances.
5. Market maturity.
6. Compatibility with the political, legislative, and administrative situation.

The first two criteria are of a quantitative character. The third, fourth, and sixth criteria are of a qualitative character and can be evaluated through the set of three normalized fuzzy values $S(F) = \{$*small, middle, large*$\}$ shown in Figure 7.35. The fifth criterion is also of a qualitative character and can be evaluated through the set of five normalized fuzzy values $S(F) = \{$*very small, small, middle, large, very large*$\}$ shown in Figure 7.35 as well.

The following four options of energy sources are considered as alternatives:

1. Solar (domestic solar water heaters).
2. Wind (wind turbines of grid-connected type).
3. Hydraulic (hydro plants in derivation schemes).
4. Biomass (combined heat and power plants fed by agricultural wastes or energy crops).

Given the evaluation matrix of the alternatives (Table 7.7), it is necessary to transform the quantitative measures related to criteria $F_1(X_k)$ and $F_2(X_k)$ and the qualitative estimates related to criteria $F_p(X_k)$, $p = 3, 4, \ldots, 6$, into the corresponding nonstrict preference relations.

To normalize the values of $F_1(X_k)$, we can apply (4.6) by taking into account that $F_1(X_k)$ is to be maximized, with $\max F_1(X_k) = 3000$ and $\min F_1(X_k) = 500$. This gives the following

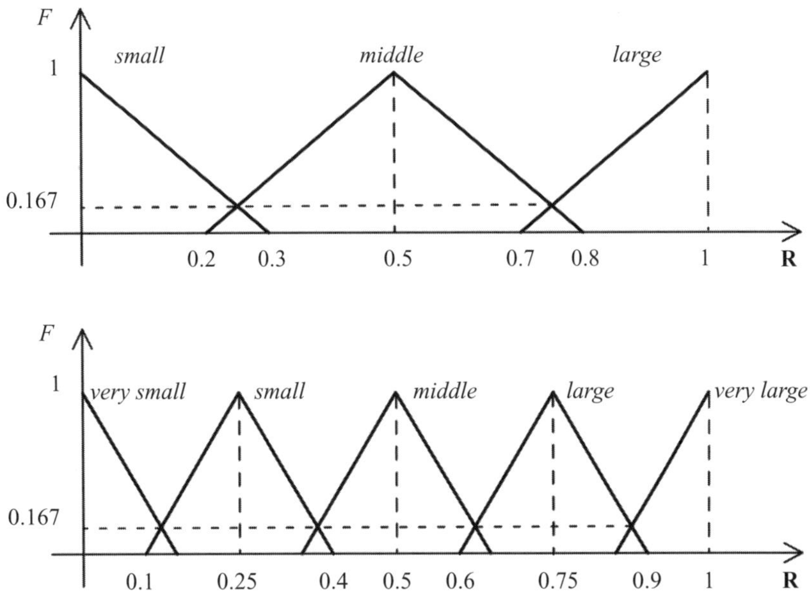

Figure 7.35 Membership functions of normalized fuzzy values.

results: 0.30, 0.92, 0.03, and 0.55 for X_1, X_2, X_3, and X_4, respectively. Applying the transformation function (6.58), we obtain the following fuzzy nonstrict preference relation for the first criterion:

$$\boldsymbol{R}_1 = \begin{bmatrix} 1 & 0.106 & 1 & 0.297 \\ 1 & 1 & 1 & 1 \\ 0.01 & 0.001 & 1 & 0.003 \\ 1 & 0.357 & 1 & 1 \end{bmatrix} \tag{7.185}$$

In a similar manner, to normalize the values of $F_2(X_k)$, we can apply (4.5) by taking into account that $F_2(X_k)$ is to be minimized, with $\max F_2(X_k) = 70$ and $\min F_2(X_k) = 40$. This gives the following results: 0.70, 0.73, 0.73, and 0.44 for X_1, X_2, X_3, and X_4, respectively. Also applying the transformation function (6.58), we obtain the following nonstrict preference

Table 7.7 Evaluation matrix of the alternatives

Alternative	$F_1(X_k)$ (TJ/year)	$F_2(X_k)$ (g CO_2/MJ)	$F_3(X_k)$	$F_4(X_k)$	$F_5(X_k)$	$F_6(X_k)$
X_1	1255	49	*large*	*small*	*very large*	*small*
X_2	2790	48	*small*	*small*	*large*	*small*
X_3	574	48	*middle*	*large*	*very large*	*large*
X_4	1884	56.7	*small*	*large*	*large*	*middle*

relation for the second criterion:

$$\boldsymbol{R}_2 = \begin{bmatrix} 1 & 0.919 & 0.919 & 1 \\ 1 & 1 & 1 & 1 \\ 1 & 1 & 1 & 1 \\ 0.395 & 0.363 & 0.363 & 1 \end{bmatrix} \tag{7.186}$$

The fuzzy nonstrict preference relations for criteria of a qualitative character are as follows:

$$\boldsymbol{R}_3 = \begin{bmatrix} 1 & 1 & 1 & 1 \\ 0 & 1 & 0.167 & 1 \\ 0.167 & 1 & 1 & 1 \\ 0 & 1 & 0.167 & 1 \end{bmatrix} \tag{7.187}$$

$$\boldsymbol{R}_4 = \begin{bmatrix} 1 & 1 & 0 & 0 \\ 1 & 1 & 0 & 0 \\ 1 & 1 & 1 & 1 \\ 1 & 1 & 1 & 1 \end{bmatrix} \tag{7.188}$$

$$\boldsymbol{R}_5 = \begin{bmatrix} 1 & 1 & 1 & 1 \\ 0.167 & 1 & 0.167 & 1 \\ 1 & 1 & 1 & 1 \\ 0.167 & 1 & 0.167 & 1 \end{bmatrix} \tag{7.189}$$

$$\boldsymbol{R}_6 = \begin{bmatrix} 1 & 1 & 0 & 0.167 \\ 1 & 1 & 0 & 0.167 \\ 1 & 1 & 1 & 1 \\ 1 & 1 & 0.167 & 1 \end{bmatrix} \tag{7.190}$$

The global nonstrict preference relation, obtained on the basis of (7.113) and considering the fuzzy quantifier "most", is given by

$$\boldsymbol{R} = \begin{bmatrix} 1 & 0.978 & 0.706 & 0.543 \\ 0.778 & 1 & 0.178 & 0.778 \\ 0.778 & 1 & 1 & 1 \\ 0.576 & 0.830 & 0.288 & 1 \end{bmatrix} \tag{7.191}$$

The guided quantifier nondominance degree is obtained here also using the fuzzy quantifier "most":

$$\text{QGNDD}(X_k) = [0.959 \quad 0.643 \quad 1 \quad 0.810] \tag{7.192}$$

Thus, the ranking of alternatives is $X_3 \succ X_1 \succ X_4 \succ X_2$. But, particularly, in a further analysis, it is interesting to observe, in Table 7.7, that although alternative X_3 has the best evaluation for the second, fourth, fifth, and sixth criteria, it has the worst evaluation for the first criterion. Does X_3 remain in first position if the DM requires all criteria to be highly satisfied and utilizes the fuzzy quantifier "all" instead of "most"? The reader is invited to analyze this question in Problem 7.10 of the Exercises section below.

7.9 Conclusions

We have discussed the questions of the emergence and importance of problems of multicriteria evaluation, comparison, choice, prioritization, and/or ordering of alternatives. Diverse techniques for the multicriteria analysis of alternatives in a fuzzy environment developed on the basis of fuzzy preference modeling have been considered. The *first technique* is related to constructing and analyzing the membership function of the set of nondominated alternatives simultaneously considering all criteria. The *second technique* is of a lexicographic character and consists of a step-by-step analysis of preference relations. The *third technique* is based on building and aggregating the membership functions of the sets of nondominated alternatives for each preference relation. The *fourth technique* provides a way for fuzzy ordering of criteria, considering the information on their importance provided by a DM in the form of a nonreciprocal fuzzy preference relation. The *fifth technique* for the analysis of $\langle \mathbf{X}, \boldsymbol{R} \rangle$ models uses the concept of fuzzy majority as the aggregation rule for considering the criteria. Finally, a method from the French School, that is, the fuzzy version of Promethee, offers an approach for analyzing $\langle \mathbf{X}, \boldsymbol{R} \rangle$ models based on the exploitation of a specific type of fuzzy preference relation, named the outranking relation. The net flow observed in the weighted graph of outranking relations provides a measure of the effective outranking quality of each alternative. The application of the proposed techniques has been illustrated by practical examples.

The discussed techniques can lead to different solutions. However, this is to be considered natural and intuitively appealing. We must emphasize here that the choice of a specific technique is a prerogative of the DM. This selection has to be based on the essence of the problem and the possible sources of information and its uncertainty.

Exercises

Problem 7.1. Construct the membership function of the fuzzy set of nondominated alternatives for the following fuzzy nonstrict preference relation:

$$\boldsymbol{R} = \begin{bmatrix} 1 & 0.5 & 1 & 0.3 \\ 1 & 1 & 0.8 & 0.7 \\ 0.6 & 1 & 1 & 0.6 \\ 1 & 0.9 & 0.4 & 1 \end{bmatrix}$$

Problem 7.2. Verify the existence of the nonfuzzy solution in the decision-making problem described by the following fuzzy nonstrict preference relation:

$$\boldsymbol{R} = \begin{bmatrix} 1 & 0.6 & 1 & 0.5 \\ 0.4 & 1 & 0.8 & 0.2 \\ 1 & 0.8 & 1 & 0.3 \\ 0.9 & 0.9 & 0.7 & 1 \end{bmatrix}$$

Problem 7.3. Apply the *first technique* for analyzing $\langle \mathbf{X}, \boldsymbol{R} \rangle$ models to solve the problem which includes the following fuzzy nonstrict preference relations:

$$\boldsymbol{R}_1 = \begin{bmatrix} 1 & 1 & 1 \\ 1 & 1 & 1 \\ 0.8 & 0.8 & 1 \end{bmatrix}$$

$$\boldsymbol{R}_2 = \begin{bmatrix} 1 & 0.8 & 0.6 \\ 1 & 1 & 0.8 \\ 1 & 1 & 1 \end{bmatrix}$$

$$\boldsymbol{R}_3 = \begin{bmatrix} 1 & 1 & 1 \\ 0.8 & 1 & 1 \\ 0.8 & 1 & 1 \end{bmatrix}$$

considering, if necessary, the following: $\lambda_1 = 0.4$, $\lambda_2 = 0.3$, and $\lambda_3 = 0.3$.

Problem 7.4. Apply the *third technique* for analyzing $\langle \mathbf{X}, \boldsymbol{R} \rangle$ models to solve the problem which includes the fuzzy nonstrict preference relations given in Problem 7.3 by considering, if necessary, $\lambda_1 = 0.5$, $\lambda_2 = 0.3$, and $\lambda_3 = 0.2$.

Problem 7.5. Apply the *second technique* for analyzing $\langle \mathbf{X}, \boldsymbol{R} \rangle$models to solve the problem which includes the fuzzy nonstrict preference relations (7.149), (7.152), and (7.153) with the following order of their importance: $p = 3$, $p = 2$, and $p = 1$.

Problem 7.6. Solve the problem formulated in Example 7.12 if the alternatives have the estimates

$$\begin{array}{lll} F_1(X_1) = \textit{middle}, & F_1(X_2) = \textit{middle}, & \text{and } F_1(X_3) = \textit{large} \\ F_2(X_1) = \textit{small}, & F_2(X_2) = \textit{middle}, & \text{and } F_2(X_3) = \textit{large} \\ F_3(X_1) = \textit{large}, & F_3(X_2) = \textit{middle}, & \text{and } F_3(X_3) = \textit{middle} \end{array}$$

from the normalized fuzzy values of Figure 7.7.

Problem 7.7. Verify the possibility of changing the solution (alternative X_1) of the problem defined by Example 7.12, if the information related to the importance of criteria is presented in the following form:

$$\Lambda = \begin{bmatrix} 1 & 0.5 & 0.2 \\ 0.8 & 1 & 0.1 \\ 0.6 & 0.7 & 1 \end{bmatrix}$$

Problem 7.8. Given the set of normalized values $\{0.5, 0.1, 0.2, 0.7\}$, obtain the set of weights associated with the fuzzy quantifier "at least three" (it is illustrated in Figure 7.22) to be utilized by OWA. Calculate the aggregated value.

Table 7.8 Evaluation of $F_4(X_k)$

	$F_4(X_k)$ (min)
X_1	(18, 22, 24)
X_2	(12, 17, 21)
X_3	(8, 12, 16)
X_4	(8, 10, 16)
X_5	(4, 7, 10)
X_6	(2, 5, 6)

Problem 7.9. Examine Examples 7.5 and 7.10 and explain why (7.129) coincides with (7.81).

Problem 7.10. Solve the problem analyzed in Example 7.15, but now use the fuzzy quantifier "all" instead of "most". Does X_3 remain in first position in the new ranking? Try to explain why it remains (or not) in the same position, taking into account the evaluation matrix in Table 7.7.

Problem 7.11. Solve the problem considered in Example 7.11, but now consider the fuzzy estimates for the criterion $F_4(X_k)$ given in Table 7.8.

References

Banerjee, A. (1993) Rational choice under fuzzy preferences: the Orlovsky choice function. *Fuzzy Sets and Systems*, **54** (3), 295–299.

Barrett, C.R., Patanalk, P.K., and Salles, M. (1990) On choosing rationally when preferences are fuzzy. *Fuzzy Sets and Systems*, **34** (2), 197–212.

Beccali, M., Cellura, M., and Ardente, D. (1998) decision making in energy planning: the ELECTRE multicriteria analysis approach compared to a fuzzy-sets methodology. *Energy Conversion and Management*, **39** (16–18), 1869–1881.

Berredo, R.C., Ekel, P.Ya., Galperin, E.A., and Sant'anna, A.S. (2005) Fuzzy preference modeling and its management applications. Proceedings of the International Conference on Industrial Logistics, Montevideo, pp. 41–50.

Botter, R.C. and Ekel, P.Ya. (2005) Fuzzy preference relations and their naval engineering applications. Proceedings of the XIX Congress of Pan-American Institute of Naval Engineering, Guayaquil, Paper 7–8.

Bouyssou, D. (1997) Acyclic fuzzy preference and the Orlovski choice function: a note. *Fuzzy Sets and Systems*, **89** (1), 107–111.

Brans, J.P. and Vincke, Ph. (1985) A preference ranking organization method: the Promethee method for multiple criteria decision making. *Management Science*, **31** (6), 647–656.

Canha, L., Ekel, P., Queiroz, J., and Schuffner Neto, F. (2007) Models and methods of decision making in fuzzy environment and their applications to power engineering problems. *Numerical Linear Algebra with Applications*, **14** (3), 369–390.

Chiclana, F., Herrera, F., and Herrera-Viedma, E. (1998) Integrating three representation models in fuzzy multipurpose decision making based on fuzzy preference relations. *Fuzzy Sets and Systems*, **97** (1), 33–48.

Chiclana, F., Herrera, F., Herrera-Viedma, E., and Poyatos, M.C. (1996) A classification method of alternatives for multiple preference ordering criteria based on fuzzy majority. *Journal of Fuzzy Mathematics*, **4** (4), 801–813.

Delgado, M., Kacprzyk, J., Verdegay, J.L., and Vila, M.A. (eds) (1994) *Fuzzy Optimization: Recent Advances*, Physica-Verlag, Heidelberg.

Dubois, D. and Prade, H. (1980) *Fuzzy Sets and Systems: Theory and Applications*, Academic Press, New York.

Ekel, P., Pedrycz, W., and Schinzinger, R. (1998) A general approach to solving a wide class of fuzzy optimization problems. *Fuzzy Sets and Systems*, **97** (1), 49–66.

Ekel, P.Ya. (2001) Methods of decision making in fuzzy environment and their applications. *Nonlinear Analysis: Theory, Methods and Applications*, **47** (5), 979–990.

Ekel, P.Ya. (2002) Fuzzy sets and models of decision making. *Computers and Mathematics with Applications*, **44** (7), 863–875.

Ekel, P.Ya. and Schuffner Neto, F.H. (2006) Algorithms of discrete optimization and their application to problems with fuzzy coefficients. *Information Sciences*, **176** (19), 2846–2868.

Ekel, P.Ya., Martini, J.S.C., and Palhares, R.M. (2008) Multicriteria analysis in decision making under information uncertainty. *Applied Mathematics and Computation*, **200** (2), 501–516.

Fodor, J. and Roubens, M. (1994) *Fuzzy Preference Modelling and Multicriteria Decision Support*, Kluwer, Boston, MA.

Fontoura Filho, R.N., Ales, J.C.O., and Tortelly, D.L.S. (1994) Uncertainty models applied to the substation planning. Technical Papers of the 4th International Conference on Probabilistic Methods Applied to Power Systems, Rio de Janeiro.

Galperin, E.A. and Ekel, P.Ya. (2005) Synthetic realization approach to fuzzy global optimization via gamma algorithm. *Mathematical and Computer Modelling*, **41** (13), 1457–1468.

Geldermann, J., Spengler, T., and Rentz, O. (2000) Fuzzy outranking for environmental assessment. Case study: iron and steel making industry. *Fuzzy Sets and Systems*, **115** (1), 45–65.

Goumas, M. and Lygerou, V. (2000) An extension of the Promethee method for decision making in fuzzy environment: ranking of alternative energy exploitation projects. *European Journal of Operational Research*, **123** (3), 606–613.

Grabisch, M., Orlovski, S.A., and Yager, R.R. (1998) Fuzzy aggregation of numerical preferences, in *Fuzzy Sets in Decision Analysis, Operations Research and Statistics*, The Handbooks of Fuzzy Sets Series, Vol. **4**, Kluwer, Boston, MA, pp. 31–68.

Halouani, N., Chabchoub, H., and Martel, J.M. (2009) Promethee-MD-2T method for project selection. *European Journal of Operational Research*, **195** (3), 841–849.

Negoita, C.V. and Ralescu, D.A. (1975) *Application of Fuzzy Sets to Systems Analysis*, Birkhäuser, Basle.

Orlovski, S.A. (1978) Decision making with a fuzzy preference relation. *Fuzzy Sets and Systems*, **1** (3), 155–167.

Orlovsky, S.A. (1981) *Problems of Decision Making with Fuzzy Information* , Nauka, Moscow (in Russian).

Orudjev, F.D. (1983) Expert estimates and fuzzy set theory in investigating electrical systems. *Elektrichestvo*, **85** (4), 7–11 (in Russian).

Pedrycz, W. and Gomide, F. (1998) *An Introduction to Fuzzy Sets: Analysis and Design*, MIT Press, Cambridge, MA.

Rao, S. (1996) *Engineering Optimization: Theory and Practice*, John Wiley & Sons, Inc., New York.

Roy, B. (1968) Classement et choi en presence de points de vue multiples (la method ELECTRE). *RIRO*, **8** (1), 57–75.

Roy, B. (1991) The outranking approach and the foundations of Electre methods. *Theory and Decision*, **31** (1), 49–73.

Sengupta, K. (1998) Fuzzy preference and Orlovsky choice procedure. *Fuzzy Sets and Systems*, **93** (2), 231–234.

Vahidnia, M.H., Alesheikh, A.A., and Alimohammadi, A. (2009) Hospital site selection using fuzzy AHP and its derivatives. *Journal of Environmental Management*, **90** (10), 3048–3056.

Wei-xiang, L. and Bang-yi, L. (2010) An extension of the Promethee II method based on generalized fuzzy numbers. *Expert Systems with Applications*, **37** (7), 5314–5319.

Yager, R.R. (1988) On ordered weighted averaging operators in multicriteria decision making. *IEEE Transactions on Systems, Man, and Cybernetics*, **18** (3), 183–190.

Yager, R.R. (1995) Multicriteria decision making using fuzzy quantifiers. Proceedings of the IEEE Conference on Computational Intelligence for Financial Engineering, New York, pp. 42–47.

Zimmermann, H.J. (1996) *Fuzzy Set Theory and Its Application*, Kluwer, Boston, MA.

Zimmermann, H.J. (2008) *Fuzzy Sets, Decision Making, and Expert Systems*, Kluwer, Boston, MA.

Zorin, V.V. and Ekel, P.Ya. (1980) Discrete-optimization methods for electrical supply systems. *Power Engineering*, **18** (5), 19–30.

8

Generalization of a Classic Approach to Dealing with Uncertainty of Information for Multicriteria Decision Problems

This chapter focuses on the generalization of the classic approach to dealing with uncertainty of information (based on constructing and analyzing payoff matrices defining effects, which result from the occurrence of different combinations of solution alternatives and different states of nature) in monocriteria decision-making for multicriteria problems. The questions of constructing aggregated payoff matrices, modifying the choice criteria, and evaluating particular and aggregated risks (regrets) in decision-making are discussed. The generalization of the classic approach as well as the use of the analysis of $\langle \mathbf{X}, \mathbf{M} \rangle$ and $\langle \mathbf{X}, \boldsymbol{R} \rangle$ models allows us to construct a general scheme of multicriteria decision-making in the presence of uncertainty of information. This scheme is directed at using the available quantitative information to the highest extent to reduce the decision uncertainty regions. If a resolving capability of the processing of quantitative information does not lead to unique solutions, the scheme presumes the application of qualitative information based on the knowledge, experience, and intuition of experts involved in the decision-making process.

8.1 Classic Approach to Dealing with Uncertainty of Information

The classic approach (Luce and Raiffa, 1957; Raiffa, 1968; Webster, 2003) encountered when considering uncertainty, which comes with a broad range of practical applications (Kaufman, 1961; Belyaev, 1977), is based on the assumption that the underlying analysis is carried out for the given solution alternatives (strategies) X_k, $k = 1, 2, \ldots, K$, and the given representative combinations of initial data or states of nature (scenarios) Y_s, $s = 1, 2, \ldots, S$. Making use of alternatives and scenarios, we associate with them the corresponding payoff matrix in the form shown in Table 8.1.

Fuzzy Multicriteria Decision-Making: Models, Methods and Applications Witold Pedrycz, Petr Ekel and Roberta Parreiras

Table 8.1 Payoff matrix

	Y_1	...	Y_s	...	Y_S
X_1	$F(X_1, Y_1)$	...	$F(X_1, Y_s)$	...	$F(X_1, Y_S)$
...	...	...	...	...	...
X_k	$F(X_k, Y_1)$	...	$F(X_k, Y_s)$	...	$F(X_k, Y_S)$
...	...	...	...	...	...
X_K	$F(X_K, Y_1)$	...	$F(X_K, Y_s)$	...	$F(X_K, Y_S)$

The payoff matrix quantifies the effects (or consequences) of actions X_k, $i = 1, 2, \ldots, K$, for the corresponding states of nature Y_s, $s = 1, 2, \ldots, S$.

Belyaev (1977) presented the following basic phases to support the application of the classic approach when dealing with the uncertainty factor:

- mathematical formulation of the problem;
- selection of the representative combinations of initial data (selection of the states of nature);
- determination and preliminary analysis of solution alternatives;
- construction of the payoff matrix;
- analysis of the payoff matrix and the choice of the rational solution alternatives;
- selection of the final solution.

Without discussing in detail the phases identified above, it is worth emphasizing that an analysis of the payoff matrices and the choice of the rational solution alternatives are based on the use of the corresponding choice criteria (Luce and Raiffa, 1957; Raiffa, 1968; Webster, 2003). Those utilized most frequently and exhibiting a general character are the criteria of Wald, Laplace, Savage, and Hurwicz. To better understand the nature of these criteria, the matrix presented in Table 8.1 is extended and shown in Table 8.2, in which we take into account recommendations presented by Belyaev (1977). This extension is associated with the incorporation of the following estimates:

- The objective function maximum level described as

$$F^{\max}(X_k) = \max_{1 \leq s \leq S} F(X_k, Y_s) \tag{8.1}$$

This level is determined for the given solution alternative and, as the name stipulates, is the most optimistic estimate when the objective function is to be maximized or the

Table 8.2 Payoff matrix with characteristic estimates

	Y_1	...	Y_s	...	Y_S	$F^{\max}(X_k)$	$F^{\min}(X_k)$	$\overline{F}(X_k)$	$r^{\max}(X_k)$
X_1	$F(X_1, Y_1)$	...	$F(X_1, Y_s)$	...	$F(X_1, Y_S)$	$F^{\max}(X_1)$	$F^{\min}(X_1)$	$\overline{F}(X_1)$	$r^{\max}(X_1)$
...	...	...	...	...	...	...	...	...	...
X_k	$F(X_k, Y_1)$	...	$F(X_k, Y_s)$	...	$F(X_k, Y_S)$	$F^{\max}(X_k)$	$F^{\min}(X_k)$	$\overline{F}(X_k)$	$r^{\max}(X_k)$
...	...	...	...	...	...	...	...	...	...
X_K	$F(X_K, Y_1)$	...	$F(X_K, Y_s)$	...	$F(X_K, Y_S)$	$F^{\max}(X_K)$	$F^{\min}(X_K)$	$\overline{F}(X_K)$	$r^{\max}(X_K)$
$F^{\max}(Y_s)$	$F^{\max}(Y_1)$	...	$F^{\max}(Y_s)$	...	$F^{\max}(Y_S)$				

most pessimistic estimate if the objective function is to be minimized for the considered solution alternative.

- The objective function minimum level

$$F^{\min}(X_k) = \min_{1 \le s \le S} F(X_k, Y_s) \tag{8.2}$$

computed for the given solution alternative. This is the most pessimistic estimate when the objective function is to be maximized or is treated as the most optimistic estimate if the objective function is to be minimized for the considered solution alternative.

- The objective function average level

$$\overline{F}(X_k) = \frac{1}{S} \sum_{s=1}^{S} F(X_k, Y_s) \tag{8.3}$$

determined for the given solution alternative.

- The risk (regret) maximum level

$$r^{\max}(X_k) = \max_{1 \le s \le S} r(X_k, Y_s) \tag{8.4}$$

where $r(X_k, Y_s)$ is an over-expenditure which takes place under a combination of the state of nature Y_s and the choice of the solution alternative X_k instead of the solution alternative that is locally optimal for the given Y_s. The estimates of over-expenditures provide a certain description of the situation as they show a relative difference in the objective function values under the choice of one solution alternative in place of another. In fact, the over-expenditures characterize a damage level associated with the uncertainty of the situation itself.

To determine risks (regrets) $r(X_k, Y_s)$, it is necessary to define the maximum value of the objective function (if it is to be maximized, as considered in Table 8.2) for each combination of the state of nature Y_s (for each column of the payoff matrix):

$$F^{\max}(Y_s) = \max_{1 \le k \le K} F(X_k, Y_s) \tag{8.5}$$

It is evident that if the objective function is to be minimized, it is necessary to define its minimum for each combination of the state of nature Y_s (for each column of the payoff matrix):

$$F^{\min}(Y_s) = \min_{1 \le k \le K} F(X_k, Y_s) \tag{8.6}$$

The risk associated with any solution alternative X_k and any state of nature Y_s can be evaluated as

$$r(X_k, Y_s) = F^{\max}(Y_s) - F(X_k, Y_s) \tag{8.7}$$

if the objective function is to be maximized or

$$r(X_k, Y_s) = F(X_k, Y_s) - F^{\min}(Y_s) \tag{8.8}$$

if the objective function is to be minimized.

Table 8.3 Risk matrix

	Y_1	...	Y_s	...	Y_S	$r^{\max}(X_k)$
X_1	$r(X_1, Y_1)$	...	$r(X_1, Y_s)$	...	$r(X_1, Y_S)$	$r^{\max}(X_1)$
...	...	...	...	...	...	...
X_k	$r(X_k, Y_1)$	...	$r(X_k, Y_s)$	...	$r(X_k, Y_S)$	$r^{\max}(X_k)$
...	...	...	...	...	...	...
X_K	$r(X_K, Y_1)$	...	$r(X_K, Y_s)$	...	$r(X_K, Y_S)$	$r^{\max}(X_K)$

Carrying out calculations based on (8.7) or (8.8) for all X_k, $k = 1, 2, \ldots, K$, and Y_s, $s = 1, 2, \ldots, S$, we obtain the risk (regret) matrix shown in Table 8.3. Note that any column of this matrix includes at least a single zero element, $r(X_k, Y_s) = 0$.

8.2 Choice Criteria

The choice criteria of Wald, Laplace, Savage, and Hurwicz are based on the use of the characteristic estimates $F^{\max}(X_k)$, $F^{\min}(X_k)$, $\overline{F}(X_k)$, and $r^{\max}(X_k)$, defined by (8.1)–(8.4). To focus our discussion, the choice criteria are represented under the assumption that the objective function is to be maximized.

The Wald criterion utilizes the estimate $F^{\min}(X_k)$ and stipulates that one has to choose the solution alternative X^W, for which this estimate attains a maximum, that is,

$$\max_{1\le k\le K} F^{\min}(X_k) = \max_{1\le k\le K}\min_{1\le s\le S} F(X_k, Y_s) \tag{8.9}$$

The Laplace criterion uses the estimate $\overline{F}(X_k)$ and is oriented to choose the solution alternative X^L, for which this estimate attains its maximum:

$$\max_{1\le k\le K} \overline{F}(X_k) = \max_{1\le k\le K}\frac{1}{S}\sum_{s=1}^{S} F(X_k, Y_s) \tag{8.10}$$

On the other hand, the Savage criterion is associated with the use of the estimate $r^{\max}(X_k)$ and allows one to choose the solution alternative X^S, for which this estimate reaches a minimum:

$$\min_{1\le k\le K} r^{\max}(X_k) = \min_{1\le k\le K}\max_{1\le s\le S} r(X_k, Y_s) \tag{8.11}$$

Finally, the Hurwicz criterion utilizes a linear combination of the estimates $F^{\min}(X_k)$ and $F^{\max}(X_k)$ and chooses the solution alternative X^H for which this combination is a maximum:

$$\begin{aligned}&\max_{1\le k\le K}[\alpha F^{\min}(X_k) + (1-\alpha)F^{\max}(X_k)]\\ &= \max_{1\le k\le K}\left[\alpha \min_{1\le s\le S} F(X_k, Y_s) + (1-\alpha)\max_{1\le s\le S} F(X_k, Y_s)\right]\end{aligned} \tag{8.12}$$

Table 8.4 Modified payoff matrix

	Y_1	...	Y_s	...	Y_S
X_1	$A_p(X_1, Y_1)$	...	$A_p(X_1, Y_s)$	...	$A_p(X_1, Y_S)$
...	...	...	...	...	...
X_k	$A_p(X_k, Y_1)$	...	$A_p(X_k, Y_s)$	...	$A_p(X_k, Y_S)$
...	...	...	...	...	...
X_K	$A_p(X_K, Y_1)$	...	$A_p(X_K, Y_s)$	...	$A_p(X_K, Y_S)$

where $\alpha \in [0, 1]$ is the "pessimism–optimism" index whose magnitude is defined in advance by a DM.

We will not proceed with a discussion of the main features of applying the considered criteria (their advantages and shortcomings are discussed, for example, in Belyaev, 1977), but instead we will try to apply them when more than one objective function is to be considered.

Before proceeding with the generalization of the classic approach to dealing with the uncertainty component in decision-making problems, it is worth noting that some models have addressed this problem (see, for instance, Yager, 1996; Kuchta, 2007; Wen and Iwamura, 2008), where fuzzy sets were discussed as a viable alternative yet the studies reported were focused on monocriteria problems.

8.3 Generalization of the Classic Approach

As shown in Chapter 4, the application of the Bellman–Zadeh approach to solve multiobjective decision-making problems provides a constructive way to derive harmonious solutions on the basis of analyzing the associated problem of a *max–min* nature. Taking this into consideration, it is possible to talk about the generalization of the classic approach on the basis of its use for maximizing (4.28) or solving the problem expressed by (4.29) (see Ekel, Martini, and Palhares, 2008; Ekel *et al.*, 2010). It is also evident that, when dealing with q objective functions, q payoff matrices are to be constructed and analyzed.

Applying (4.30) to the maximized objective functions or using (4.31) on the maximized objective functions, we can construct the modified (normalized) payoff matrix for the pth criterion in the form shown in Table 8.4.

The availability of q modified payoff matrices allows us, by applying (4.28), to construct the aggregated payoff matrix as given in Table 8.5.

Table 8.5 Aggregated payoff matrix with characteristic estimates

	Y_1	...	Y_s	...	Y_S	$D^{\max}(X_k)$	$D^{\min}(X_k)$	$\overline{D}(X_k)$	$r^{\max}(X_k)$
X_1	$D(X_1, Y_1)$	...	$D(X_1, Y_s)$	...	$D(X_1, Y_S)$	$D^{\max}(X_1)$	$D^{\min}(X_1)$	$\overline{D}(X_1)$	$r^{\max}(X_1)$
...	...	...	...	...	...	...	...	...	...
X_k	$D(X_k, Y_1)$	...	$D(X_k, Y_s)$	...	$D(X_k, Y_S)$	$D^{\max}(X_k)$	$D^{\min}(X_k)$	$\overline{D}(X_k)$	$r^{\max}(X_k)$
...	...	...	...	...	...	...	...	...	...
X_K	$D(X_K, Y_1)$	...	$D(X_K, Y_s)$	...	$D(X_K, Y_S)$	$D^{\max}(X_K)$	$D^{\min}(X_K)$	$\overline{D}(X_K)$	$r^{\max}(X_K)$
$D^{\max}(Y_s)$	$D^{\max}(Y_1)$	...	$D^{\max}(Y_s)$	...	$D^{\max}(Y_S)$				

The characteristic estimates shown there are formed as follows:

- The membership function maximum level (optimistic estimate)

$$D^{\max}(X_k) = \max_{1\le s\le S} D(X_k, Y_s) \tag{8.13}$$

- The membership function minimum level (corresponding to the pessimistic estimate)

$$D^{\min}(X_k) = \min_{1\le s\le S} D(X_k, Y_s) \tag{8.14}$$

- The membership function average level

$$\overline{D}(X_k) = \frac{1}{S}\sum_{s=1}^{S} D(X_k, Y_s) \tag{8.15}$$

- The risk maximum level, which is defined as (8.4) with $r(X_i, Y_s) = D^{\max}(Y_s) - \mu_D(X_i, Y_s)$ where $D^{\max}(Y_s) = \max_{1\le k\le K} D(X_k, Y_s)$.

In this case, it is possible to construct the aggregated risk matrix (similar to the risk matrix given in Table 8.3) as well. We observe that if the risk matrices constructed for each objective function reflect the particular risks (monocriteria risk estimates), the aggregated risk matrix reflects the aggregated risks (multicriteria risk) encountered in decision-making.

8.4 Modification of the Choice Criteria

The characteristic estimates $D^{\max}(X_k)$, $D^{\min}(X_k)$, $\overline{D}(X_k)$, and $r^{\max}(X_k)$ considered above can serve as the basis for the choice criteria which are to be used under the generalization of the classic approach (Ekel, Martini, and Palhares, 2008; Ekel *et al.*, 2010).

In particular, the modified Wald criterion assumes the following form:

$$\max_{1\le k\le K} D(X_k) = \max_{1\le k\le K}\min_{1\le s\le S}\min_{1\le p\le q} A_p(X_k, Y_s) \tag{8.16}$$

The Laplace criterion can be expressed as follows:

$$\max_{1\le k\le K} D(X_i) = \max_{1\le k\le K}\frac{1}{S}\sum_{s=1}^{S}\min_{1\le p\le q} A_p(X_k, Y_s) \tag{8.17}$$

The Savage criterion comes in the following form:

$$\min_{1\le k\le K} r^{\max}(X_k) = \min_{1\le k\le K}\max_{1\le s\le S}\left[\max_{1\le k\le K}\min_{1\le p\le q} A_p(X_k, Y_s) - \min_{1\le p\le q} A_p(X_k, Y_s)\right] \tag{8.18}$$

Finally, the Hurwicz criterion takes on the following form:

$$\begin{aligned}&\max_{1\leq k\leq K}\left[\alpha\min_{1\leq k\leq K}D(X_k)+(1-\alpha)\max_{1\leq k\leq K}D(X_k)\right]\\&=\max_{1\leq k\leq K}\left[\alpha\min_{1\leq k\leq K}\min_{1\leq p\leq q}D(X_k,Y_s)+(1-\alpha)\max_{1\leq k\leq K}\min_{1\leq p\leq q}D(X_k,Y_s)\right]\end{aligned}\tag{8.19}$$

Although the generalization of the classic approach (Luce and Raiffa, 1957; Raiffa, 1968; Webster, 2003) to consider uncertainty of information in multicriteria decision-making is concerned here with the modification of the criteria of Wald, Laplace, Savage, and Hurwicz, the same line of thought can be extended to other types of choice criteria available in the literature, for example, the criteria of Hodges and Lehmann, Bayes, maximal probability, and so on (Hodges and Lehmann, 1952; Trukhaev, 1981). However, the use of these criteria presumes the availability of a certain type of information (usually in a probabilistic format) about the states of nature.

8.5 General Scheme of Multicriteria Decision-Making under Uncertainty

Following our brief presentation above, we can introduce a general scheme of multicriteria decision-making under uncertainty. The scheme comprises three main phases.

First, we construct q payoff matrices (completed according to the number of objective functions under consideration) for all combinations of the given solution alternatives X_k, $k = 1, 2, \ldots, K$, and the given representative states of nature Y_s, $s = 1, 2, \ldots, S$. We do not consider in detail the techniques that can be used to select representative states of nature (scenarios); those are represented and discussed, for example, in Belyaev (1977) and Meristo (1989). Here the use of so-called LP_τ-sequences (Sobol', 1979; Sobol' and Statnikov, 2006) can be beneficial in generating the representative states of nature as well. These sequences are used in Example 8.1. To construct the payoff matrices, it is necessary to solve S multiobjective problems formalized within the framework of the $\langle\mathbf{X}, \mathbf{M}\rangle$ models as discussed in Chapter 4. Their resulting solutions produce the solution alternatives $X_k, k = 1, 2, \ldots, K \leq S$ (naturally, some solutions can be added by a DM and all solution alternatives can be defined by a DM). Hereafter the solution alternatives X_k, $k = 1, 2, \ldots, K$, are substituted into $F_p(X)$, $p = 1, 2, \ldots, q$, for Y_s, $s = 1, 2, \ldots, S$. These substitutions generate q payoff matrices.

The second phase of the scheme is associated with the analysis of the obtained payoff matrices. This phase is based on the use of the generalization of the classic approach to dealing with uncertainty of information in multicriteria decision problems. It might happen, however, that the solution obtained in this way is not unique. In this case we have to proceed with further processing (which forms the third phase of the scheme). Nevertheless, it is important to note here that the second stage helps us to evaluate the particular risks (in the case of monocriteria risk estimates) as well as the aggregated risks (multicriteria risk estimates) in decision-making for any particular solution alternative.

The third phase is associated with the construction and analysis of the $\langle\mathbf{X}, \boldsymbol{R}\rangle$ models for the subsequent contraction of the decision uncertainty regions (the construction and analysis of such models were covered in Chapter 7; the results related to group decision-making presented in Chapter 9 are applicable here as well). The use of the $\langle \mathbf{X}, \boldsymbol{R} \rangle$ models allows us to take into consideration quantitative as well as qualitative indices, whose formation is based on the knowledge, experience, and intuition of experts.

In this way, the application of the general scheme of multicriteria decision-making under uncertainty is crucial for utilizing information of a formal nature (quantitative information) to the greatest extent and only then does one exploit qualitative information to reduce the corresponding decision uncertainty regions.

8.6 Application Example

The example presented below demonstrates the ways of realizing the two first phases of the general scheme of multicriteria decision-making under uncertainty. The third phase is not considered as Chapter 7 included detailed and convincing examples of applying the $\langle \mathbf{X}, \boldsymbol{R} \rangle$ models.

Example 8.1. Let us consider the following multiobjective problem with interval coefficients present in the objective functions:

$$F_1(x) = [2.70, 3.30]x_1 + [11.70, 14.30]x_2 + [7.20, 8.80]x_3 \to \min \tag{8.20}$$

$$F_2(x) = [5.40, 6.60]x_1 + [3.60, 4.40]x_2 + [4.50, 5.50]x_3 \to \min \tag{8.21}$$

subject to the following constraints:

$$0 \le x_1 \le 10 \tag{8.22}$$

$$0 \le x_2 \le 12 \tag{8.23}$$

$$0 \le x_3 \le 14 \tag{8.24}$$

$$x_1 + x_2 + x_3 = 36 \tag{8.25}$$

The first phase in the decision-making process is concerned with the construction of two payoff matrices for all combinations of the solution alternatives X_k, $k = 1, 2, \ldots, K$, and the representative states of nature Y_s, $s = 1, 2, \ldots, S$. Without going into detail on how the LP_τ-sequences (which have superior characteristics of uniformity among other uniformly distributed sequences, Sobol', 1979) are constructed, we will apply them to generate the representative states of nature.

The results of Sobol' (1979) allow us to determine points Q_s, $s = 1, 2, \ldots, S$, with coordinates q_{st}, $t = 1, 2, \ldots, T$, in the corresponding unit hypercube Q^T. In particular, taking into account that here $T = 6$ (we have six coefficients in (8.20) and (8.21)) and setting $S = 7$, Table 8.6 includes coordinates of points Q_s, $s = 1, 2, \ldots, 7$, for $t = 1, 2, \ldots, 6$ determined on the basis of recommendations of Sobol' (1979) and Sobol' and Statnikov (2006).

In essence, the selection of representative states of nature is reduced to the formation of points of a uniformly distributed sequence in Q^6 and their transformation to the hypercube C^6 defined by the lower c_t' and upper c_t'' bounds of the corresponding coefficients of (8.20) and (8.21). Taking this into account, if points Q_s, $s = 1, 2, \ldots, 7$, with coordinates q_{st},

Table 8.6 Points of the LP_τ-sequences in Q^6

s	$t=1$	$t=2$	$t=3$	$t=4$	$t=5$	$t=6$
1	0.500	0.500	0.500	0.500	0.500	0.500
2	0.250	0.750	0.250	0.750	0.250	0.750
3	0.750	0.250	0.750	0.250	0.750	0.250
4	0.125	0.625	0.875	0.875	0.625	0.125
5	0.625	0.125	0.375	0.375	0.125	0.625
6	0.375	0.375	0.625	0.125	0.875	0.875
7	0.875	0.875	0.125	0.625	0.375	0.375

$t = 1, 2, \ldots, 6$, form a uniformly distributed sequence in Q^6, then points C_s, $s = 1, 2, \ldots, 7$, with the coordinates expressed as

$$c_{st} = c_t' + (c_t'' - c_t')q_{st}, \quad t = 1, 2, \ldots, 6 \tag{8.26}$$

form a uniformly distributed sequence in C^6 which is included in Table 8.7.

The coordinates of points in Table 8.7 serve as a basis for constructing seven (in accordance with the number of states of nature) multiobjective optimization problems with deterministic coefficients

$$F_1(x) = 3.00x_1 + 13.00x_2 + 8.00x_3 \to \min \tag{8.27}$$
$$F_2(x) = 6.00x_1 + 4.00x_2 + 5.00x_3 \to \min \tag{8.28}$$
$$F_1(x) = 2.85x_1 + 13.65x_2 + 7.60x_3 \to \min \tag{8.29}$$
$$F_2(x) = 6.30x_1 + 3.80x_2 + 5.25x_3 \to \min \tag{8.30}$$
$$F_1(x) = 3.15x_1 + 12.35x_2 + 8.40x_3 \to \min \tag{8.31}$$
$$F_2(x) = 5.70x_1 + 4.20x_2 + 4.75x_3 \to \min \tag{8.32}$$
$$F_1(x) = 2.93x_1 + 12.68x_2 + 8.20x_3 \to \min \tag{8.33}$$
$$F_2(x) = 5.55x_1 + 4.30x_2 + 5.38x_3 \to \min \tag{8.34}$$
$$F_1(x) = 2.78x_1 + 13.33x_2 + 8.60x_3 \to \min \tag{8.35}$$
$$F_2(x) = 6.45x_1 + 4.10x_2 + 4.63x_3 \to \min \tag{8.36}$$
$$F_1(x) = 3.08x_1 + 12.03x_2 + 7.80x_3 \to \min \tag{8.37}$$

Table 8.7 Representative states of nature

s	$t=1$	$t=2$	$t=3$	$t=4$	$t=5$	$t=6$
1	3.00	13.00	8.00	6.00	4.00	5.00
2	2.85	13.65	7.60	6.30	3.80	5.25
3	3.15	12.35	8.40	5.70	4.20	4.75
4	2.93	12.68	8.20	5.55	4.30	5.38
5	2.78	13.33	8.60	6.45	4.10	4.63
6	3.08	12.03	7.80	5.85	3.70	5.13
7	3.23	13.98	7.40	6.15	3.90	4.88

$$F_2(x) = 5.85x_1 + 3.70x_2 + 5.13x_3 \rightarrow \min \tag{8.38}$$

$$F_1(x) = 3.23x_1 + 13.98x_2 + 7.40x_3 \rightarrow \min \tag{8.39}$$

$$F_2(x) = 6.15x_1 + 3.90x_2 + 4.88x_3 \rightarrow \min \tag{8.40}$$

which are subject to the same constraints (8.22)–(8.25).

The solutions to the above multiobjective problems, obtained with the use of the AIDMS1 described in Chapter 4, are as follows:

$$s = 1\text{: } x_1^0 = 7.00,\ x_2^0 = 9.00,\ x_3^0 = 14.00 \text{ for (8.27) and (8.28)}$$

$$s = 2\text{: } x_1^0 = 8.95,\ x_2^0 = 10.50,\ x_3^0 = 10.55 \text{ for (8.29) and (8.30)}$$

$$s = 3\text{: } x_1^0 = 7.00,\ x_2^0 = 9.00,\ x_3^0 = 14.00 \text{ for (8.31) and (8.32)}$$

$$s = 4\text{: } x_1^0 = 9.95,\ x_2^0 = 10.50,\ x_3^0 = 9.55 \text{ for (8.33) and (8.34)}$$

$$s = 5\text{: } x_1^0 = 7.00,\ x_2^0 = 9.00,\ x_3^0 = 14.00 \text{ for (8.35) and (8.36)}$$

$$s = 6\text{: } x_1^0 = 9.93,\ x_2^0 = 11.35,\ x_3^0 = 8.72 \text{ for (8.37) and (8.38)}$$

$$s = 7\text{: } x_1^0 = 7.00,\ x_2^0 = 9.00,\ x_3^0 = 14.00 \text{ for (8.39) and (8.40)}$$

In such a way, we can form the following four solution alternatives for the problem (8.20)–(8.25):

$$X_1 = (7.00, 9.00, 14.00)$$

$$X_2 = (8.95, 10.50, 10.55)$$

$$X_3 = (9.95, 10.50, 9.55)$$

$$X_4 = (9.93, 11.35, 8.72)$$

Substituting these solutions into (8.27), (8.29), (8.31), (8.33), (8.35), (8.37), and (8.39), we construct a payoff matrix for the first objective function (Table 8.8). When substituting them into (8.28), (8.30), (8.32), (8.34), (8.36), (8.38), and (8.40), we construct a payoff matrix for the second objective function (Table 8.9).

Let us consider the solution of the single-criterion problem (8.20) subject to the constraints (8.22)–(8.25) when analyzing the payoff matrix given in Table 8.8.

First of all, we construct the payoff matrix with the characteristic estimates represented in Table 8.10. In particular, this table includes the estimates $F^{\max}(X_k)$, $F^{\min}(X_k)$, and $\overline{F}(X_k)$

Table 8.8 Payoff matrix for the first criterion

	Y_1	Y_2	Y_3	Y_4	Y_5	Y_6	Y_7
X_1	250.00	249.20	250.80	249.43	259.83	239.03	252.03
X_2	243.75	249.01	246.49	245.87	255.58	236.17	253.77
X_3	242.75	244.26	241.24	240.60	249.76	231.45	249.60
X_4	247.10	249.50	244.70	244.52	253.89	235.14	255.27

Table 8.9 Payoff matrix for the second criterion

	Y_1	Y_2	Y_3	Y_4	Y_5	Y_6	Y_7
X_1	148.00	151.80	144.20	152.87	146.87	146.07	146.47
X_2	148.45	151.67	145.23	151.58	149.62	145.33	147.48
X_3	149.45	152.72	146.18	151.75	151.44	146.05	148.75
X_4	148.58	151.47	145.69	150.83	150.96	144.82	147.89

defined on the basis of (8.1)–(8.3). We note that the last line in Table 8.10 is $F^{\min}(Y_s)$ because the first criterion is to be minimized. Its elements are defined by the relationship described by (8.6). Further, the elements of the risk matrix (Table 8.11) have been calculated on the basis of (8.8). The estimates $r^{\max}(X_k)$, given in Tables 8.10 and 8.11, have been obtained by making use of (8.4).

Taking into account that the first objective function is to be minimized, the Wald criterion (8.9) using which we choose the solution alternative X^W, assumes the following form:

$$\min_{1\le k\le K} F^{\min}(X_k) = \min_{1\le k\le K} \max_{1\le s\le S} F(X_k, Y_s) \tag{8.41}$$

Thus $X^W = \{X_3\}$. The Laplace criterion (8.10), which leads to the formation of X^L, reads

$$\min_{1\le k\le K} \overline{F}(X_k) = \min_{1\le k\le K} \frac{1}{S}\sum_{s=1}^{S} F(X_k, Y_s) \tag{8.42}$$

and yields$X^L = \{X_3\}$. The Savage criterion (8.11) returns here $X^S = \{X_3\}$. Finally, the Hurwicz criterion (8.12) is modified as follows:

$$\begin{aligned}&\min_{1\le k\le K} [\alpha F^{\max}(X_k) + (1-\alpha)F^{\min}(X_k)]\\ &= \min_{1\le k\le K} \left[\alpha \max_{1\le s\le S} F(X_k, Y_s) + (1-\alpha) \min_{1\le s\le S} F(X_k, Y_s)\right]\end{aligned} \tag{8.43}$$

The use of (8.43) with $\alpha = 0.75$ (as recommended in Belyaev, 1977) generates $X^H = \{X_3\}$ as well.

Table 8.10 Payoff matrix with characteristic estimates for the first criterion

	Y_1	Y_2	Y_3	Y_4	Y_5	Y_6	Y_7	$F^{\max}(X_k)$	$F^{\min}(X_k)$	$\overline{F}(X_k)$	$r^{\max}(X_k)$
X_1	250.00	249.20	250.80	249.43	259.83	239.03	252.03	259.83	239.03	250.05	10.07
X_2	243.75	249.01	246.49	245.87	255.58	236.17	253.77	255.58	236.17	247.81	5.82
X_3	242.75	244.26	241.24	240.60	249.76	231.45	249.60	249.76	231.45	242.81	0
X_4	247.10	249.50	244.70	244.52	253.89	235.14	255.27	255.27	235.14	247.16	5.67
$F^{\min}(Y_s)$	242.75	244.26	241.24	240.60	249.76	231.45	249.60				

Table 8.11 Risk matrix for the first criterion

	Y_1	Y_2	Y_3	Y_4	Y_5	Y_6	Y_7	$r^{\max}(X_k)$
X_1	7.25	4.94	9.56	8.83	10.07	7.58	2.43	10.07
X_2	5.00	4.75	5.25	5.27	5.82	4.72	4.17	5.82
X_3	0	0	0	0	0	0	0	0
X_4	4.35	5.24	3.46	3.92	4.13	3.69	5.67	5.67

In such a way, as the solution to the single-criterion problem (8.20) with the constraints (8.22)–(8.25), alternativeX_3 is considered with a high degree of confidence.

Let us consider the solution of the single-criterion problem (8.21) subject to the constraints (8.22)–(8.25) by analyzing the payoff matrix given in Table 8.9.

The corresponding payoff matrix with characteristic estimates for the second criterion is presented in Table 8.12. The risk matrix, constructed by applying (8.8), is presented in Table 8.13.

The use of the Wald criterion, applied on the basis of (8.41), leads to $X^W = \{X_4\}$. The Laplace criterion, applied as given by (8.42), produces the selection $X^L = \{X_1\}$. The Savage criterion (8.11) results in $X^S = \{X_1\}$. Finally, the Hurwicz criterion, based on the use of (8.43) with $\alpha = 0.75$, generates $X^H = \{X_4\}$.

In such a way, as the solution to the single-criterion problem (8.21) with the constraints (8.22)–(8.25), the alternatives X_1 and X_4 are to be considered. Formally, these alternatives cannot be distinguished.

Let us return to the problem described by (8.20)–(8.25).

Taking into account that $\max_{1\le k\le 4} F^{\max}(X_k) = 259.87$ and $\min_{1\le k\le 4} F^{\min}(X_k) = 231.45$ for the first objective function (see Table 8.10), and applying (4.31), we construct the modified (normalized) payoff matrix for the first criterion presented in Table 8.14. In a similar manner, considering that $\max_{1\le k\le 4} F^{\max}(X_k) = 152.87$ and $\min_{1\le k\le 4} F^{\min}(X_k) = 144.20$ for the second objective function (see Table 8.12), and applying (4.32), we construct the modified (normalized) payoff matrix for the second criterion, which is presented in Table 8.15.

The two modified payoff matrices result in the construction of the aggregated payoff matrix presented in Table 8.16; it has been generated with the use of (4.28). Table 8.16 also includes the characteristic estimates $D^{\max}(X_k)$, $D^{\min}(X_k)$, and $\overline{D}(X_k)$, defined on the basis of (8.13)–(8.15), as well as the estimates $D^{\max}(Y_s)$. The last ones serve for the construction of the risk matrix shown in Table 8.17. This matrix has been used to obtain the estimates $r^{\max}(X_k)$ presented in Table 8.16 and Table 8.17.

Table 8.12 Payoff matrix with characteristic estimates for the second criterion

	Y_1	Y_2	Y_3	Y_4	Y_5	Y_6	Y_7	$F^{\max}(X_k)$	$F^{\min}(X_k)$	$\overline{F}(X_k)$	$r^{\max}(X_k)$
X_1	148.00	151.80	144.20	152.87	146.87	146.07	146.47	152.87	144.20	148.04	2.04
X_2	148.45	151.67	145.23	151.58	149.62	145.33	147.48	151.67	145.23	148.48	2.75
X_3	149.45	152.72	146.18	151.75	151.44	146.05	148.75	152.72	146.05	149.48	4.57
X_4	148.58	151.47	145.69	150.83	150.96	144.82	147.89	151.47	144.82	148.61	4.09
$F^{\min}(Y_s)$	148.00	151.80	144.20	152.87	146.87	146.07	146.47				

Table 8.13 Risk matrix for the first criterion

	Y_1	Y_2	Y_3	Y_4	Y_5	Y_6	Y_7	$r^{\max}(X_k)$
X_1	0	0.33	0	2.04	0	1.25	0	2.04
X_2	0.45	0.20	0.93	0.75	2.75	0.51	1.01	2.75
X_3	1.45	1.25	1.98	0.92	4.57	1.23	2.28	4.57
X_4	0.58	0	1.49	0	4.09	0	1.42	4.09

Table 8.14 Modified payoff matrix for the first criterion

	Y_1	Y_2	Y_3	Y_4	Y_5	Y_6	Y_7
X_1	0.35	0.37	0.32	0.37	0	0.73	0.27
X_2	0.57	0.38	0.47	0.49	0.15	0.83	0.21
X_3	0.60	0.55	0.66	0.68	0.35	1	0.36
X_4	0.45	0.36	0.53	0.54	0.21	0.87	0.16

Table 8.15 Modified payoff matrix for the second criterion

	Y_1	Y_2	Y_3	Y_4	Y_5	Y_6	Y_7
X_1	0.56	0.12	1	0	0.69	0.78	0.74
X_2	0.51	0.14	0.88	0.15	0.37	0.87	0.62
X_3	0.39	0.02	0.77	0.13	0.16	0.79	0.48
X_4	0.49	0.16	0.83	0.24	0.22	0.93	0.57

Table 8.16 Aggregated payoff matrix with characteristic estimates

	Y_1	Y_2	Y_3	Y_4	Y_5	Y_6	Y_7	$D^{\max}(X_k)$	$D^{\min}(X_k)$	$\overline{D}(X_k)$	$R^{\max}(X_k)$
X_1	0.35	0.12	0.32	0	0	0.73	0.27	0.73	0	0.26	0.34
X_2	0.51	0.14	0.47	0.15	0.15	0.83	0.21	0.83	0.14	0.34	0.19
X_3	0.39	0.02	0.66	0.13	0.16	0.79	0.36	0.79	0.02	0.36	0.14
X_4	0.45	0.16	0.53	0.24	0.21	0.87	0.16	0.87	0.16	0.37	0.20
$D^{\max}(Y_s)$	0.51	0.16	0.66	0.24	0.21	0.87	0.36				

Table 8.17 Aggregated risk matrix

	Y_1	Y_2	Y_3	Y_4	Y_5	Y_6	Y_7	$R^{\max}(X_k)$
X_1	0.16	0.04	0.34	0.24	0.21	0.14	0.09	0.34
X_2	0.00	0.02	0.19	0.09	0.06	0.04	0.15	0.19
X_3	0.12	0.14	0	0.11	0.05	0.08	0	0.14
X_4	0.06	0.00	0.13	0	0	0	0.20	0.20

It is not difficult to work out that the use of the modified Wald criterion (8.16) leads to $X^W = \{X_4\}$. The application of the modified Laplace criterion (8.17) shows that $X^L = \{X_4\}$ as well. The use of the modified Savage criterion produces $X^S = \{X_3\}$. Finally, the application of the modified Hurwicz criterion, based on (8.19) with $\alpha = 0.75$, generates $X^H = \{X_4\}$.

It is evident that the obtained results require the application of the third phase of the general scheme by considering the solution alternatives X_3 and X_4 realized on the basis of constructing and analyzing the corresponding $\langle \mathbf{X}, \boldsymbol{R} \rangle$ model.

8.7 Conclusions

We have discussed the general scheme of multicriteria decision-making under uncertainty of information. This scheme is based on constructing and analyzing $\langle \mathbf{X}, \mathbf{M} \rangle$ as well as $\langle \mathbf{X}, \boldsymbol{R} \rangle$ models and also on the generalization of the classic approach to dealing with uncertainty of information in monocriteria decision-making for multicriteria problems. The aspects of constructing aggregated payoff matrices and modifying the choice criteria of the classic approach have been considered.

The remarkable feature of the general scheme is associated with the fact that it is directed at using the available quantitative information to the greatest extent to reduce the decision uncertainty regions. If the resolving capacity of the processing of the available quantitative information does not allow unique solutions to be obtained, the scheme presumes the application of qualitative information based on the knowledge, experience, and intuition of the involved experts. Further, the application of the general scheme permits the evaluation of not only particular (monocriteria) but also aggregated (multicriteria) risks (regrets) in decision-making.

Exercises

Problem 8.1. Apply the classic approach to dealing with uncertainty of information to analyze the payoff matrix given in Table 8.18. The objective function is to be minimized.

Problem 8.2. Apply the classic approach to considering the uncertainty of information to analyze the payoff matrix given in Table 8.19. Here the objective function is to be maximized.

Problem 8.3. Apply the generalization of the classic approach to considering the uncertainty of information to analyze the multicriteria problem reflected by the payoff matrix (Table 8.18)

Table 8.18 Payoff matrix

	Y_1	Y_2	Y_3	Y_4
X_1	10	11	12	11
X_2	8	14	11	13
X_3	9	11	12	13
X_4	12	13	12	13

Table 8.19 Payoff matrix

	Y_1	Y_2	Y_3	Y_4
X_1	90	119	96	110
X_2	100	110	98	112
X_3	110	95	110	120
X_4	110	125	112	100

for the minimized objective function and by the payoff matrix (Table 8.19) for the maximized objective function.

Problem 8.4. Verify the possibility of changing the solution (alternatives X_3 and X_4) of the problem defined by Example 8.1, in the case of increasing the importance of the first objective function by two times.

Problem 8.5. Demonstrate the possibility of changing the solution (alternatives X_3 and X_4) of the problem discussed in Example 8.1, in the case of decreasing the importance of the second objective function by two times.

References

Belyaev, L.S. (1977) *A Practical Approach to Choosing Alternative Solutions to Complex Optimization Problems under Uncertainty*, IIASA, Laxenburg.

Ekel, P., Kokshenev, I., Palhares, R., *et al.* (2010) Multicriteria analysis based on constructing payoff matrices and applying methods of decision making in fuzzy environment. *Optmization and Engineering*, http://www.springerlink.com/content/150263x887328465/, Published online 4 March 2010.

Ekel, P.Ya., Martini, J.S.C., and Palhares, R.M. (2008) Multicriteria analysis in decision making under information uncertainty. *Applied Mathematics and Computation*, **200** (2), 501–516.

Hodges, J.L. and Lehmann, E.L. (1952) The use of previous experience in reaching statistical decision. *Annals of Mathematics Studies*, **23** (3), 396–407.

Kaufman, G.M. (1961) *Statistical Decision and Related Techniques in Oil and Gas Exploration*, Prentice Hall, Englewood Cliffs, NJ.

Kuchta, D. (2007) Choice of the best alternative in case of a continuous set of states of nature – application of fuzzy numbers. *Fuzzy Optimization and Decision Making*, **6** (3), 173–178.

Luce, R.D. and Raiffa, H. (1957) *Games and Decisions*, John Wiley & Sons, Inc., New York.

Meristo, T. (1989) Not forecasts but multiple scenarios when coping with uncertainties in the competitive environment. *European Journal of Operational Research*, **38** (3), 350–357.

Raiffa, H. (1968) *Decision Analysis*, Addison-Wesley, Reading, MA.

Sobol', I.M. (1979) On the systematic search in a hypercube. *SIAM Journal on Numerical Analysis*, **16** (5), 790–793.

Sobol', I.M. and Statnikov, R.B. (2006) *Choice of Optimal Parameters in Problems with Many Criteria*, Drofa, Moscow (in Russian).

Trukhaev, R. (1981) *Models of Decision-Making in Conditions of Uncertainty*, Nauka, Moscow (in Russian).

Webster, T.J. (2003) *Managerial Economics: Theory and Practice*, Academic Press, London.

Wen, M. and Iwamura, K. (2008) Fuzzy facility location-allocation problem under the Hurwicz criterion. *European Journal of Operational Research*, **184** (2), 627–635.

Yager, R.R. (1996) Fuzzy set methods for uncertainty representation in risky financial decisions. Proceedings of the IEEE/IAFE Conference on Computational Intelligence for Financial Engineering, New York, pp. 59–65.

9

Group Decision-Making: Fuzzy Models

This chapter is concerned with discrete multiattribute decision-making problems, in a group environment. One possible approach for solving this class of problems is to consider some aggregation procedures as the exclusive arbitration scheme to arrive at a collective decision. This type of approach can be considered to be dictatorial, as it does not require a consensus to be achieved within the group members. The chapter presents three strategies, based on different aggregation procedures, which can be utilized for extending multiattribute decision methods, related to the analysis of $\langle \mathbf{X}, \boldsymbol{R} \rangle$ models, to group settings. Among the main differences between these strategies, we can highlight the following: (1) the time at which the aggregation of the opinions becomes realized; (2) the way the experts are considered in the decision process (being treated as mutually dependent or independent individuals); and (3) the character of numerical values being aggregated, say fuzzy estimates, fuzzy preference relations, and fuzzy nondominance degrees. We include some examples to illustrate how these strategies are utilized to solve group decision problems by means of different multiattribute decision methods.

9.1 Group Decision-Making Problem and its Characteristics

The group decision problem involves the following main elements:

- The set of alternatives $\mathbf{X} = \{X_1, X_2, \ldots, X_n\}$, which is finite, discrete, and contains two or more alternatives.
- The set of criteria $\mathbf{F} = \{F_1, F_2, \ldots, F_q\}$, with two or more criteria of a quantitative or qualitative nature.
- The team of experts involved in the decision process $\mathbf{E} = \{E_1, E_2, \ldots, E_v\}$, which contains two or more qualified professionals.

Fuzzy Multicriteria Decision-Making: Models, Methods and Applications Witold Pedrycz, Petr Ekel and Roberta Parreiras

The group decision environment can have different characteristics concerning attributes such as the spatial and temporal distances among experts; the structure of the decision process; the commonality of goals; and the level of mutual cooperation within the group (Bui and Jarke, 1986; Matsatsinis and Samaras, 2001). Next, we briefly analyze these attributes, except for the last one, which is left for discussion in Section 9.2.

With respect to the spatial and temporal distances, the group members can be located nearby or positioned quite distantly, which means that they can be in different places and working at different points in time. Several organizations have implemented conferencing systems by means of replacing face-to-face meetings among geographically distributed groups, being motivated by both time and money savings. The communication media habitually utilized in group decision-making include face-to-face meetings, teleconferencing, videoconferencing, instant messages (chat), and electronic mail, ranked in accordance with their respective synchronization degree (that is, the degree of temporal distance each medium permits among group members), from the most to the least synchronous media (Baltes *et al.*, 2002). However, it should be stressed that some negative phenomena, which may occur in a group environment, such as miscommunication between members, lack of involvement, insufficient time spent in the analysis, and group polarization, may be strengthened by the distance factor. In spite of the logistic and economic appeal of using electronic communications instead of face-to-face meetings, several works conclude that, when we take into account group satisfaction and time spent in analyzing a problem, face-to-face meetings remain the most fruitful medium for group decision-making (Baltes *et al.*, 2002; Thompson and Coovert, 2002).

Concerning the structure of the process, as indicated in Bui and Jarke (1986) it can be hierarchical, in the sense that more power can be delegated to a group leader, or, in contrast, it can be democratic, when authority is equally distributed among all members. Whereas in the latter case all members are supposed to influence the decision directly and actively, in an extreme case of the hierarchical structure, the leader can make the decision on his/her own, with just some assistance from other members.

With regard to the commonality of goals, the group can work in a cooperative or noncooperative fashion. As Ng and Abramson (1990) point out, cooperative work is more frequent in medical as well as scientific and engineering areas, where experts are regularly invited to pool their knowledge with the intention of finding a better solution than any of them could obtain without help. In contrast, this tends to be less common in political and economical fields.

In cooperative decision-making, all experts are supposed to work together toward the same goals, in order to achieve a common decision for which they must share the responsibility. Hence, the experts do not play the role of disputants, as in noncooperative decision-making, which, on the other hand, usually involves bargaining among members over some common interest and requires the use of negotiation methodologies (Lu *et al.*, 2007). However, it is important to emphasize that, even in such a friendly environment as the one observed in cooperative work, the occurrence of disagreements across the group is inevitable. In practice, it has been observed that each expert often has a distinct perception of the problem and different information at hand (some of them may even have privileged access to restricted information). Further, although the experts are supposed to have similar fundamental goals, they may just partially share all the aspirations of the other members. Therefore, even under the conditions of cooperative work, achieving a perfect consensus among group members on a final solution is almost an impossible ideal.

9.2 Strategies for the Analysis of Group Decision-Making Problems: Multiperson and Multiattribute Aggregation Modes

The current literature contains several strategies for extending multiattribute decision methods to group settings, in order to obtain solutions that reflect the collective vision of a problem. Here, we concentrate on the group decision methods designed for working specifically in a cooperative environment, without distinguishing whether or not the group is spatially and temporarily distributed, as well as whether or not the structure of the process is democratic or hierarchical. Next, we present three different strategies and utilize them to enhance the multiattribute decision-making methods based on $\langle \mathbf{X}, \boldsymbol{R} \rangle$ models for dealing with the input of multiple experts (Ekel *et al.*, 2009):

- *Aggregation of individual evaluations* (AIE). As represented in Figure 9.1, the experts are supposed to evaluate each alternative by forming fuzzy or linguistic estimates. Afterward, the

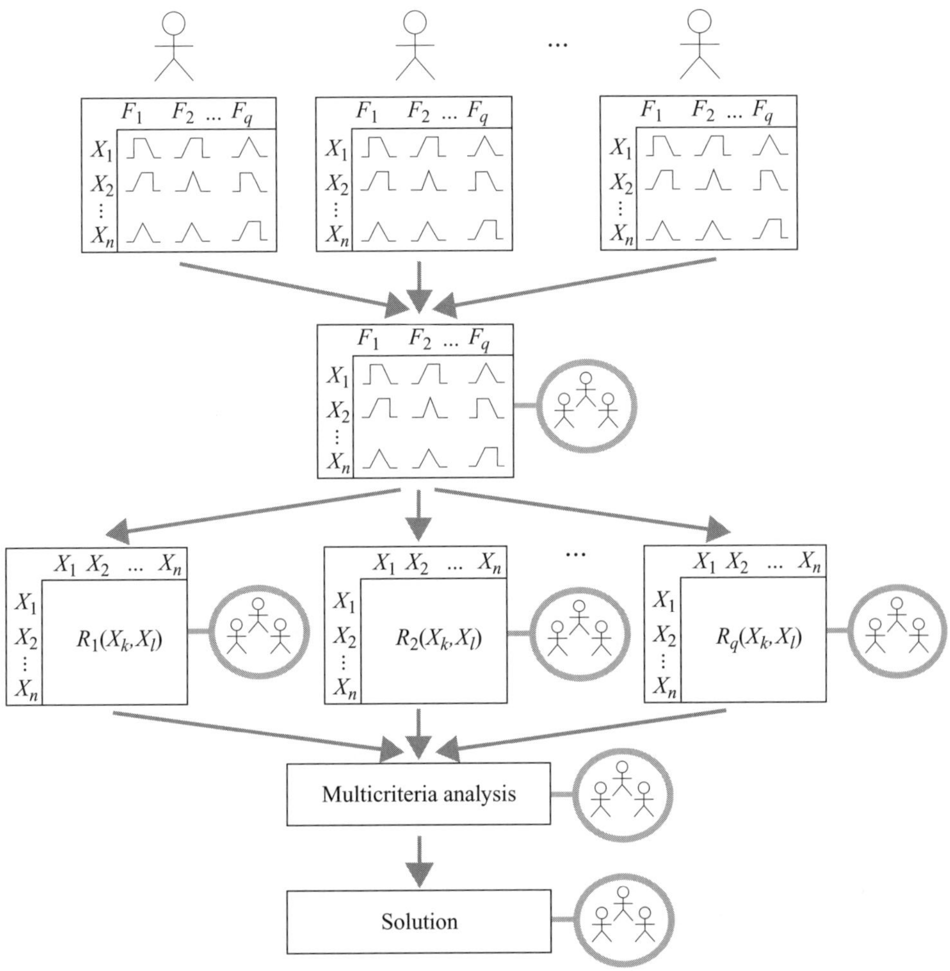

Figure 9.1 Aggregation of individual evaluations.

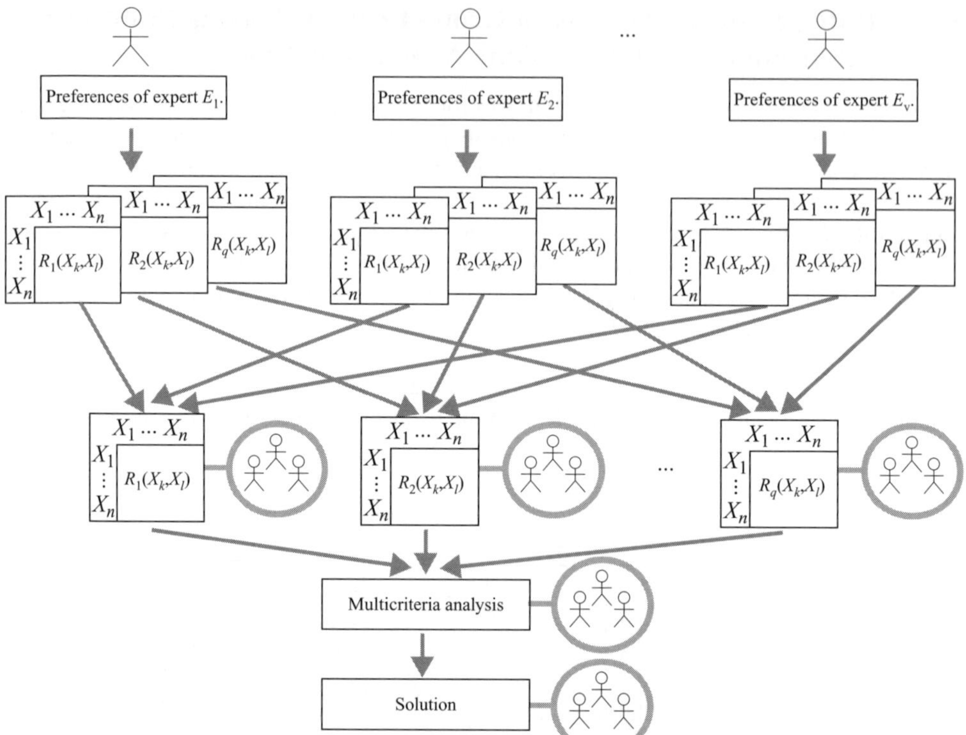

Figure 9.2 Aggregation of individual preferences per criterion.

estimates provided by each expert for each alternative, and taking into account each criterion, are aggregated into some collective estimates. Having at hand an evaluation matrix of the alternatives, it is possible to construct fuzzy preference relations per criterion and, then, different methods for multiattribute analysis, such as those described in Chapter 7, can be utilized.

- *Aggregation of preferences per criterion* (AIC). As can be seen in Figure 9.2, the experts can provide their preferences for each criterion, using any preference format considered in Chapter 6. After the information is made uniform, being converted into fuzzy preference relations, the resulting matrices, obtained for each expert, are aggregated into a collective fuzzy preference relation per criterion. Having at hand a collective fuzzy preference relation for each criterion, it is possible to apply a multiattribute method based on $\langle \mathbf{X}, \boldsymbol{R} \rangle$ models to analyze the problem and obtain a ranking of the alternatives.
- *Aggregation of individual results* (AIR). The multiattribute decision-making problem is solved by each member of the group and, then, the individual results are combined into a collective result, as illustrated in Figure 9.3. When AIR is used, a priori, each DM is allowed to select a different multiattribute decision-making method to solve the problem. However, as each method has its own fundaments and underlying principles for addressing the problem, it is expected that each of them may produce different results for the same problem and the same input of preferences. In this context, the use of different methods may

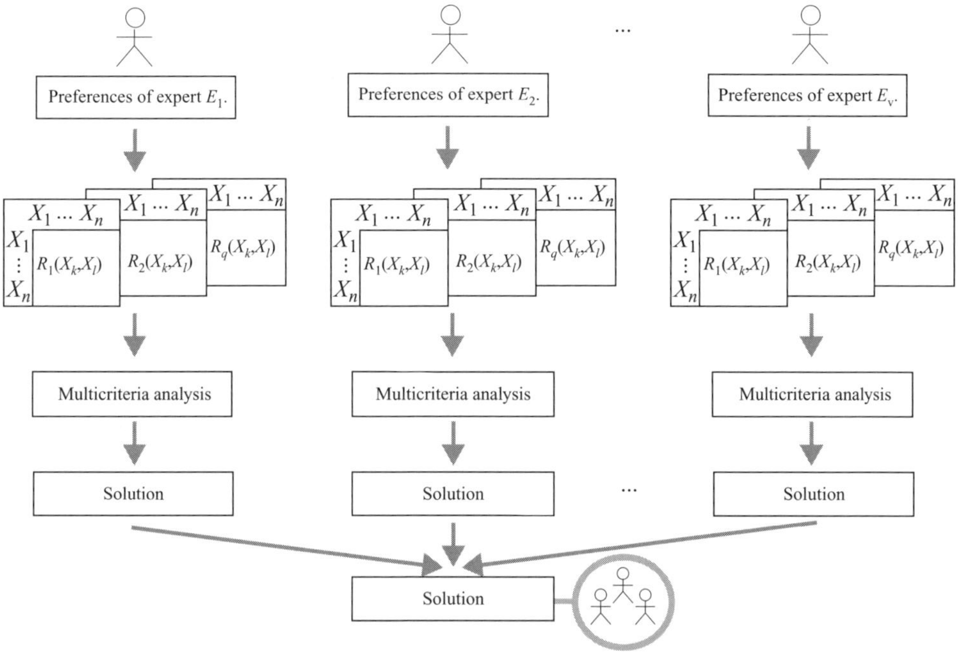

Figure 9.3 Aggregation of individual results.

increase the dissimilarities among results and, consequently, can make it harder to construct satisfactory results of aggregation.

By comparing the three strategies, their main differences are associated with: the different points along the process of the multicriteria analysis in which the aggregation of the opinions of the multiple experts is realized; the character of numerical values being aggregated: fuzzy sets, fuzzy preference relations, or fuzzy nondominance degrees; the way the group members are handled, as a synergetic unique individual or a collection of individuals. Table 9.1 summarizes the main differences among the three strategies, taking into account these three aspects.

Table 9.1 Summary of the main differences among AIE, AIC, and AIR

	Moment of aggregation	Aggregated units	Group management
AIE	Subsequent to the input of the opinions of all experts	Fuzzy sets	Unique individual
AIC	Subsequent to the transformation of the opinions or preferences provided by all experts to fuzzy preference relations	Fuzzy preference relations	Unique individual
AIR	After the problem has been solved by each expert	Fuzzy nondominance degrees	Collection of individuals

Two fundamental aspects should be analyzed in the selection of a suitable strategy for a specific application. The first one concerns the observed fact that each expert frequently prefers a different format to express their opinions. No DM should be pressed to use a specific preference structure against his/her will; otherwise, the input of preference information can become a very critical step in the group decision analysis. Due to the difficulty in assessing preference levels or of understanding the particularities of a preference format, a DM may provide data that do not correspond to the true state of affairs, which, obviously, reduces the soundness of results.

Does any expert want to use a different preference structure to analyze each criterion? Do any of them disagree on using fuzzy or linguistic estimates to evaluate each alternative? An affirmative answer to any one of these questions suggests discarding AIE as an admissible strategy, but does not provide enough reasons or incentives to decide between AIC and AIR.

The second aspect can be briefly summarized as the following question: can the group be considered as a synergistic unit or can it be managed as a collection of independent individuals (Forman and Peniwati, 1998)? In practice, if it can be assumed that each member has perfect knowledge of the problem, it may be interesting to allow each member to completely solve the problem using AIR. But if each member has only partial knowledge of the problem, it is more reasonable to manage the group as a unique individual, using AIE or AIC. Particularly, when the decision-making is performed on the basis of the analysis of $\langle \mathbf{X}, \boldsymbol{R} \rangle$ models, AIC may be the most natural choice when a different expert (or a different set of experts) is supposed to analyze each criterion.

Having defined which strategy is to be utilized, there are two other interrelated aspects that still need to be analyzed. One aspect corresponds to the selection of an adequate operator to combine the individual opinions, judgments, or results into collective ones. As will be discussed in Section 9.4, this aspect depends on the strategy selected to be implemented, as well as on the correspondence between the group requirements and the mathematical properties of each aggregation operator. Finally, the other aspect to be considered can be briefly stated as follows: the relative importance of the opinion provided by each DM may be assumed to be equal or not. If the opinions of each expert are not equally important, how can their respective importance factors be specified? This question will be discussed in Section 9.3.

9.3 The Different Levels of Influence of Each Expert in the Construction of the Collective Opinion

In the context of group decision-making, sometimes it is relevant to differentiate the level of influence of each expert in determining the collective opinion, in an attempt to correlate influence with expertise. The most common way to implement levels of influence is by considering weighted aggregation operators, which admit as input parameter a weighting vector. In this way, a weight w_y is specified for the yth member of the group and then it is associated with the information provided by the corresponding DM, by means of the weighted aggregation operator. The larger the weight assigned to an expert, the greater his/her influence on the final result. However, when specifying unequal weights, it is important to be aware that inappropriate assignments may produce biased results.

It is possible to distinguish three sources of knowledge to determine the values of such unequal weights:

- Each expert, who is supposed to specify a weight w_y representing their self-confidence. Afterward, these weights are usually normalized in such a way that $\sum_{y=1}^{v} w_y = 1$, before being used (Ng, 1992).
- The manager, who assigns a weight to each expert. In this specific case, the influence level of each expert has been frequently evaluated and quantified with the use of AHP (which was outlined in Chapter 3). In the analysis for determining the importance weight of each DM, Saaty (1980) suggests considering factors such as expertise, experience, previous performance, and persuasive abilities. Particularly, when strategy AIC is being used, a criterion that can be utilized by the manager to set these weights is the level of consistency within the judgments of each expert. As will be discussed in the next section, it may be useful to reduce the weights of those experts who give inconsistent judgments, in order to improve the consistency of the aggregated fuzzy preference relation.
- Mathematical models. Among the available models for estimating the weight of each expert, it is worth mentioning those based on fuzzy indices which reflect the concordance level between two opinions. For instance, in several studies (Hsu and Chen, 1996; Lu, Lan, and Wang, 2006; Bernandes *et al.* 2008; Bernandes, Ekel, and Parreiras, 2009) the experts express their opinions using fuzzy estimates and the importance weight of each opinion is calculated taking into account two factors: the importance weight of each expert (which can be provided by the manager as described above) and, predominantly, the mean concordance level among each expert and the others. Lowest weights are assigned to the opinions of the most discordant experts.

Another type of mathematical model that can be utilized to determine the weight of each expert corresponds to the maximization of an index that reflects the consensus level among all experts. For instance, in Ben-Arieh and Chen (2006), the scalar weights associated to each individual result are derived from a mathematical model, in order to increase the measure of consensus until a certain minimum level. We leave this type of procedure, whose main focus is to search for a satisfactory level of consensus across the group, until Chapter 10.

9.4 Aggregation Operators for Constructing Collective Opinions on the Basis of Fuzzy Models and their Properties

As we know, in the process of group decision-making, at some point there is a need to aggregate some numerical values in order to make it possible to construct a meaningful solution for the group. Depending on the selected strategy, AIE, AIC, or AIR, these numerical values are associated with fuzzy estimates, fuzzy preference relations, or the cardinal rating of each alternative, reflected by their respective fuzzy nondominance levels. Obviously each strategy involves peculiarities that must be considered in the choice of an aggregation operator.

The literature contains several aggregation operators which can be applied to group decision-making, with the aim of summarizing different points of view in a unique opinion. Particularly, when AIC and AIR are considered, more than one aggregation operator has been traditionally utilized. Considering that each aggregation operator reflects individual information in a different manner, it is assumed as part of the group decision problem that some imposed requirements are to be satisfied by the aggregation operator, in order to guarantee reasonable results. Next, we analyze the properties of some common aggregation operators in the context of each strategy.

First, let us consider the AIE strategy, which requires each DM to express their opinion using fuzzy or linguistic estimates. Given a particular alternative $X_k \in \mathbf{X}$, all evaluations for a specific criterion F_p are represented as fuzzy sets $F_p^y(X_k)$, $y = 1, 2, \ldots, v$, in the same universe of discourse, but possibly having different membership functions. The collective opinion $F_p^C(X_k)$ is commonly obtained by applying the weighted arithmetic mean to combine the estimates provided by each expert $F_p^y(X_k)$, $y = 1, 2, \ldots, v$, into a collective estimate as follows:

$$F_p^C(X_k) = \sum_{y=1}^{v} w_y F_p^y(X_k) \tag{9.1}$$

where $X_k \in \mathbf{X}$, $0 \leq w_y \leq 1$, for $y = 1, 2, \ldots, v$, $\sum_{y=1}^{v} w_y = 1$. In expression (9.1), the sum and the multiplication are implemented in accordance with the addition operation (refer to expression (3.42)) and the multiplication operation (refer to expression (3.44)) described in Chapter 3, with each scalar weight w_y represented as a fuzzy singleton.

Example 9.1. Consider that three experts, E_1, E_2, and E_3, evaluated an alternative X_1 from the point of view of the same criterion using the following trapezoidal fuzzy estimates: $F^1(X_1) = \{2.1, 2.2, 2.7, 2.8\}$, $F^2(X_1) = \{1, 1.2, 1.4, 1.6\}$, $F^3(X_1) = \{2, 3, 3.5, 4\}$. If all experts are supposed to have the same level of influence on the results, their corresponding weights can be set as: $w_1 = w_2 = w_3 = 0.33$. In this way, the collective estimate, obtained using (9.1), is given by

$$\begin{aligned} F^C(X_1) &= \{0.33, 0.33, 0.33, 0.33\}.\{2.1, 2.2, 2.7, 2.8\} \\ &\quad + \{0.33, 0.33, 0.33, 0.33\}.\{1, 1.2, 1.4, 1.6\} \\ &\quad + \{0.33, 0.33, 0.33, 0.33\}.\{2, 3, 3.5, 4\} \\ &= \{1.68, 2.11, 2.5, 2.77\} \end{aligned} \tag{9.2}$$

Now, let us concentrate on AIC, which deals with the aggregation of fuzzy preference relations rather than fuzzy estimates. However, before going any further, we want to call attention to the fact that, in this chapter, we use a notation coherent with that utilized in Chapter 6, that is, we use $\boldsymbol{R}$ for fuzzy preference relations in general, **RR** for the additive reciprocal fuzzy preference relation (ARFPR), and **NR** for the nonreciprocal fuzzy preference relation (NRFPR). Further, it must be apparent to the reader that we make a distinction between ARFPR and NRFPR only in the stage of preference input of the decision-making process. Once the preferences have been supplied, we handle both types of fuzzy preference relations as the general fuzzy preference relation denoted by $\boldsymbol{R}$.

In AIC, the operation to generate collective information is performed over the fuzzy preference relations per criterion (before the aggregation across all criteria has been completed). Hence, given the pth criterion, the operation of aggregation makes use of a function that maps a vector of fuzzy nonstrict preference relations $\boldsymbol{R}_p = [\boldsymbol{R}_p^1 \quad \boldsymbol{R}_p^2 \quad \ldots \quad \boldsymbol{R}_p^v]$, to another fuzzy nonstrict preference relation $\boldsymbol{R}_p^C$. Among the operators commonly utilized in this context, we can name the OWA operator, the weighted arithmetic mean (WAM), the weighted geometric mean (WGM), and the min operator. When OWA is utilized, the collective fuzzy preference relation is given by (Herrera-Viedma, Herrera, and Chiclana, 2002)

$$\boldsymbol{R}_p^C(X_k, X_l) = \sum_{y=1}^{v} w_y b_y \tag{9.3}$$

where $(X_k, X_l) \in \mathbf{X} \times \mathbf{X}$, b_y is the yth largest value in the collection $\boldsymbol{R}_p^1(X_k, X_l)$, $\boldsymbol{R}_p^2(X_k, X_l), \ldots, \boldsymbol{R}_p^v(X_k, X_l)$, and the weights $w_1, w_2, \ldots, w_v$ satisfy conditions $0 \leq w_y \leq 1$, $y = 1, 2, \ldots, v$, and $\sum_{y=1}^{v} w_y = 1$.

On the other hand, when WAM is utilized, we have (Peneva and Popchev, 2003)

$$\boldsymbol{R}_p^C(X_k, X_l) = \sum_{y=1}^{v} w_y \boldsymbol{R}_p^y(X_k, X_l) \tag{9.4}$$

where $(X_k, X_l) \in X \times X$, $0 \leq w_y \leq 1$, $y = 1, 2, \ldots, v$, and $\sum_{y=1}^{v} w_y = 1$.

The use of WGM as the selected aggregation operator leads to (Peneva and Popchev, 2003)

$$\boldsymbol{R}_p^C(X_k, X_l) = \prod_{y=1}^{v} \left(\boldsymbol{R}_p^y(X_k, X_l)\right)^{w_y} \tag{9.5}$$

where $(X_k, X_l) \in X \times X$, $0 \leq w_y \leq 1$, for $y = 1, 2, \ldots, v$, and $\sum_{y=1}^{v} w_y = 1$.

Finally, if min is the selected aggregation operator, the collective fuzzy preference relation is obtained as (Peneva and Popchev, 2003)

$$\boldsymbol{R}_p^C(X_k, X_l) = \min_{1 \leq y \leq v} \boldsymbol{R}_p^y(X_k, X_l) \tag{9.6}$$

In AIC, the collective fuzzy preference relation $\boldsymbol{R}_p^C$ is supposed to provide the possibility of deciding by means of a multiattribute decision method among the ones based on the Orlovsky choice function, which were described in Chapter 7. It implies that the resultant aggregated relations $\boldsymbol{R}_p^C$, $p = 1, 2, \ldots, q$, should satisfy at least weak transitivity (Sengupta, 1998); otherwise, unreasonable outcomes may be derived from their analysis. Thus, taking into account that the aggregated fuzzy preference relation may inherit inconsistencies from the individual ones, each member of the group is supposed to cooperate, providing consistent judgments. If an expert cannot provide consistent judgments, some action can be taken in order to reduce the negative impact of the inconsistent judgments in the construction of collective preferences, as will be shown in the next section. But, for now, let us consider that all experts provided consistent judgments. In this case, one aspect to be considered is: does the selected aggregation operator have the capability to preserve the consistency of individual fuzzy preference relations? To what extent is the consistency preserved? We note that other authors also have observed the importance of considering the fact that, in spite of the good properties of some aggregation operators, they may not assure consistent decisions in the framework of fuzzy preferences (García-Lapresta and Meneses, 2005). A very comprehensive treatment of this subject by Peneva and Popchev (2003) addresses the capability of several aggregation operators to transmit the mathematical properties of individual fuzzy preference relations (reflexivity, symmetry, transitivity, and others) to the aggregated one. However, as far as we are aware, the current literature has not included a relevant property such as additive transitivity, or an important aggregation operator such as OWA, in this kind of inquiry. Furthermore, at present, the literature still lacks a consensus on an adequate consistency condition to be satisfied by each expert, as well as by the whole group, when it is handled as a unique individual. For these reasons, it has so far remained impossible to fairly select an aggregation operator taking this criterion into account.

On the other hand, an important factor that should be (and can be) considered in the choice of an operator corresponds to the set of requirements imposed by the group, when taking into account the expectations of each expert. For instance, if the group agrees that the content of the opinion is more important than its author, it may be more helpful to utilize the min operator or else the OWA operator, whose weights do not depend on the source, but on the position of each element in the rank of all individual elements. Conversely, if certain experts need to be privileged (rather than certain opinions), then it is interesting to utilize WAM or WGM.

It also should be taken into account that, whereas OWA may have (or not) a compensatory behavior (that is, it allows a bad evaluation given by a DM to be compensated by a good one from another DM), depending on the fuzzy linguistic quantifier selected, WAM and WGM always have a compensatory character (the compensatory character of WGM is weaker than that of WAM). The min operator, however, always has a noncompensatory behavior. Particularly, its use is helpful when the group agrees that the collective decision should be pessimistic, in the sense that an alternative which was badly evaluated by any expert should be badly evaluated by the group in a noncompensatory way.

Finally, another helpful property of OWA lies in the fact that it can be utilized to consider only part of the opinions, by a proper setting of the linguistic operator. When the opinions of experts are very discordant, if we aggregate all of them by means of an averaging operator, the result may be an intermediate one which does not satisfy any expert in the group. On the other hand, if we consider, for instance, "most" opinions rather than "all" opinions, this undesired outcome can be avoided.

Example 9.2. Consider that three experts, E_1, E_2, and E_3, compared three alternatives, X_1, X_2, and X_3. The provided judgments are expressed in terms of the nonreciprocal fuzzy nonstrict preference relations

$$\mathbf{RN}^1 = \begin{bmatrix} 1 & 0.92 & 0.92 \\ 1 & 1 & 1 \\ 1 & 1 & 1 \end{bmatrix} \tag{9.7}$$

$$\mathbf{RN}^2 = \begin{bmatrix} 1 & 1 & 0 \\ 1 & 1 & 0 \\ 1 & 1 & 1 \end{bmatrix} \tag{9.8}$$

$$\mathbf{RN}^3 = \begin{bmatrix} 1 & 0.6 & 0.6 \\ 1 & 1 & 0.6 \\ 1 & 1 & 1 \end{bmatrix} \tag{9.9}$$

OWA, WAM, WGM, and min are to be utilized to aggregate the fuzzy preference relations $\mathbf{RN}^1$, $\mathbf{RN}^2$, and $\mathbf{RN}^3$ into collective relations. In the case of OWA, the linguistic quantifier "most" defined, as shown in Figure 7.22, is utilized. The resulting weights are $w_1 = 0.066$, $w_2 = 0.667$, $w_3 = 0.266$. For the sake of comparison among all operators, the same weights are also considered for WAM and WGM.

In particular, the application of OWA generates

$$\boldsymbol{R}^C = \begin{bmatrix} 1 & 0.84 & 0.46 \\ 1 & 1 & 0.46 \\ 1 & 1 & 1 \end{bmatrix} \tag{9.10}$$

The use of WAM leads to

$$\boldsymbol{R}^C = \begin{bmatrix} 1 & 0.89 & 0.22 \\ 1 & 1 & 0.23 \\ 1 & 1 & 1 \end{bmatrix} \tag{9.11}$$

By using WGM, we have

$$\boldsymbol{R}^C = \begin{bmatrix} 1 & 0.87 & 0 \\ 1 & 1 & 0 \\ 1 & 1 & 1 \end{bmatrix} \tag{9.12}$$

Finally, when the minimum operator is applied, the collective fuzzy preference relation is the matrix

$$\boldsymbol{R}^C = \begin{bmatrix} 1 & 0.6 & 0 \\ 1 & 1 & 0 \\ 1 & 1 & 1 \end{bmatrix} \tag{9.13}$$

By observing these results, it is interesting to note that the low values of $\boldsymbol{R}^C(X_1, X_3)$ and $\boldsymbol{R}^C(X_2, X_3)$ in both (9.11) and (9.12) are due to the low values observed in the positions $\mathbf{RN}^2(X_1, X_3)$ and $\mathbf{RN}^2(X_2, X_3)$ of the NRFPR (9.8), which has a high importance. Further, another interesting aspect to be noted by comparing (9.11) and (9.12) is that WGM tends to penalize judgments with low values more than WAM does. In (9.13), we have a pessimistic aggregated preference relation where the most severe judgments from (9.7), (9.8), and (9.9) prevailed. On the other hand, when we used OWA (which does not fix an association between the weights and the relations) and allowed a compensatory behavior, we obtained higher values for $\boldsymbol{R}^C(X_1, X_3)$ and $\boldsymbol{R}^C(X_2, X_3)$, as in (9.10).

In AIR, regardless of the multiattribute decision method selected by each expert to solve the problem, the individual results invariably correspond to the fuzzy nondominance degree associated with each alternative, which can be taken as cardinal ratings of each alternative. Again, the aggregation operators commonly utilized in this context are OWA, WAM, WGM, and min.

The use of OWA leads to

$$\mathbf{ND}^C(X_k) = \sum_{y=1}^{v} w_y b_y \tag{9.14}$$

where $X_k \in \mathbf{X}$, the degree of fuzzy nondominance b_y is the yth largest value among $\mathbf{ND}^1(X_k), \ldots, \mathbf{ND}^v(X_k)$, and the weights $w_1, w_2, \ldots, w_y$ satisfy conditions $0 \leq w_y \leq 1$, for $y = 1, 2, \ldots, v$ and $\sum_{y=1}^{v} w_y = 1$.

The use of WAM leads to

$$\mathbf{ND}^C(X_k) = \sum_{y=1}^{v} w_y \mathbf{ND}^y(X_k) \tag{9.15}$$

where $X_k \in \mathbf{X}$, $0 \leq w_y \leq 1$, for $y = 1, 2, \ldots, v$, and $\sum_{y=1}^{v} w_y = 1$.

The application of WGM generates

$$\mathbf{ND}^C(X_k) = \prod_{y=1}^{v} (\mathbf{ND}^y(X_k))^{w_y} \tag{9.16}$$

where $X_k \in \mathbf{X}$, $0 \leq w_y \leq 1$, for $y = 1, 2, \ldots, v$, and $\sum_{y=1}^{v} w_y = 1$.

Finally, with the application of the min operator, we have

$$\mathbf{ND}^C(X_k) = \min_{1 \leq y \leq v} \mathbf{ND}^y(X_k) \tag{9.17}$$

In AIR, the aggregation operator should be selected by considering the correspondence between the requirements of the group and the mathematical properties of the operator. As already indicated, OWA is the best option when it is necessary to associate each weight to the value of each score, rather than to the author responsible for providing the score. Conversely, the use of the averaging operators WAM and WGM may be particularly attractive because of their intuitive appeal: certainly all experts know both WAM and WGM and have already utilized them to aggregate some kind of scores during their lives. The choice between WAM and WGM should consider the fact that WGM, more than WAM, tends to penalize alternatives with at least a low score. Finally, min has a useful capability of generating pessimistic results, in the sense that it takes as the final cardinal ratings the worst rating obtained by each alternative. In this way, it does not allow any alternative to assume a good position in the collective ranking if at least one expert does not agree with that position.

Example 9.3. Consider that the fuzzy nondominance levels of alternatives X_1, X_2, and X_3 are calculated taking into account the preferences of experts E_1, E_2, and E_3, separately. Then, by applying OWA, WAM, WGM, and min, four different collective results are obtained. Again, the same weights utilized in Example 9.2 are utilized here. Table 9.2 shows the individual as well as the collective results. As can be seen, different operators can lead to different results: the use of WAM leads to $X_1 \sim X_2 \succ X_3$; the ranking provided by both min and WGM is $X_2 \succ X_1 \succ X_3$; the use of OWA yields $X_1 \succ X_2 \succ X_3$. With this simple example we intend

Table 9.2 Fuzzy nondominance levels of alternatives

	E_1	E_2	E_3	OWA	WAM	WGM	min
X_1	1	1	0.6	0.89	0.89	0.87	0.6
X_2	0.8	0.9	0.9	0.87	0.89	0.89	0.8
X_3	0.9	0.5	1	0.80	0.66	0.62	0.5

to call attention to the importance of selecting an aggregation operator compatible with the requirements of the group decision problem.

9.5 Consistency of Pairwise Judgments in Group Decision-Making

When preferences are expressed in terms of pairwise comparisons, it is important to verify whether the supplied judgments are consistent, because the occurrence of inconsistencies may be a sign of erroneous judgments and, principally, inconsistencies may lead to incoherent results. Indeed, as indicated in Chapter 7, the multiattribute decision-making methods based on the Orlovsky choice function require the fuzzy preference relation to satisfy at least weak transitivity to characterize a rational decision. Therefore, in group decision-making, we can distinguish two distinct problems associated with the issue of consistency (Herrera-Viedma *et al.*, 2004):

- the problem of individual inconsistency, which refers to the fact that the preferences of each expert, considered separately, should be consistent;
- the problem of collective inconsistency, which refers to the fact that the preferences of the whole group of experts, considered as a unique individual, should be consistent.

When we focus on the specific problem of collective inconsistency, it is important to indicate that it may be raised by two factors:

- A member of the group may supply an inconsistent fuzzy preference relation or an inconsistent multiplicative preference relation, which may cause inconsistencies in the collective preferences.
- As already mentioned, in spite of the good properties of some aggregation operators, they do not necessarily assure consistent aggregated fuzzy preference relations. In this way, even if all individual preferences are considered consistent, the occurrence of different preferences across the group may result in an inconsistent collective fuzzy preference relation.

As discussed in Chapter 5 and Chapter 6, the transitivity property has been conventionally utilized as a condition for attesting to the consistency of the pairwise judgments provided by an expert. However, when we concentrate on the consistency of fuzzy preference relations, no transitivity property has been assumed to be the ideal consistency condition to be applied in all circumstances. Among the different transitivity properties, three of them receive special attention here: the additive transitivity, which has been applied to additive reciprocal fuzzy preference relations (refer to expressions (6.7) or (6.33)); and the min-transitivity (refer to expression (5.23)) and the weak transitivity (refer to expression (5.59)), which have been applied to both reciprocal and nonreciprocal fuzzy preference relations.

The problem of individual inconsistency may take place when AIC and AIR are in use and at least one expert expresses his/her opinions in terms of fuzzy preference relations or multiplicative preference relations. On the other hand, the problem of collective inconsistency is specifically associated with AIC, which is based on the analysis of collective fuzzy preference relations. In the execution of the AIC strategy, the transitivity property can also be used to verify whether the collective fuzzy preference relation is within a minimally acceptable level of consistency.

Example 9.4. Consider that three experts supplied their preferences over three alternatives, X_1, X_2, and X_3, in terms of the ARFPRs $\mathbf{RR}^1$, $\mathbf{RR}^2$, and $\mathbf{RR}^3$, which perfectly satisfy the additive transitivity condition:

$$\mathbf{RR}^1 = \begin{bmatrix} 0.5 & 0.1 & 0 \\ 0.9 & 0.5 & 0.4 \\ 1 & 0.6 & 0.5 \end{bmatrix} \tag{9.18}$$

$$\mathbf{RR}^2 = \begin{bmatrix} 0.5 & 0.2 & 0.6 \\ 0.8 & 0.5 & 0.9 \\ 0.4 & 0.1 & 0.5 \end{bmatrix} \tag{9.19}$$

$$\mathbf{RR}^3 = \begin{bmatrix} 0.5 & 0.9 & 0.7 \\ 0.1 & 0.5 & 0.3 \\ 0.3 & 0.7 & 0.5 \end{bmatrix} \tag{9.20}$$

It is interesting to observe that (9.18)–(9.20) are coherent with the rankings $X_3 \succ X_2 \succ X_1$, $X_2 \succ X_1 \succ X_3$, and $X_1 \succ X_3 \succ X_2$, respectively. Next, we aggregate (9.18)–(9.20) using four different operators: OWA (with the quantifier "most"), min, and the averaging operators WAM and WGM, both with weights $w_1 = 0.2$, $w_2 = 0.3$, $w_3 = 0.5$. The corresponding collective fuzzy preference relations are given by (9.21), (9.22), (9.23), and (9.24), respectively:

$$\boldsymbol{R}^C = \begin{bmatrix} 0.500 & 0.220 & 0.447 \\ 0.620 & 0.50 & 0.407 \\ 0.413 & 0.473 & 0.500 \end{bmatrix} \tag{9.21}$$

$$\boldsymbol{R}^C = \begin{bmatrix} 0.5 & 0.1 & 0 \\ 0.1 & 0.50 & 0.3 \\ 0.3 & 0.1 & 0.50 \end{bmatrix} \tag{9.22}$$

$$\boldsymbol{R}^C = \begin{bmatrix} 0.50 & 0.53 & 0.53 \\ 0.47 & 0.50 & 0.50 \\ 0.47 & 0.50 & 0.50 \end{bmatrix} \tag{9.23}$$

$$\boldsymbol{R}^C = \begin{bmatrix} 0.500 & 0.369 & 0 \\ 0.289 & 0.500 & 0.442 \\ 0.416 & 0.378 & 0.500 \end{bmatrix} \tag{9.24}$$

Note that, except for (9.23), the obtained fuzzy preference relations do not satisfy the reciprocity condition, as well as the additive transitivity condition. For instance, according to the additive transitivity property, we should have $\boldsymbol{R}^C(X_1, X_2) + \boldsymbol{R}^C(X_2, X_3) + \boldsymbol{R}^C(X_3, X_1) = 3/2$, but in (9.21) we obtain $0.220 + 0.407 + 0.413 = 1.04$; in (9.22) we have $0.1 + 0.3 + 0.3 = 0.7$; and, finally, in (9.24) we have $0.369 + 0.442 + 0.416 = 1.227$. On the other hand, it should be indicated that, in this example, except for (9.27), the other aggregated fuzzy preference relations satisfy weak transitivity. For instance, in (9.24) we have that $\boldsymbol{R}^C(X_2, X_3) > \boldsymbol{R}^C(X_3, X_2)$ and $\boldsymbol{R}^C(X_3, X_1) > \boldsymbol{R}^C(X_1, X_3)$. Consequently, we should have $\boldsymbol{R}^C(X_2, X_1) > \boldsymbol{R}^C(X_1, X_2)$. But, instead of this, we have $\boldsymbol{R}^C(X_2, X_1) = 0.289$ which is lower than $\boldsymbol{R}^C(X_1, X_2) = 0.369$.

The current literature still lacks deep considerations of the problem of individual, as well as of collective, inconsistency. Particularly, up to now, researchers have not agreed on a fair consistency condition to be satisfied by the collective judgments, represented in terms of fuzzy preference relations. One could argue: is it reasonable to impose the same consistency condition on each individual and on the whole group? Example 9.3 confirms that, even when all experts supply their preferences as reciprocal fuzzy preference relations satisfying the additive transitivity property, the aggregated fuzzy preference relation may not satisfy this condition. In this example, the additive and the min-transitivity properties seem to be excessively hard conditions for the case of collective preferences. In this chapter, we considered that it suffices to require the collective fuzzy preference relation to satisfy weak transitivity, which is an acceptable level of consistency for the application of the decision methods based on the Orlovsky choice function (Sengupta, 1998).

Considering that the current literature does not present specific procedures for improving the consistency of collective fuzzy preference relations, it is recommended to take some preventive actions against this problem. Obviously, first of all, it is important to minimize the problem of individual inconsistency as much as possible. As indicated in Chapter 6, when an expert is not capable of revising or adjusting his/her inconsistent preference relations, this task can be delegated to an analyst. This procedure may be considered a little dictatorial, but, in practice, it may be useful, considering that at least a minimum level of consistency should be observed within all individual judgments in order to promote the consistency of the aggregated judgments. Further, it is also valuable to make use of automated procedures to improve the consistency of the individual preferences till weak transitivity, such as the one introduced in Ma *et al.* (2006) (refer to Chapter 6 for its description). Alternatively, it is also possible to ask the expert to consider using another preference format to express his/her opinions.

Nevertheless, if the aggregated preferences contain inconsistent judgments, there are still some other possible actions to alleviate this problem, as follows (they are not listed in order of priority):

- Verify empirically whether another admissible aggregation operator can produce more consistent collective fuzzy preference relations.
- Measure the degree of inconsistency of the individual preference relations and reduce the weight associated with the most inconsistent ones, in order to decrease their influence in the collective fuzzy preference relation. Obviously, this action should be pondered before being made, as it involves neglecting the opinion of the corresponding experts.
- Ask the expert to consider using another preference format to express his/her opinions.

9.6 Fuzzy Group Decision-Making Methods

Some group decision-making methods are more suitable than others for a particular group or a particular problem. With the accessibility to various methods, it becomes possible for the group to deal more effectively with diverse types of problems (Zhang and Lu, 2009). Here, we give further details on how the three strategies, AIE, AIC, and AIR, can extend the methods for multiattribute analysis based on $\langle \mathbf{X}, \boldsymbol{R} \rangle$ models, including the ones described in Chapter 7, to deal with the group environment. The beginning of the group decision process, regardless of the selected strategy, always involves the following actions:

- Definition of the group standards. It is desirable that all members agree upon the strategy they are going to utilize and, if AIE or AIC is selected by the group, a multiattribute decision technique to be utilized must be selected as well. It is also important that the group defines how the members are going to communicate: that is, which media they are going to use and how the interactive conversations are going to be organized.
- Weight assignment for each expert. The moderator or the group should select an approach for specifying the importance weights for each expert. As discussed in Section 9.3, each expert can specify a weight to him/herself; the moderator (or manager) can assign a weight to each expert with the use of an elicitation tool or a mathematical model can be utilized to estimate these weights.
- Identification of the problem. Certain aspects of the problem are specified with the help of all members: that is, the goals to be achieved and the requirements to be satisfied by the possible alternatives.

After a consensual statement of the problem and specification of group standards have been obtained, the group should focus on the design of a shared decision space, that is, the set of alternatives and the set of criteria to be considered by the group, in the decision process. Thus, the following actions should be carried out, although not necessarily in a chronological sequence:

- Brainstorming alternatives. In the process of creating the set of alternatives, each DM lists a few possible alternatives they have in mind. The infeasible ones must be excluded from further analysis and similar ones can be merged (Lu *et al.*, 2007).
- Brainstorming criteria. Each DM proposes some criteria for the evaluation of the alternatives. They must be correlated to the previously defined goals, in the sense that they must reflect the level of achievement of those goals and, at the same time, allow discrimination among alternatives. Some of the criteria proposed by the group can be eliminated from further analysis for several reasons: two or more criteria may be sufficiently similar to be merged; a certain criterion may be considered too ambiguous or too hard to be assessed, so that its inclusion can reduce the validity of the results; a certain criterion may not be sufficiently significant to be considered, as its inclusion can increase the costs of the decision process in a unjustifiable manner.

More practical rules such as "a candidate must be named by at least three people" can also be utilized in the brainstorming of alternatives and of criteria, in order to avoid spending

Table 9.3 Evaluation matrix of the alternatives

	$F_1(X_k)$	$F_2(X_k)$	...	$F_q(X_k)$
X_1	$F_1(X_1)$	$F_2(X_1)$	...	$F_q(X_1)$
X_2	$F_1(X_2)$	$F_2(X_2)$	...	$F_q(X_2)$
...	...	...	...	...
X_n	$F_1(X_n)$	$F_2(X_n)$	...	$F_q(X_n)$

excessive amounts of time in these phases (Bui and Jarke, 1986). The use of such rules may be particularly effective when the group is so large that it is difficult to achieve a perfect consensus on the definition of a shared decision space.

Next, each strategy is presented as sequential actions. The steps related to the multiattribute decision method are omitted here. However, it should be indicated that if, during the execution of the selected multiattribute decision method, it becomes necessary to specify the importance weight of each criterion, it is possible to use a group version of AHP to estimate them (Forman and Peniwati, 1998).

Let us begin with AIE, which can be outlined as follows:

Step 1. The experts individually evaluate each alternative, taking into account each criterion F_p, $p = 1, 2, \ldots, q$, using fuzzy or linguistic estimates $F_p^y(X_k)$, $p = 1, 2, \ldots, q$, $y = 1, 2, \ldots, \upsilon$, $X_k \in \mathbf{X}$.

Step 2. The fuzzy estimates provided by each expert for each alternative are aggregated into collective estimates $F_p^C(X_k)$, $p = 1, 2, \ldots, q$, $X_k \in \mathbf{X}$, using the weighted sum given by (9.1). At the end of the current step, an evaluation matrix similar to the one shown in Table 9.3 is obtained.

Step 3. Afterward, with the collective evaluation matrix at hand, the fuzzy preference relations that represent each criterion are constructed on the basis of (6.12) and (6.13) if F is a quantitative criterion (attribute) or on the basis of (6.14) and (6.15) if F is a qualitative criterion (attribute).

Step 4. Finally, the procedure selected by the group to perform the multiattribute analysis is executed.

Example 9.5. Strategy AIE is utilized here jointly with the *first technique* (refer to Chapter 7 for its description) to analyze $\langle \mathbf{X}, \boldsymbol{R} \rangle$ models to solve a group decision problem, where a set of five experts $\mathbf{E} = \{E_1, E_2, \ldots, E_5\}$ must rank four alternatives $\mathbf{X} = \{X_1, X_2, X_3, X_4\}$, taking into account three criteria $\mathbf{F} = \{F_1, F_2, F_3\}$. As the professionals are considered to be of the same importance, the weights are set as $w_y = 0.2$, $y = 1, 2, \ldots, 5$.

In Step 1, the specialists are asked to give their opinion relative to each alternative in terms of fuzzy estimates, using the linguistic values shown in Figure 9.4. Table 9.4 presents all the linguistic estimates provided by the experts.

In Step 2, a collective fuzzy estimate is obtained for each alternative taking into account each criterion, using (9.1). The aggregated fuzzy estimates are shown in Table 9.5 and represented in Figure 9.5.

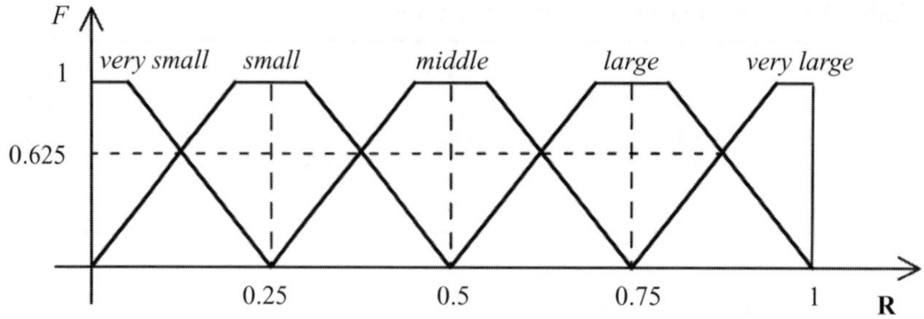

Figure 9.4 Membership functions for normalized fuzzy values.

Table 9.4 Evaluation matrix of the alternatives

F_1					
	E_1	E_2	E_3	E_4	E_5
X_1	*very large*	*very large*	*very large*	*large*	*very large*
X_2	*very small*	*very small*	*small*	*small*	*small*
X_3	*small*	*middle*	*middle*	*small*	*small*
X_4	*large*	*large*	*large*	*middle*	*large*
F_2					
	E_1	E_2	E_3	E_4	E_5
X_1	*middle*	*middle*	*small*	*middle*	*middle*
X_2	*very large*	*very large*	*large*	*very large*	*large*
X_3	*large*	*very large*	*very large*	*very large*	*very large*
X_4	*very small*	*very small*	*very small*	*small*	*very small*
F_3					
	E_1	E_2	E_3	E_4	E_5
X_1	*small*	*very small*	*very small*	*small*	*very small*
X_2	*middle*	*small*	*small*	*middle*	*small*
X_3	*large*	*middle*	*middle*	*middle*	*middle*
X_4	*large*	*large*	*very large*	*large*	*very large*

Table 9.5 Collective fuzzy estimates.

	F_1	F_2	F_3
X_1	{0.7, 0.9, 0.96, 1}	{0.2, 0.4, 0.5, 0.7}	{0, 0.08, 0.15, 0.35}
X_2	{0, 0.12, 0.2, 0.4}	{0.65, 0.85, 0.92, 1}	{0.05, 0.21, 0.3, 0.5}
X_3	{0.1, 0.3, 0.4, 0.6}	{0.7, 0.9, 0.96, 1}	{0.3, 0.5, 0.6, 0.8}
X_4	{0.45, 0.65, 0.75, 0.95}	{0, 0.04, 0.1, 0.3}	{0.6, 0.8, 0.88, 1}

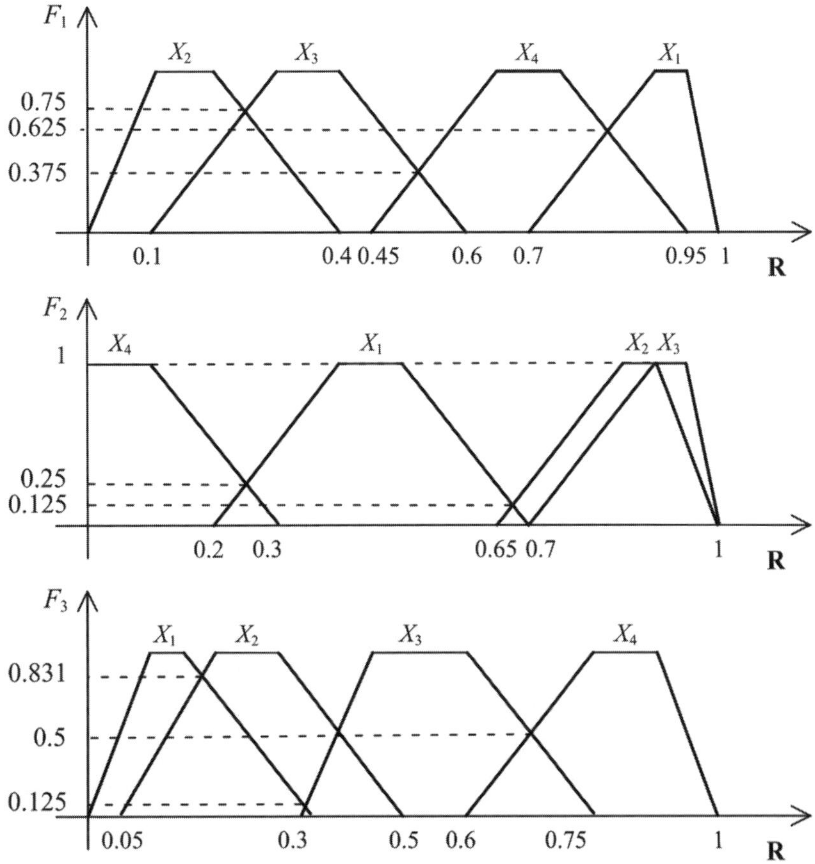

Figure 9.5 Collective fuzzy estimates.

The following fuzzy nonstrict preference relations are derived from the collective fuzzy estimates with the use of (6.14) and (6.15):

$$\boldsymbol{R}_1 = \begin{bmatrix} 1 & 1 & 1 & 1 \\ 0 & 1 & 0.75 & 0 \\ 0 & 1 & 1 & 0.375 \\ 0.625 & 1 & 1 & 1 \end{bmatrix} \tag{9.25}$$

$$\boldsymbol{R}_2 = \begin{bmatrix} 1 & 0.125 & 0 & 1 \\ 1 & 1 & 1 & 1 \\ 1 & 1 & 1 & 1 \\ 0.25 & 0 & 0 & 1 \end{bmatrix} \tag{9.26}$$

$$\boldsymbol{R}_3 = \begin{bmatrix} 1 & 0.831 & 0.125 & 0 \\ 1 & 1 & 0.5 & 0 \\ 1 & 1 & 1 & 0.5 \\ 1 & 1 & 1 & 1 \end{bmatrix} \tag{9.27}$$

By applying the *first technique* (see its description in Chapter 7), the intersection of (9.25)–(9.27) leads to

$$\boldsymbol{R} = \begin{bmatrix} 1 & 0.125 & 0 & 0 \\ 0 & 1 & 0.5 & 0 \\ 0 & 1 & 1 & 0.375 \\ 0.25 & 0 & 0 & 1 \end{bmatrix} \tag{9.28}$$

Then, applying (5.35) to (9.28), the fuzzy strict preference relation is obtained

$$\mathbf{P} = \begin{bmatrix} 0 & 0.125 & 0 & 0 \\ 0 & 0 & 0 & 0 \\ 0 & 0.5 & 0 & 0.375 \\ 0.25 & 0 & 0 & 0 \end{bmatrix} \tag{9.29}$$

Finally, by applying (7.54) to (9.29), we obtain the fuzzy nondominance degree of each alternative

$$\mathbf{ND} = [0.75 \quad 0.5 \quad 1 \quad 0.625] \tag{9.30}$$

which corresponds to the following ranking of the alternatives: $X_3 \succ X_1 \succ X_4 \succ X_2$.

If AIC is the selected strategy to work in the group environment, the following steps should be executed:

Step 1. The experts provide their preferences, taking into account each criterion, using any preference format among the ones considered in Chapter 6. It is necessary to check the consistency of the individual preference judgments, which are provided in terms of fuzzy preference relations or multiplicative preference relations. As a general rule, the experts are supposed to provide judgments that satisfy at least weak transitivity. But if any of them cannot perform this task, it may be helpful to call an analyst to do so, as in the process of consistency improvement outlined in Chapter 6.

Step 2. All information is converted into nonreciprocal fuzzy preference relations, which are taken as a uniform base to construct the aggregated fuzzy preference relations per criterion and, subsequently, to perform the multiattribute analysis.

Step 3. A collective fuzzy preference relation per criterion is obtained using an aggregation operator such as OWA, min, WAM, or WGM.

Step 4. It is necessary to check whether each collective fuzzy preference relation satisfies at least the weak transitivity condition; otherwise, one of the actions to deal with the problem of collective inconsistency, listed at the end of Section 9.6, should be implemented.

Step 5. Having at hand a consistent collective fuzzy preference relation per criterion, it is possible to apply any one among the multiattribute techniques for the analysis of $\langle \mathbf{X}, \boldsymbol{R} \rangle$ models described in Chapter 7.

Example 9.6. The same problem considered in Example 9.5 is now solved using the AIC strategy jointly with the *third technique* for analyzing the $\langle \mathbf{X}, \boldsymbol{R} \rangle$ model. In this way, first, the experts provided their preferences using different formats. For the sake of simplicity, we omit here the details concerning both Step 1 and Step 2 of this strategy and go straight to Step 3, assuming that all information provided by the experts has already been translated into the format of nonreciprocal fuzzy preference relations. In this way, we have:

- Preferences of E_1:

$$\boldsymbol{R}_1^1 = \begin{bmatrix} 1 & 1 & 1 & 1 \\ 0 & 1 & 1 & 0 \\ 0 & 1 & 1 & 0 \\ 1 & 1 & 1 & 1 \end{bmatrix}, \quad \boldsymbol{R}_2^1 = \begin{bmatrix} 1 & 0 & 0.5 & 1 \\ 1 & 1 & 1 & 1 \\ 1 & 0.44 & 1 & 1 \\ 0.53 & 0 & 0 & 1 \end{bmatrix}, \quad \boldsymbol{R}_3^1 = \begin{bmatrix} 1 & 1 & 0 & 0 \\ 1 & 1 & 0 & 0 \\ 1 & 1 & 1 & 1 \\ 1 & 1 & 1 & 1 \end{bmatrix} \tag{9.31}$$

- Preferences of E_2:

$$\boldsymbol{R}_1^2 = \begin{bmatrix} 1 & 0.33 & 0.4 & 1 \\ 1 & 1 & 1 & 1 \\ 1 & 1 & 1 & 1 \\ 1 & 0.36 & 0.44 & 1 \end{bmatrix}, \quad \boldsymbol{R}_2^2 = \begin{bmatrix} 1 & 0.53 & 0.54 & 1 \\ 1 & 1 & 1 & 1 \\ 1 & 1 & 1 & 1 \\ 1 & 0.61 & 0.51 & 1 \end{bmatrix},$$

$$\boldsymbol{R}_3^2 = \begin{bmatrix} 1 & 1 & 1 & 0.63 \\ 0.63 & 1 & 0.63 & 0 \\ 1 & 1 & 1 & 0.63 \\ 1 & 1 & 1 & 1 \end{bmatrix} \tag{9.32}$$

- Preferences of E_3:

$$\boldsymbol{R}_1^3 = \begin{bmatrix} 1 & 1 & 1 & 1 \\ 0 & 1 & 1 & 0.5 \\ 0 & 1 & 1 & 0.35 \\ 1 & 1 & 1 & 1 \end{bmatrix}, \quad \boldsymbol{R}_2^3 = \begin{bmatrix} 1 & 0 & 0.64 & 1 \\ 1 & 1 & 1 & 1 \\ 1 & 0.72 & 1 & 1 \\ 1 & 0 & 0.53 & 1 \end{bmatrix}, \quad \boldsymbol{R}_3^3 = \begin{bmatrix} 1 & 0.72 & 0.84 & 0 \\ 1 & 1 & 1 & 0 \\ 1 & 1 & 1 & 0 \\ 1 & 1 & 1 & 1 \end{bmatrix} \tag{9.33}$$

- Preferences of E_4:

$$\boldsymbol{R}_1^4 = \begin{bmatrix} 1 & 1 & 1 & 1 \\ 0 & 1 & 1 & 0.23 \\ 0 & 1 & 1 & 0.35 \\ 0.44 & 1 & 1 & 1 \end{bmatrix}, \quad \boldsymbol{R}_2^4 = \begin{bmatrix} 1 & 0.63 & 0.63 & 1 \\ 1 & 1 & 1 & 1 \\ 1 & 1 & 1 & 1 \\ 0.63 & 0 & 0 & 1 \end{bmatrix}, \quad \boldsymbol{R}_3^4 = \begin{bmatrix} 1 & 1 & 0.46 & 0 \\ 1 & 1 & 0.64 & 0 \\ 1 & 1 & 1 & 0.55 \\ 1 & 1 & 1 & 1 \end{bmatrix} \tag{9.34}$$

- Preferences of E_5:

$$\boldsymbol{R}_1^5 = \begin{bmatrix} 1 & 1 & 1 & 1 \\ 0 & 1 & 1 & 0 \\ 0 & 1 & 1 & 0 \\ 1 & 1 & 1 & 1 \end{bmatrix}, \boldsymbol{R}_2^5 = \begin{bmatrix} 1 & 0.41 & 0 & 1 \\ 1 & 1 & 1 & 1 \\ 1 & 0.32 & 1 & 1 \\ 0.6 & 0 & 0.64 & 1 \end{bmatrix}, \boldsymbol{R}_3^5 = \begin{bmatrix} 1 & 1 & 1 & 0.63 \\ 0.63 & 1 & 0.63 & 0 \\ 1 & 1 & 1 & 0.63 \\ 1 & 1 & 1 & 1 \end{bmatrix} \tag{9.35}$$

As in Example 9.5, we consider again the professionals with the same importance, which results in $w_y = 0.2$, $y = 1, 2, \ldots, 5$. In Step 3, the individual fuzzy preference relations are aggregated using WAM in the following collective fuzzy preference relations:

$$\boldsymbol{R}_1^C = \begin{bmatrix} 1 & 0.87 & 0.88 & 1 \\ 0.2 & 1 & 1 & 0.35 \\ 0.2 & 1 & 1 & 0.34 \\ 0.89 & 0.87 & 0.89 & 1 \end{bmatrix} \tag{9.36}$$

$$\boldsymbol{R}_2^C = \begin{bmatrix} 1 & 0.31 & 0.46 & 0.8 \\ 1 & 1 & 1 & 1 \\ 1 & 0.70 & 1 & 1 \\ 0.75 & 0.12 & 0.34 & 1 \end{bmatrix} \tag{9.37}$$

$$\boldsymbol{R}_3^C = \begin{bmatrix} 1 & 0.94 & 0.66 & 0.25 \\ 0.85 & 1 & 0.58 & 0 \\ 1 & 1 & 1 & 0.56 \\ 1 & 1 & 1 & 1 \end{bmatrix} \tag{9.38}$$

In Step 4, a test confirms that (9.36)–(9.38) satisfy the weak-transitivity condition (refer to Chapter 7 for a test of weak-transitivity based on the Orlovsky choice function) and, therefore, can be exploited in the subsequent multiattribute analysis.

In Step 5, the *third technique*, which is described in Chapter 7, is applied. Therefore, first, the fuzzy strict preference relation is calculated for each criterion by means of (5.35), which results in

$$\boldsymbol{P}_1 = \begin{bmatrix} 0 & 0.67 & 0.68 & 0.11 \\ 0 & 0 & 0 & 0 \\ 0 & 0 & 0 & 0 \\ 0 & 0.53 & 0.55 & 0 \end{bmatrix} \tag{9.39}$$

$$\boldsymbol{P}_2 = \begin{bmatrix} 0 & 0 & 0 & 0.25 \\ 0.69 & 0 & 0.3 & 0.88 \\ 0.54 & 0 & 0 & 0.66 \\ 0 & 0 & 0 & 0 \end{bmatrix} \tag{9.40}$$

$$\boldsymbol{P}_3 = \begin{bmatrix} 0 & 0.09 & 0 & 0 \\ 0 & 0 & 0 & 0 \\ 0.34 & 0.46 & 0 & 0 \\ 0.75 & 1 & 0.44 & 0 \end{bmatrix} \tag{9.41}$$

Then, by applying (7.54) to (9.39)–(9.41), the fuzzy sets of nondominated alternatives are calculated for each criterion separately as follows:

$$\mathbf{ND}_1 = [1 \quad 0.33 \quad 0.32 \quad 0.89] \tag{9.42}$$

$$\mathbf{ND}_2 = [0.31 \quad 1 \quad 0.7 \quad 0.12] \tag{9.43}$$

$$\mathbf{ND}_3 = [0.25 \quad 0 \quad 0.56 \quad 1] \tag{9.44}$$

The intersection of (9.42)–(9.44) with the use of (7.76) results in

$$\mathbf{ND} = [0.25 \quad 0 \quad 0.32 \quad 0.12] \tag{9.45}$$

which corresponds to the following ranking of the alternatives: $X_3 \succ X_1 \succ X_4 \succ X_2$.

Finally, AIR involves the execution of the following specific steps to allow the use of the multiattribute decision methods for analyzing the $\langle \mathbf{X}, \boldsymbol{R} \rangle$ model in a group environment:

> ***Step 1.*** The experts provide their preferences taking into account each criterion using any preference format considered in Chapter 6. As in AIC, it is important to assure that all experts supply consistent judgments.
>
> ***Step 2.*** All information is translated to nonreciprocal fuzzy preference relations, which is the preference format utilized to perform the multiattribute analysis.
>
> ***Step 3.*** The problem is solved taking into account, separately, the information provided by each expert. As already mentioned, each expert is free to use a different decision method.
>
> ***Step 4.*** The fuzzy nondominance degrees of each alternative, obtained by dealing with the preferences of each expert, independently, are aggregated using OWA, min, WAM, or WGM.

Example 9.7. The same problem considered in both Examples 9.5 and 9.6 is now solved using strategy AIR. As in Example 9.6, the experts are allowed to provide their preferences using different formats. But, for the sake of simplicity, we assume that each expert provides the same preferences as in Example 9.6. The details concerning both Step 1 and Step 2 of this strategy are omitted: we go directly to Step 3, assuming that all preference information provided by the experts has already been converted to nonreciprocal fuzzy preference relations. Further, we also assume that, in the opinion of all experts, all criteria are equally important and their corresponding weights should not be differentiated. Regarding the decision method selected by each expert, we consider that all of them agreed to use the *first technique* for analyzing the $\langle \mathbf{X}, \boldsymbol{R} \rangle$ model.

We begin execution of the *first technique* by performing the intersection of $\boldsymbol{R}_1$, $\boldsymbol{R}_2$ and $\boldsymbol{R}_3$, with the use of the min operator (obviously, the preferences of each expert are considered separately). Thus we have:

- Expert E_1:

$$\boldsymbol{R}^1 = \begin{bmatrix} 1 & 0 & 0 & 0 \\ 0 & 1 & 0 & 0 \\ 0 & 0.44 & 1 & 0 \\ 0.53 & 0 & 0 & 1 \end{bmatrix} \tag{9.46}$$

- Expert E_2:

$$\boldsymbol{R}^2 = \begin{bmatrix} 1 & 0.33 & 0.4 & 0 \\ 0.63 & 1 & 0.63 & 0 \\ 1 & 1 & 1 & 0.63 \\ 1 & 0.36 & 0.44 & 1 \end{bmatrix} \tag{9.47}$$

- Expert E_3:

$$\boldsymbol{R}^3 = \begin{bmatrix} 1 & 0 & 0.64 & 0 \\ 0 & 1 & 1 & 0 \\ 0 & 0.72 & 1 & 0 \\ 1 & 0 & 0.53 & 1 \end{bmatrix} \tag{9.48}$$

- Expert E_4:

$$\boldsymbol{R}^4 = \begin{bmatrix} 1 & 0.63 & 0.46 & 0 \\ 0 & 1 & 0.64 & 0 \\ 0 & 1 & 1 & 0.35 \\ 0.44 & 0 & 0 & 1 \end{bmatrix} \tag{9.49}$$

- Expert E_5:

$$\boldsymbol{R}^5 = \begin{bmatrix} 1 & 0.41 & 0 & 0.63 \\ 0 & 1 & 0.63 & 0 \\ 0 & 0.32 & 1 & 0 \\ 0.6 & 0 & 0.64 & 1 \end{bmatrix} \tag{9.50}$$

Then, by applying (5.35) to the fuzzy nonstrict preference relations (9.46)–(9.50), separately, the corresponding strict fuzzy preference relations are obtained:

- Expert E_1:

$$\boldsymbol{P}^1 = \begin{bmatrix} 0 & 0 & 0 & 0 \\ 0 & 0 & 0 & 0 \\ 0 & 0.44 & 0 & 0 \\ 0.53 & 0 & 0 & 0 \end{bmatrix} \tag{9.51}$$

- Expert E_2:

$$\boldsymbol{P}^2 = \begin{bmatrix} 0 & 0 & 0 & 0 \\ 0.3 & 0 & 0 & 0 \\ 0.6 & 0.37 & 0 & 0.19 \\ 1 & 0.36 & 0 & 0 \end{bmatrix} \tag{9.52}$$

- Expert E_3:

$$\boldsymbol{P}^3 = \begin{bmatrix} 0 & 0 & 0.64 & 0 \\ 0 & 0 & 0.28 & 0 \\ 0 & 0 & 0 & 0 \\ 1 & 0 & 0.53 & 0 \end{bmatrix} \tag{9.53}$$

- Expert E_4:

$$\boldsymbol{P}^4 = \begin{bmatrix} 0 & 0.63 & 0.46 & 0 \\ 0 & 0 & 0 & 0 \\ 0 & 0.36 & 0 & 0.35 \\ 0.44 & 0 & 0 & 0 \end{bmatrix} \tag{9.54}$$

- Expert E_5:

$$\boldsymbol{P}^5 = \begin{bmatrix} 0 & 0.41 & 1 & 0.03 \\ 0 & 0 & 0.31 & 0 \\ 0 & 0 & 0 & 0 \\ 0 & 0 & 0.64 & 0 \end{bmatrix} \tag{9.55}$$

Finally, by applying (7.54) to (9.51)–(9.55), we obtain the fuzzy nondominance degree of each alternative:

- Expert E_1:

$$\mathbf{ND}^1 = [0.47 \quad 0.56 \quad 1 \quad 1] \tag{9.56}$$

- Expert E_2:

$$\mathbf{ND}^2 = [0.4 \quad 0.63 \quad 1 \quad 0.81] \tag{9.57}$$

- Expert E_3:

$$\mathbf{ND}^3 = [0 \quad 1 \quad 0.36 \quad 1] \tag{9.58}$$

- Expert E_4:

$$\mathbf{ND}^4 = [0.56 \quad 0.37 \quad 0.54 \quad 0.65] \tag{9.59}$$

- Expert E_5:

$$\mathbf{ND}^5 = [1 \quad 0.59 \quad 0.36 \quad 0.97] \tag{9.60}$$

It should be mentioned that, in individual decision-making, the fuzzy nondominance values obtained for experts E_1 and E_3 have not permitted us to distinguish the alternatives X_3 and X_4 and X_2 and X_4, respectively.

As the group agreed that all criteria are equally important, we can go straight to Step 4 and aggregate all individual results with the use of WAM, which gives rise to the following:

$$\mathbf{ND}^C = [0.49 \quad 0.63 \quad 0.65 \quad 0.89] \tag{9.61}$$

All alternatives can be distinguished on the basis of the collective results, which lead to the following ranking: $X_4 \succ X_3 \succ X_2 \succ X_1$.

9.7 Conclusions

In this chapter, we have presented three strategies for dealing with the input of multiple experts in the analysis of $\langle \mathbf{X}, \boldsymbol{R} \rangle$ models, for multiattribute decision-making, namely AIE, AIC, and AIR.

Each strategy allows for the more or less flexible participation of each expert in the decision process. For instance, in AIR, the experts can select any preference format among the ones presented in Chapter 6 to express their preferences, as well as choose any decision technique among the ones described in Chapter 7 to complete the multiattribute analysis. On the other hand, AIE is the least flexible strategy in the sense that it requires all experts to use fuzzy or linguistic estimates to express their respective opinions and the group of experts is handled as a unique individual since the beginning of the decision process, which is supposed to use a unique decision method to solve the problem. AIC can be considered an intermediate strategy: it allows a flexible preference input as in the AIR strategy. But all experts are handled as a unique individual, who is supposed to use a unique decision method as in the AIE strategy.

In practical applications, as discussed in Section 9.2, the group can select the most appropriate strategy in accordance with the requirements of each member and the characteristics of the group as a whole. However, it is important to mention that each strategy may lead to a different result. In Problems 9.2, 9.3, and 9.4 below, the reader is invited to solve the same decision problem, considering the same input of each expert and the same method for multiattribute analysis, but using strategies AIE, AIC, and AIR each time. The results obtained using different strategies do not necessarily coincide. A priori, we cannot identify an approach producing the best results as a general rule, but some aspects may help us to answer this question for each specific application. For instance, when AIC is being used, we should consider that inconsistent collective fuzzy preference relations may reduce the validity of the results. When AIE is being used, very discrepant opinions may reduce the validity of its results if the weighted sum is utilized to aggregate the individual estimates (we should remember that the weighted sum may generate an intermediate collective fuzzy estimate that does not satisfy any expert among the group). However, if none of these aspects affects the results, it becomes more difficult to identify which is the most reasonable result, unless we consider the level of satisfaction of group members.

Exercises

Problem 9.1. Consider the fuzzy preference relations $\boldsymbol{R}^1$and $\boldsymbol{R}^2$ shown below. Check whether they satisfy weak transitivity (hint: verify whether the nonfuzzy, nondominated set $\mathbf{X}^{NFND}$ is nonempty, as described in Chapter 7). Aggregate them in a collective fuzzy preference relation using WAM and the collections of weights given by ($w_1 = 0.4$, $w_2 = 0.6$) and ($w_1 = 0.6$, $w_2 = 0.4$). Verify whether each aggregated fuzzy preference relation satisfies weak transitivity.

$$\boldsymbol{R}^1 = \begin{bmatrix} 1 & 0.8 & 1 \\ 1 & 1 & 1 \\ 0.8 & 0.2 & 1 \end{bmatrix}, \quad \boldsymbol{R}^2 = \begin{bmatrix} 1 & 1 & 0.4 \\ 0.8 & 1 & 1 \\ 1 & 0.8 & 1 \end{bmatrix}$$

Problem 9.2. Consider a decision problem, which involves the ranking of alternatives X_1, X_2, X_3, and X_4 taking into account the criteria F_1, F_2, and F_3, under group settings. Suppose that three experts supplied their opinions by means of the linguistic estimates shown in Figure 9.4. Table 9.6 presents all the linguistic estimates provided by the experts. Obtain the ranking of all alternatives from the best to the worst, using the AIE strategy and the *second technique* for the analysis of $\langle \mathbf{X}, \boldsymbol{R} \rangle$models. Assume that all experts have the same level of influence and that the criteria are arranged in the following order of importance:$p = 1$, $p = 2$, and $p = 3$.

Table 9.6 Evaluation matrix of the alternatives

F_1			
	E_1	E_2	E_3
X_1	*large*	*middle*	*middle*
X_2	*middle*	*middle*	*small*
X_3	*small*	*middle*	*middle*
X_4	*large*	*small*	*middle*
F_2			
	E_1	E_2	E_3
X_1	*middle*	*middle*	*small*
X_2	*very large*	*very large*	*large*
X_3	*large*	*very large*	*very large*
X_4	*very small*	*very small*	*very small*
F_3			
	E_1	E_2	E_3
X_1	*small*	*very small*	*very small*
X_2	*middle*	*small*	*small*
X_3	*large*	*middle*	*middle*
X_4	*large*	*large*	*very large*

Problem 9.3. Consider a decision problem which involves the ranking of alternatives X_1, X_2, X_3, and X_4 taking into account the criteria F_1, F_2, and F_3, under group settings. Suppose that three experts supplied their opinions by means of the nonreciprocal fuzzy preference relations shown next (you can see that these fuzzy preference relations coincide with the ones constructed on the basis of the linguistic estimates provided by each expert in the previous exercise):

- Preferences of E_1:

$$\boldsymbol{R}_1^1 = \begin{bmatrix} 1 & 1 & 1 & 1 \\ 0.62 & 1 & 1 & 0.62 \\ 0 & 0.62 & 1 & 0 \\ 1 & 1 & 1 & 1 \end{bmatrix}, \quad \boldsymbol{R}_2^1 = \begin{bmatrix} 1 & 0 & 0.62 & 1 \\ 1 & 1 & 1 & 1 \\ 1 & 0.62 & 1 & 1 \\ 0 & 0 & 0 & 1 \end{bmatrix}, \quad \boldsymbol{R}_3^1 = \begin{bmatrix} 1 & 0.62 & 0 & 0 \\ 1 & 1 & 0.62 & 0.62 \\ 1 & 1 & 1 & 1 \\ 1 & 1 & 1 & 1 \end{bmatrix}$$

- Preferences of E_2:

$$\boldsymbol{R}_1^2 = \begin{bmatrix} 1 & 1 & 1 & 1 \\ 1 & 1 & 1 & 1 \\ 1 & 1 & 1 & 1 \\ 0.62 & 0.62 & 0.62 & 1 \end{bmatrix}, \quad \boldsymbol{R}_2^2 = \begin{bmatrix} 1 & 0 & 0 & 1 \\ 1 & 1 & 1 & 1 \\ 1 & 1 & 1 & 1 \\ 0 & 0 & 0 & 1 \end{bmatrix}, \quad \boldsymbol{R}_3^2 = \begin{bmatrix} 1 & 0.62 & 0 & 0 \\ 1 & 1 & 0.62 & 0 \\ 1 & 1 & 1 & 0.62 \\ 1 & 1 & 1 & 1 \end{bmatrix}$$

- Preferences of E_3:

$$\boldsymbol{R}_1^3 = \begin{bmatrix} 1 & 1 & 1 & 1 \\ 0.62 & 1 & 0.62 & 0.62 \\ 1 & 1 & 1 & 1 \\ 1 & 1 & 1 & 1 \end{bmatrix}, \quad \boldsymbol{R}_2^3 = \begin{bmatrix} 1 & 0 & 0 & 1 \\ 1 & 1 & 0.62 & 1 \\ 1 & 1 & 1 & 1 \\ 1 & 0 & 0 & 1 \end{bmatrix}, \quad \boldsymbol{R}_3^3 = \begin{bmatrix} 1 & 0.62 & 0 & 0 \\ 1 & 1 & 0.62 & 0 \\ 1 & 1 & 1 & 0 \\ 1 & 1 & 1 & 1 \end{bmatrix}$$

Solve the multiattribute decision problem with the use of the AIC strategy combined with the *second technique* for analyzing the $\langle \mathbf{X}, \boldsymbol{R} \rangle$model. Consider that all experts exhibit the same level of importance and use the WAM aggregation operator in order to obtain the collective fuzzy preference relations. In the multiattribute analysis, suppose that the criteria are arranged in the following order of importance: $p = 1$, $p = 2$, and $p = 3$.

Problem 9.4. Solve the group decision problem from the previous exercise. Now, use strategy AIR, combined with the second technique for analyzing the $\langle \mathbf{X}, \boldsymbol{R} \rangle$ model. In the multicriteria analysis, suppose that the criteria are arranged in the following order of importance: $p = 1$, $p = 2$, and $p = 3$. Consider that all experts have the same importance and use the WAM aggregation operator to generate a collective result.

Problem 9.5. Do the results obtained in Problems 9.2, 9.3 and/or 9.4 coincide? Analyze the results from each exercise and, considering the discussion presented in Section 9.7, indicate which, in your opinion, is the most satisfactory result.

References

Baltes, B.B., Dickson, M.W., Sherman, M.P., *et al.* (2002) Computer-mediated communication and group decision-making: a meta-analysis. *Organizational Behavior and Human Decision Processes*, **87** (1), 156–179.

Ben-Arieh, D. and Chen, Z. (2006) Linguistic aggregation and consensus measure for autocratic decision-making using group recommendations. *IEEE Transactions on Systems, Man, and Cybernetics – Part A: Systems and Humans*, **36** (3), 558–568.

Bernardes, P., Ekel, P., and Parreiras, R. (2009) A new consensus scheme for multicriteria group decision-making under linguistic assessments, in *Mathematics and Mathematical Logic: New Research* (eds P. Miloslav and I. Ercegovaca), Nova Science, New York, pp. 67–86.

Bernardes, P., Ekel, P., Kotlarewski, J., and Parreiras, R. (2008) Fuzzy set based multicriteria decision-making and its applications, in *Progress on Nonlinear Analysis* (ed. E. T. Hoffman), Nova Science, Hauppauge, pp. 247–272.

Bui, T.X. and Jarke, M. (1986) Communications design for Co-oP: a group decision support system. *ACM Transactions on Office Information Systems*, **4** (2), 81–103.

Ekel, P., Queiroz, J., Parreiras, R., and Palhares, R. (2009) Fuzzy set based models and methods of multicriteria group decision-making. *Nonlinear Analysis: Theory, Methods & Applications*, **71** (12), e409–e419.

Forman, E. and Peniwati, K. (1998) Aggregating individual judgments and priorities with the Analytic Hierarchy Process. *European Journal of Operational Research*, **108** (1), 165–169.

García-Lapresta, J.L. and Meneses, L.C. (2005) Individual-valued preferences and their aggregation: consistency analysis in a real case. *Fuzzy Sets and Systems*, **151** (2), 269–284.

Herrera-Viedma, E., Herrera, F., and Chiclana, F. (2002) A consensus model for multiperson decision-making with different preference structures. *IEEE Transactions on Systems, Man, and Cybernetics – Part A: Systems and Humans*, **32** (3), 394–402.

Herrera-Viedma, E., Herrera, F., Chiclana, F., and Luque, M. (2004) Some issues on consistency of fuzzy preference relations. *European Journal of Operational Research*, **154** (1), 98–109.

Hsu, H.M. and Chen, C.T. (1996) Aggregation of fuzzy opinions under group decision-making. *Fuzzy Sets and Systems*, **79** (3), 279–285.

Lu, C., Lan, J., and Wang, Z. (2006) Aggregation of fuzzy opinions under group decision-making based on similarity and distance. *Journal of Systems Science and Complexity*, **19** (1), 63–71.

Lu, J., Zhang, G., Ruan, D., and Wu, F. (2007) *Multi-objective Group Decision-Making: Methods, Software and Applications with Fuzzy Set Techniques*, Imperial College Press, London.

Ma, J., Fan, Z.-P., Jiang, Y.-P., *et al.* (2006) A method for repairing the inconsistency of fuzzy preference relations. *Fuzzy Sets and Systems*, **157** (1), 20–33.

Matsatsinis, N.F. and Samaras, A.P. (2001) MCDA and preference disaggregation in group decision support systems. *European Journal of Operational Research*, **130** (2), 414–429.

Ng, K.-C. (1992) Consensus diagnosis: a simulation study. *IEEE Transactions on Systems, Man, And Cybernetics*, **22** (5), 916–928.

Ng, K.-C. and Abramson, B. (1990) Consensus in a multi-expert system. Proceedings of the 1990 ACM Annual Conference on Cooperation, Washington, DC, pp. 351–357.

Peneva, V. and Popchev, I. (2003) Properties of the aggregation operators related with fuzzy relations. *Fuzzy Sets and Systems*, **139** (3), 615–633.

Saaty, T. (1980) *The Analytic Hierarchy Process*, McGraw-Hill, New York.

Sengupta, K. (1998) Fuzzy preference and Orlovsky choice procedure. *Fuzzy Sets and Systems*, **93** (2), 231–234.

Thompson, L.F. and Coovert, M.D. (2002) Stepping up to the challenge: a critical examination of face-to-face and computer-mediated team decision-making. *Group Dynamics: Theory, Research, and Practice*, **6** (1), 52–64.

Zhang, G. and Lu, J. (2009) A linguistic intelligent user guide for method selection in multi-objective decision support systems. *Information Sciences*, **179** (14), 2299–2308.

10

Use of Consensus Schemes in Group Decision-Making

In this chapter, we present a suite of procedures for achieving a consensus in the analysis of discrete multicriteria decision-making problems, which involves the evaluation, comparison, choice, prioritization, and/or ordering of alternatives, in a group environment. The chapter brings together two different approaches for the construction of collective opinions under a rubric of satisfactory consensus: the consensus schemes and the procedures for the formation of an optimized consensus. Whereas the former approach requires experts to review and update their respective opinions within the process of an iterative discussion, the latter approach represents an attempt to automate the process of constructing and improving the collective opinion, in such a way that the level of consensus in the group becomes elevated. Each approach has its own advantages and drawbacks. The selection of the most suitable method for a specific application depends mostly on the available time and on the cost of facilitating meetings among the members of the group.

10.1 Consensus in Group Decision-Making

As discussed in Chapter 9, when dealing with multicriteria decision problems under group settings, conflicting opinions among group members are very likely to occur, even in a cooperative environment. As Shanteau (2001) points out, disagreements between domain experts are inevitable and should not be taken as evidence of the incompetence of any expert, but a reflection of the way that experts think and a consequence of the type of work they do. For instance, in medicine, there may be diverse treatments for an illness. If one treatment does not work as expected, the physician seeks another one (Shanteau, 2001). It should not be surprising to find experts disagreeing about which course of action should be taken. However, as in practice only one strategy can be implemented, finding a unique solution is crucial, in spite of the discordances among the experts.

Fuzzy Multicriteria Decision-Making: Models, Methods and Applications Witold Pedrycz, Petr Ekel and Roberta Parreiras

The importance of achieving a satisfactory level of concordance among the experts has motivated several researchers to develop procedures for increasing the rationality of the collective solution, as well as the efficiency of the discussion among experts. If we concentrate our attention on the available procedures, we can note that they follow two types of approaches (Ben-Arieh and Chen, 2006; Ekel *et al.*, 2009):

- The consensus schemes. These consist of a systematic and iterative discussion process, implemented under the supervision of a moderator, with the intention of reducing the discordance among opinions. The consensus is achieved as long as the experts move from their original positions toward a predominant opinion.
- Procedures for constructing an optimized consensus. These consist of a method for obtaining a suitable adjustment of the weight associated with the opinion of each expert, in order to improve a consensus index. In this approach, the experts are not supposed to modify their opinions toward a consensus. The weighted aggregation of individual opinions, with a suitable adjustment of weights, is the exclusive arbitration scheme for defining the collective decision.

Both approaches have some advantages and disadvantages, which can make one or the other approach most suitable for a certain application. As Ekel *et al.* (2009) points out, the main disadvantage of automatically constructing an optimized consensus lies in the fact that the opinion of a discordant expert, having deep knowledge of the problem, can be easily neglected. This may occur because the process of forming an optimal set of weights may demand an excessive reduction of the weight associated with his/her opinion. Further, the process of obtaining an adequate set of weights may demand significant computational effort. On the other hand, the consensus scheme also has some drawbacks. In order to achieve an adequate level of consensus, a discordant expert may have to change drastically his/her initial position and, maybe, in an unjustified way. Moreover, the experts may be repeatedly invited to review their respective opinions, making the discussion last too long and, as a consequence, being too ineffective and expensive, if not frustrating.

However, it should be mentioned that, in spite of the drawbacks of consensus schemes, the current literature contains several valuable arguments in favor of its use. As has been recognized in several studies (see for instance Bui and Jarke, 1986; Madu and Kuei, 1995; Salo, 1995; Jiang and Klein, 2000), human preferences are not rigid, in the sense that they may be formed (when they do not even exist at the beginning of the decision process) or changed during the discussion. Further, it is also true that discordant members may have access to exclusive information which could influence their respective opinions. Hence, we cannot neglect the fact that, by promoting further discussions among the participants, it is possible to modify their opinions, in such a way that their discordances are minimized. Furthermore, it is also possible to gather vital information for the decision process, which can lead the group toward more rational and better justified outcomes (Madu and Kuei, 1995; Salo, 1995).

In this way, as a general rule, when it is desirable to achieve a consensus among the experts but it is impossible to allow group members to discuss their conflicting opinions, due to the large size of the group or to logistic, timing, or monetary restrictions, the first approach should be utilized. Otherwise, whenever it is feasible to give all experts the opportunity to discuss their discordances, the consensus scheme should be applied.

10.2 Consensus Schemes: Definition and Motivation

A consensus scheme can be defined as an interactive multistage process, in which the experts discuss the problem in a systematic way, toward a consensual decision (Herrera-Viedma, Herrera, and Chiclana, 2002). The discussion is conducted by a human or artificial moderator, playing the role of an impartial arbiter who has authority to indirectly interfere in the process, with the purpose of helping the group to achieve better solutions.

Intuitively, we know that discussions on concordant assignments are meaningless, in the sense that they usually have no power to change the collective decision. Thus, in practice, it is more fruitful to rapidly bypass the concordant opinions and intensify efforts for minimizing the discordances.

A priori, discordant opinions should not be neglected, as they can help the group to identify sources of crucial information for the decision (Madu and Kuei, 1995). Actually, in practice, a discordant expert with persuasive abilities can convince other members of the group to modify their own positions and, as a result, the predominant position in the group can move toward a more justifiable decision.

Taking all of this into consideration, the guiding principle of the discussions among experts, under the context of group decision-making, should be to gather more information at each round in an effective way. Inadequate interactions among members affect the efficiency of the discussions, resulting in unsatisfactory outcomes and/or time-consuming and unfruitful debates. In this context, the participation of a human or artificial moderator becomes essential, in order to conduct the discussion and reduce time losses inherent in group meetings. The basic dynamic iteration among the experts and the moderator should be as follows: at each cycle, the moderator should identify the least concordant specialist from the group in order to invite him/her to review his/her opinion. The invited expert is supposed to acquire more information on the problem in order to review his/her opinion, which can be changed or maintained. As already mentioned, in the last case, it may be helpful to invite the expert to let the group know the reason for his/her assignment, because that explanation possibly contains original information that can change the opinions of the other experts (Madu and Kuei, 1995; Eklund, Rusinowska, and De Swart, 2007). If the expert refuses to change his/her opinion, then the moderator should identify and invite the second least concordant expert (and so forth), in order to allow the other members of the group to review their opinions, taking into account the arguments of the least concordant expert. This cyclic iteration is repeated until a stop condition has been satisfied.

Ideally, the condition for terminating the discussion should be the achievement of a perfect concordance among all experts. However, in reality, it is implemented by means of verifying whether the current level of an index of consensus is higher than a minimum threshold value. Other conditions are also admissible, but we leave this subject till Section 10.6. In the next section, we focus on some basic tools utilized by the moderator to control the discussion: the consensus and the concordance indices.

10.3 Fuzzy Concordance and Fuzzy Consensus Measures

Let us focus on methods designed for constructing a consensus within a group of experts $\mathbf{E} = \{E_1, E_2, \ldots, E_v\}$ on a solution for a discrete multicriteria decision problem, which involves the ranking of a discrete set of alternatives $\mathbf{X} = \{X_1, X_2, \ldots, X_n\}$, taking into account a set of criteria $\mathbf{F} = \{F_1, F_2, \ldots, F_q\}$.

The fuzzy environment offers several types of mathematical models which can be utilized as common media for expressing, associating, and comparing the preferences of each expert. In Chapter 9, we considered several aggregation rules for associating the individual preferences expressed in terms of fuzzy estimates, fuzzy preference relations, or fuzzy sets of nondominated alternatives. Here, we move further on and study the indices of concordance and consensus, which are essential tools for measuring the degree of compatibility between the preferences reflected in those different forms.

The concordance index is a function that quantifies the level of similarity or correspondence between any pair of opinions. For practical purposes, it is supposed to satisfy some conditions such as (García-Lapresta, 2008):

- it achieves its maximum value only if both opinions are identical;
- the value of the index depends on the agreement between two opinions, regardless of which expert is responsible for each opinion.

In consensus schemes, the main use of a concordance index is associated with the identification of the least concordant expert in each cycle of the discussion. As already mentioned, this expert is supposed to review his/her opinion or to explain it to the group. With the use of such an index, it is possible to calculate the level of concordance between the current opinion of each expert and the group's temporary opinion or the level of concordance among the opinions of all experts, in order to identify who is the expert with the most discrepant opinion within the group.

The consensus index assumes values in the unit interval and is modeled as a function that quantifies how far a group of experts is from perfect unanimity. Here, the value of 1 corresponds to full and unanimous concordance, whereas 0 corresponds to nonexistent concordance. Intermediate values, between 0 and 1, are also possible to reflect levels of partial agreement among all experts.

The classical definition of the term "consensus" as a unanimous concordance among individuals has been considered excessively binary (two valued) or rigid to characterize the agreement among group members. A Boolean (two-valued) notion of consensus does not allow different levels of concordance to be distinguished among experts. Such discrimination is important due to the fact that, in practice, the discussion process toward a consensus is frequently interrupted before a perfect concordance among experts is achieved, considering that a perfect consensus is almost unachievable. In this way, with the use of a consensus index, it becomes possible to interrupt the discussion at an earlier moment, as soon as an acceptable level of concordance has been achieved.

Next, we describe different indices of concordance and consensus which have been proposed in the literature and can be utilized to deal with information expressed in three different formats: fuzzy estimates, fuzzy preference relations, and fuzzy sets of nondominance.

Let us begin by considering the comparison of opinions expressed in terms of fuzzy estimates. In this case, we can utilize a concordance index that reflects the level of similarity between a pair of fuzzy estimates and a consensus index that reflects the mean level of similarity among a collection of fuzzy estimates. The concordance index proposed in Hsu and Chen (1996) and improved in Lu, Lan, and Wang (2006), which combines both fuzzy distance and fuzzy similarity concepts, allows a fair comparison between a pair of fuzzy estimates.

The weighted similarity between the fuzzy estimates $F_p^y(X_k)$ and $F_p^z(X_k)$, which are provided by the yth expert and the zth expert, respectively, is given by (Lu, Lan, and Wang, 2006)

$$S_w\left(F_p^y(X_k), F_p^z(X_k)\right) = \frac{\int_x \min\left(F_p^y(X_k), F_p^z(X_k)\right)^2 \mathrm{d}X}{\int_x \max\left(F_p^y(X_k), F_p^z(X_k)\right)^2 \mathrm{d}X}. \tag{10.1}$$

where the ratio between the two integrals reflects the proportion of the concordant area $\int_x \min\{F_p^y(X_k), F_p^z(X_k)\}\,\mathrm{d}X$ to the total area $\int_x \max\{F_p^y(X_k), F_p^z(X_k)\}\,\mathrm{d}X$ associated with both fuzzy estimates.

The distance between $F_p^y(X_k)$ and $F_p^z(X_k)$ can be calculated in the following form (Lu, Lan, and Wang, 2006):

$$D_h\left(F_p^y(X_k), F_p^z(X_k)\right) = \frac{1}{2}\left[\int_x \left|F_p^y(X_k) - F_p^z(X_k)\right| \mathrm{d}X + d_{inf}\left(F_p^y(X_k), F_p^z(X_k)\right)\right] \tag{10.2}$$

In (10.2), the integral corresponds to the Hamming distance between $F_p^y(X_k)$ and $F_p^z(X_k)$ and the term d_{inf} is given by

$$d_{inf} = \min_{\substack{a \in \mathrm{Supp}(F_p^y(X_k)) \\ b \in \mathrm{Supp}(F_p^z(X_k))}} (|a - b|) \tag{10.3}$$

where Supp stands for the support of a fuzzy set, as given by expression (2.12).
In this way, if we suppose for instance that the membership functions of $F_p^y(X_k)$ and $F_p^z(X_k)$ correspond to the trapezoidal fuzzy numbers $\{a_1, a_2, a_3, a_4\}$ and $\{b_1, b_2, b_3, b_4\}$, respectively, then $d_{inf} = \inf\{|a - b|, a \in [a_1, a_4], b \in [b_1, b_4]\}$. The inclusion of the term d_{inf} in (10.2) is important to adequately handle fuzzy estimates with no intersection (Lu, Lan, and Wang, 2006).

Finally, the level of concordance between $F_p^y(X)$ and $F_p^z(X)$ can be calculated as follows (Lu, Lan, and Wang, 2006):

$$S_{FE}^{y,z}\left(F_p^y(X_k), F_p^z(X_k)\right) = \beta S_w\left(F_p^y(X_k), F_p^z(X_k)\right) + (1 - \beta)\left(1 - D_h\left(F_p^y(X_k), F_p^z(X_k)\right)\right) \tag{10.4}$$

where the parameter β, assuming values in the range $0 \leq \beta \leq 1$, allows one to adjust the level of influence of S_w and D_h on the concordance value $S_{FE}^{y,z}$.

As indicated in Bernardes *et al.* (2009), in certain cases it is beneficial to normalize D_h, by dividing it by a constant, in order to guarantee that $0 \leq \bar{D}_h \leq 1$. Obviously, this constant will depend on the range of the universe of discourse being considered. This normalization will facilitate empirically adjusting the value of β in (10.4).

The level of consensus across the group per alternative can be calculated as the arithmetic average

$$C_{FE}(X_k) = \frac{1}{v}\sum_{y=1}^{v} S_{FE}^{y,C}\left(F_p^y(X_k), F_p^C(X_k)\right) \tag{10.5}$$

where $F_p^C(X_k)$ represents the collective fuzzy estimate which may be obtained by means of expression (9.1).

Table 10.1 Level of concordance and consensus

	$S_{FE}^{1,C}$	$S_{FE}^{2,C}$	$S_{FE}^{3,C}$	Consensus
Concordance	0.71	0.18	0.13	0.35

Example 10.1. As a continuation of Example 9.1, let us identify the least concordant expert in the group and verify whether the level of consensus in the group is satisfactory. In this specific case, it is assumed that the level of consensus is acceptable if it exceeds 0.5. By applying (10.4), with $\beta = 0.5$, we obtain the levels of concordance between the opinion of each expert and the collective opinion as shown in Table 10.1. As can be seen, E_3 is the least concordant expert in the group. The consensus level, calculated with (10.5), is equal to 0.35 and, therefore, it is considered unacceptable.

Let us consider indices for comparing preferences expressed in terms of nonreciprocal fuzzy preference relations. Given a pair of alternatives X_k and X_l, a simple way of calculating the level of concordance between the preferences of two experts, namely E_y and E_z, consists of calculating the differences

$$S^{y,z}(X_k, X_l) = \frac{2 - (|\boldsymbol{R}^y(X_k, X_l) - \boldsymbol{R}^z(X_k, X_l)| + |\boldsymbol{R}^y(X_l, X_k) - \boldsymbol{R}^z(X_l, X_k)|)}{2} \tag{10.6}$$

where $\boldsymbol{R}^y$ corresponds to the nonreciprocal fuzzy preference relation supplied by the yth expert.

The mean level of concordance between the preferences of E_y and E_z can also be calculated for a single alternative X_k, with the use of the expression

$$SX_k^{y,z} = \frac{1}{(n-1)} \sum_{l=1;l\neq k}^{n} S^{y,z}(X_k, X_l) \tag{10.7}$$

or for the entire set of alternatives (which corresponds to the mean level of concordance per relation), with the use of the expression

$$SR^{y,z} = \frac{1}{n} \sum_{k=1}^{n} SX_k^{y,z} \tag{10.8}$$

Finally, the level of consensus in the group can be calculated for a pair of alternatives by aggregating $S^{y,C}(X_k, X_l)$, $y = 1, 2, \ldots, v$, with the use of the arithmetic mean operator

$$C(X_k, X_l) = \frac{1}{v} \sum_{y=1}^{v} S^{y,C}(X_k, X_l) \tag{10.9}$$

where $S^{y,C}(X_k, X_l)$ is the level of concordance between the preferences of E_y and the collective preferences $\boldsymbol{R}_p^C$, which can be calculated with the use of (10.6). As discussed in Chapter 9, the collective preferences $\boldsymbol{R}_p^C$ may be obtained by means of different aggregation operators as given by (9.3), (9.4), (9.5) or (9.6).

The mean level of consensus in the group can be calculated for a single alternative X_k as

$$CX_k = \frac{1}{(n-1)} \sum_{l=1, l\neq k}^{n} C(X_k, X_l) = \frac{1}{v} \sum_{y=1}^{v} SX_k^{y,C} \tag{10.10}$$

or for the entire set of alternatives (that is, the level of consensus per relation) as given by

$$CR = \frac{1}{n} \sum_{k=1}^{n} CX_k = \frac{1}{v} \sum_{y=1}^{v} SR^{y,C} \tag{10.11}$$

Example 10.2. Let us consider three fuzzy preference relations:

$$R^1 = \begin{bmatrix} 1 & 1 & 1 \\ 0.3 & 1 & 1 \\ 0 & 0.3 & 1 \end{bmatrix} \tag{10.12}$$

$$R^2 = \begin{bmatrix} 1 & 1 & 1 \\ 0.9 & 1 & 1 \\ 0.9 & 1 & 1 \end{bmatrix} \tag{10.13}$$

$$R^3 = \begin{bmatrix} 1 & 0.3 & 1 \\ 1 & 1 & 1 \\ 0.3 & 0 & 1 \end{bmatrix} \tag{10.14}$$

The aggregation of (10.12)–(10.14) by applying WAM with equal weights yields

$$R^C = \begin{bmatrix} 1 & 0.77 & 1 \\ 0.73 & 1 & 1 \\ 0.4 & 0.43 & 1 \end{bmatrix} \tag{10.15}$$

Table 10.2 shows the concordance level per alternative, calculated for each expert by using (10.7), and the concordance level per relation, obtained for each expert by using (10.8). Furthermore, it also shows the mean level of consensus per alternative, calculated by using (10.10), and the mean level of consensus per relation, obtained with the use of (10.11). If the moderator considers that a level of consensus per relation under 0.8 is unsatisfactory, then

Table 10.2 Levels of concordance

Concordance per alternative	E_1	E_2	E_3	Consensus per alternative
X_1	0.73	0.78	0.79	0.77
X_2	0.8	0.76	0.71	0.76
X_3	0.87	0.73	0.87	0.82
Concordance per relation	0.80	0.76	0.79	–
Consensus per relation	0.78			

the least concordant expert, that is, E_2, as can be seen in Table 10.2, can be invited to review his/her opinion.

Given a pair of alternatives, the level of concordance between the preferences of each expert and the collective preferences should reflect these two types of differences. One type of difference is associated with the different orderings of the alternatives determined by the preferences of an expert and by the collective preferences. The other type is associated with the situation when the preferences of an expert and the collective preferences determine the same ordering of the alternatives, but they differ in the intensity of the preference of one alternative over the other.

For practical purposes, few may agree that the first type of difference is more critical than the second type, in the sense that it is more valuable to invite an expert with a different ordering of the alternatives to review or explain his/her preferences than an expert with the same ordering of the alternatives, but with different levels of preferences. A simple procedure is proposed here in order to accentuate the effect of the former type of difference in the calculation of the level of concordance, so that the concordance index (10.8) allows one to identify the expert whose preferences are different than the collective preferences in a more critical way. This procedure implies consideration of a modified version of the collective fuzzy preference relation to calculate the concordance index. As long as the original collective fuzzy preference relation satisfies weak transitivity, the modified collective fuzzy preference relation can be obtained through a nonlinear transformation of the original collective fuzzy preference relation in accordance with the following rule: if $\boldsymbol{R}(X_k, X_l) < 1$ and $\boldsymbol{R}(X_l, X_k) < 1$, pick the maximum value between $\boldsymbol{R}(X_k, X_l)$ and $\boldsymbol{R}(X_l, X_k)$ and set it equal to one. Considering that the matrix corresponding to the collective fuzzy preference relation may present pairs of entries verifying $\boldsymbol{R}(X_k, X_l) < 1$ and $\boldsymbol{R}(X_l, X_k) < 1$, this simple modification can provide a way of accentuating the effect of differences in the ordering of the alternatives in the calculation of the indices of concordance (10.6)–(10.8). The following example demonstrates the use of this procedure.

Example 10.3. The reader can easily confirm that, according to (10.12) and (10.15), the alternatives can be ordered as $X_1 \succ X_2 \succ X_3$. On the other hand, (10.13) is associated with $X_1 \succ X_2 \sim X_3$ and (10.14) is associated with $X_2 \succ X_1 \succ X_3$. By intuition, we know that (10.14) corresponds to the least concordant preferences with the collective preferences. However, as can be seen in Table 10.2, the level of concordance per relation calculated with (10.6)–(10.8) indicates that E_2 is the least concordant expert.

On the other hand, considering that (10.15) satisfies weak transitivity, we can apply (10.6)–(10.8) to the following version of the collective fuzzy preference relation

$$\boldsymbol{R}^{MC} = \begin{bmatrix} 1 & 1 & 1 \\ 0.73 & 1 & 1 \\ 0.4 & 0.43 & 1 \end{bmatrix} \tag{10.16}$$

which is derived from (10.15) in accordance with the procedure described above (note that, in (10.15), $\boldsymbol{R}(X_1, X_2)$ is equal to 0.77 and, in (10.16), it has changed to 1). The resulting indices of concordance are shown in Table 10.3. It is worth noting that, coherently with our expectations, E_3 has become the least concordant expert in the group.

Table 10.3 Levels of concordance for the modified collective fuzzy preference relation

Concordance per alternative	E_1	E_2	E_3
X_1	0.79	0.83	0.73
X_2	0.86	0.82	0.65
X_3	0.87	0.73	0.87
Concordance per relation	0.84	0.79	0.75

Finally, let us consider indices of concordance and consensus adequate for comparing the opinions of different experts, expressed in terms of the degree of fuzzy nondominance of each alternative. As discussed in Herrera-Viedma, Herrera, and Chiclana (2002), different cardinal results (which correspond to the degree of fuzzy nondominance of each alternative) may lead to the same ranking of the alternatives. For instance, both $\mathbf{ND}^1 = [0.8\ 1\ 0.6\ 0.7]$ and $\mathbf{ND}^2 = [0.81\ 0.9\ 0.3\ 0.57]$ result in $X_2 \succ X_1 \succ X_4 \succ X_3$. Although there is no perfect concordance on the cardinal rating of each alternative, there may be a perfect consensus on their ranking. For this reason, it is more reasonable to compare the final results of the multicriteria analysis by taking into account the final ranking of the alternatives and not their respective fuzzy nondominance degrees.

Therefore, let us assume that we have at hand the ranking of all alternatives, obtained in accordance with the opinion of the group as a whole, as well as with the opinion of each expert individually. One way of measuring the concordance between the order of an alternative X_k for the yth expert and the zth expert is by using a decreasing function such as (Herrera-Viedma, Herrera, and Chiclana, 2002)

$$SO_k^{y,z} = 1 - \left(\frac{|O_k^y - O_k^z|}{n-1}\right)^b \qquad (10.17)$$

where the term O_k^y corresponds to the position of alternative X_k, taking into account the results obtained by the yth expert. The constant b in (10.17) can assume any value in the interval [0, 1]. In particular, when b is close to one, the concordance measure is less rigorous than the case when b is close to zero. Herrera-Viedma, Herrera, and Chiclana (2002) suggest the use of 0.5, 0.7, 0.9, or 1. Besides, it is important to indicate that, when two or more alternatives are considered indistinguishable because they present exactly the same level of fuzzy nondominance, the value of O_k for each of them is given by the mean value of the positions that they would have if they were ranked in descending order of their values. For instance, if alternatives X_4 and X_5 are in second position in the ranking, then $O_4 = O_5 = (2+3)/2 = 2.5$.

Expression (10.17) determines the level of concordance between the results of E_y and E_z, restricted to alternative X_k. The average level of concordance between the results of E_y and E_z, extended to the entire set of alternatives, is given by

$$SO^{y,z} = \frac{1}{n}\sum_{k=1}^{n} SO_k^{y,z} \qquad (10.18)$$

Table 10.4 Position of each alternative in the rankings obtained for each expert and for the group

	X_1	X_2	X_3
E_1	3	2	1
E_2	1	2	3
E_3	3	2	1
Group	3	1	2

The level of consensus can be determined for a single alternative X_k as (Herrera-Viedma, Herrera, and Chiclana, 2002)

$$CO_k = \frac{1}{v} \sum_{y=1}^{v} SO_k^{y,C} \tag{10.19}$$

where $SO_k^{y,C}$ is the average level of concordance between the results of E_y and the collective results calculated with the use of (10.18).

The mean level of consensus taken over all alternatives is given by

$$CO = \frac{1}{n} \sum_{k=1}^{n} CO_k \tag{10.20}$$

Example 10.4. Consider that the results obtained by three experts correspond to $\mathbf{ND}^1 = [0.4\ 0.7\ 0.8]$, $\mathbf{ND}^2 = [0.9\ 0.4\ 0.2]$, and $\mathbf{ND}^3 = [0.1\ 0.4\ 0.5]$. By using the min operator to aggregate these results, we obtain $\mathbf{ND}^C = [0.1\ 0.4\ 0.2]$. Table 10.4 shows the ranking of the alternatives derived from the individual, as well as the collective results. Table 10.5 shows the levels of concordance between the ranking obtained in accordance with the preferences of each expert and the ranking obtained for the group, calculated with the use of (10.17) and considering $b = 0.5$. This table also shows the consensus level per alternative and the mean level of consensus, calculated with the use of (10.19) and (10.20), respectively.

If the moderator considers the mean level of consensus under 0.6 being unsatisfactory, then the least concordant expert, which is clearly E_2, can be invited to review his/her results.

Now, let us reconsider the problem of identifying the least concordant fuzzy preference relation in the group. When we want to guarantee that the expert with the fuzzy preference relation

Table 10.5 Concordance and consensus levels

Concordance per alternative	X_1	X_2	X_3	Mean concordance for all alternatives
E_1	1	0.29	0.29	0.53
E_2	0	0.29	0.29	0.19
E_3	1	0.29	0.29	0.53
Consensus per alternative	0.67	0.29	0.29	–
Mean consensus for all alternatives	0.42	–		

Table 10.6 Position of each alternative in the rankings obtained for each expert and for the group

	X_1	X_2	X_3
E_1	1	2	3
E_2	1	2.5	2.5
E_3	2	1	3
Group	1	2	3

associated with the least concordant ordering of the alternatives is identified, we can consider the use of two types of concordance indices in a lexicographic way. First, we can use indices (10.17) and (10.18) to identify the least concordant ordering of the alternatives. The use of these two indices requires the application of (5.35) and (7.54), in order to rank the alternatives based on their respective nondominance degrees. If it is not possible to distinguish which one among at least two different orderings is the least concordant ordering, then it is possible to use (10.6)–(10.8) in an attempt to differentiate them. The following example demonstrates the use of (10.17) and (10.18) to identify the least concordant fuzzy preference relation based on the ordering of the alternatives associated with each of them.

Example 10.5. By applying (5.35) and (7.54) to each fuzzy preference relation (10.12), (10.13), (10.14), and (10.15), we obtain the following fuzzy sets of nondominated alternatives:

$$\mathbf{ND}^1 = \begin{bmatrix} 1 & 0.3 & 0 \end{bmatrix} \tag{10.21}$$

$$\mathbf{ND}^2 = \begin{bmatrix} 1 & 0.9 & 0.9 \end{bmatrix} \tag{10.22}$$

$$\mathbf{ND}^3 = \begin{bmatrix} 0.3 & 1 & 0 \end{bmatrix} \tag{10.23}$$

$$\mathbf{ND}^C = \begin{bmatrix} 1 & 0.97 & 0.4 \end{bmatrix} \tag{10.24}$$

Table 10.6 shows the ranking of the alternatives derived from the individual, as well as the collective results. Table 10.7 shows the levels of concordance between the ranking obtained in accordance with the preferences of each expert and the ranking obtained for the group, calculated with the use of (10.17) and considering $b = 0.5$. As can be seen, the preferences of E_3 provided the least concordant ordering of the alternatives, which is also in accordance with the results of Example 10.3.

Table 10.7 Concordance levels

Concordance per alternative	X_1	X_2	X_3	Mean concordance for all alternatives
E_1	1	1	1	1
E_2	1	0.5	0.5	0.67
E_3	0.29	0.29	1	0.53

10.4 Moderator Interventions

The moderator's (or facilitator's) role in group decision-making is one of controlling the exchange of information among the experts during their discussions. The ideal moderator is supposed to be impartial and make only neutral interferences in the process of communication among the experts. It should be emphasized that the moderator is not meant to make decisions for the group, but is supposed to enhance the ability of the group to make decisions (Griffith, Fuller, and Northcraft, 1998).

Ekel *et al.* (2009) distinguished the following three types of direct interference of the moderator in the process of constructing a consensus:

- an invitation to the least concordant expert to review his/her opinion;
- an adjustment of the weight associated with each opinion in the construction of the collective opinion (this adjustment is usually made with the purpose of reducing the influence of a discordant opinion or of an inconsistent opinion);
- the activation of a computational procedure for constructing an optimized consensus (this kind of interference will be discussed in Section 10.5).

Here, we consider an indirect type of moderator interference, which is related to a feedback mechanism for communicating with the least concordant expert, that is, the differences between his/her own opinions and the collective opinion. Theoretically, this information should be utilized by the least concordant expert only as a reference for modifying or explaining his/her discordant position. However, it is important to indicate that this feedback does not always contribute to achieving a consensus. Although all experts are expected to contribute their true opinions, when the intermediate results do not satisfy a member of the group, it is not rare for this expert to exaggerate his/her judgments in order to move the aggregated opinion toward his/her opinion. Obviously, one way of mitigating this undesirable effect is by presenting the intermediate results only to the moderator (Salo, 1995). Therefore, it should be emphasized that, in real applications, the use of a feedback mechanism should always be pondered by a human moderator, as in some cases it may have undesirable effects.

Herrera-Viedma, Herrera, and Chiclana (2002) proposed a simple and effective feedback mechanism. It requires as input the ranking of the alternatives for each expert and for the group. Based on the positions of each alternative O_k^y and O_k^C, for the yth expert and for the group, it generates three types of instructions:

- If $O_k^C - O_k^y < 0$, then increase (improve) evaluations associated with X_k.
- If $O_k^C - O_k^y = 0$, then do not modify evaluations associated with X_k.
- If $O_k^C - O_k^y > 0$, then decrease (worsen) evaluations associated with X_k.

When the opinions are expressed in terms of the fuzzy nondominance level of each alternative, the use of this feedback mechanism is straightforward. It is interesting to observe that the feedback mechanism can also be applied when the preferences are expressed in terms of fuzzy preference relations, provided that the degree of fuzzy nondominance of the alternatives is derived from the fuzzy preference relations, with the use of (5.35) and (7.54). Other relevant feedback mechanisms, which admit incomplete and/or inconsistent fuzzy preference relations as input, can be found in Herrera-Viedma *et al.* (2007).

10.5 Optimal Consensus in a Fuzzy Environment

Here we present a procedure, proposed in Ben-Arieh and Chen (2006), for constructing an optimized (and dictatorial) consensus for the results of a decision problem. By means of a systematic adjustment of the weight associated with the opinion of each expert, the procedure tries to construct a collective result in such a way that the level of consensus, reflected by a specific index, is increased.

Considering that the weights determine the level of contribution of each expert to the construction of a collective opinion, the procedure acts as a computational arbiter responsible for determining the relevance of each opinion for the decision. It tends to penalize discordant experts in favor of an improved consensus. It is important to stress that, in this approach, the experts are supposed to provide their preferences just once only and the rest of the work is left to this computational arbiter. The aggregation operator utilized to construct the collective results should be WAM. Next, the procedure is outlined in a sequence of seven steps:

Step 1. Initialize the weight of all experts in such a way that $w_y = 1/v$, $y = 1, 2, \ldots, v$.

Step 2. Calculate the current level of a weighted consensus per alternative

$$CW(X_k) = \sum_{y=1}^{v} \left(1 - \frac{|O_k^C - O_k^y|}{n-1}\right) w_y \tag{10.25}$$

and the mean level of the weighted consensus for all alternatives

$$\overline{CW} = \frac{\sum_{k=1}^{n} CW(X_k)}{n} \tag{10.26}$$

considering that O_k^C correspond to the position of the kth alternative, taking into account the collective results, and that O_k^y correspond the position of the kth alternative in the ranking derived from the results obtained by expert E_y. It is interesting to observe in (10.25) that an expert with a high weight affects more intensely the level of consensus than an expert with a low weight.

Step 3. If the current mean level of weighted consensus is higher than a minimum threshold, then interrupt the process. Otherwise, go to Step 4.

Step 4. Obtain the rankings of the alternatives for some different formations of the group of experts. Each time, one expert is eliminated from the group, as described in Table 10.8, where the index O_k^{C-z} represents the position of alternative X_k, by taking into account the collective results, obtained for a group formed by the experts, $y = 1, 2, \ldots, z-1, z+1, \ldots, v$.

Step 5. Calculate the levels of consensus for v different configurations of the group. In each configuration a particular expert is not considered as a member.

Table 10.8 Position of each alternative in the rankings obtained for the group without an expert

	X_1	X_2	...	X_n
Group E_1	O_1^{C-1}	O_2^{C-1}	...	O_n^{C-1}
Group E_2	O_1^{C-2}	O_2^{C-2}	...	O_n^{C-2}
...	...	...	...	...
Group E_v	O_1^{C-v}	O_2^{C-v}	...	O_n^{C-v}

The following expression can be utilized to calculate these levels of consensus:

$$CW^{C-z}(X_k) = \sum_{y=1 \wedge y \neq z}^{v} \left(1 - \frac{\left|O_k^{C-y} - O_k^{y}\right|}{n-1}\right) \beta_y \tag{10.27}$$

where

$$\beta_y = \frac{w_y}{\sum\limits_{z=1 \wedge z \neq y}^{v} w_z}$$

Step 6. Calculate the contribution $\bar{D}^z$, $z = 1, 2, \ldots, v$, of each expert for the consensus with the use of the following expressions:

$$D^z(X_k) = CW(X_k) - CW^{C-z}(X_k) \tag{10.28}$$

$$\bar{D}^z = \sum_{k=1}^{n} D^z(X_k) \tag{10.29}$$

Step 7. Adjust the weights w_y, $y = 1, 2, \ldots, v$, of all experts, in accordance with the following expressions:

$$w_y^{cycle+1} = \frac{t_y^{cycle+1}}{\sum\limits_{y=1}^{v} t_y^{cycle+1}} \tag{10.30}$$

$$t_y^{cycle+1} = w_y^{cycle}(1 + D_y)^b \tag{10.31}$$

In (10.31), the value of b represents the weight of the individual contribution to the construction of a consensus. The highest value of b is associated with a faster convergence toward a desired level of consensus.

Step 8. Obtain the new collective solution and go to Step 2.

Example 10.6. Consider that three experts solved a multicriteria decision problem which consists of ranking four alternatives from the most important to the least important one. Each expert solved the problem separately and the individual results are combined into a

Table 10.9 Position of each alternative in the rankings for each expert and for the group, at the first cycle

	X_1	X_2	X_3	X_4
E_1	3	4	1	2
E_2	2	3	1	4
E_3	4	3	1	2
Group	3	4	1	2

collective result with the use of WAM, considering equal weights for all experts, that is, $w_1 = w_2 = w_3 = 0.333$. The individual as well as the collective results, expressed in terms of the degree of fuzzy nondominance of each alternative, are as follows:

- Expert E_1:

$$\mathbf{ND}^1 = [0.59 \quad 0.25 \quad 1 \quad 0.85] \tag{10.32}$$

- Expert E_2:

$$\mathbf{ND}^2 = [0.89 \quad 0.41 \quad 1 \quad 0.36] \tag{10.33}$$

- Expert E_3:

$$\mathbf{ND}^3 = [0.52 \quad 0.63 \quad 1 \quad 0.85] \tag{10.34}$$

- Group:

$$\mathbf{ND}^C = [0.63 \quad 0.43 \quad 1 \quad 0.69] \tag{10.35}$$

Table 10.9 presents the rankings of the alternatives, which are derived from each individual result and from the collective results.

In Step 1, the weights are initialized as $w_1 = w_2 = w_3 = 0.333$.

In Step 2, the mean level of consensus is calculated by considering the rankings of the alternatives for each expert and for the group, by applying (10.25) and (10.26):

$$CW(X_1) = \left(1 - \frac{|3-3|}{3}\right)0.33 + \left(1 - \frac{|2-3|}{3}\right)0.33 + \left(1 - \frac{|4-3|}{3}\right)0.33 = 0.77 \tag{10.36}$$

$$CW(X_2) = \left(1 - \frac{|4-4|}{3}\right)0.33 + \left(1 - \frac{|3-4|}{3}\right)0.33 + \left(1 - \frac{|3-4|}{3}\right)0.33 = 0.77 \tag{10.37}$$

$$CW(X_3) = \left(1 - \frac{|1-1|}{3}\right)0.33 + \left(1 - \frac{|1-1|}{3}\right)0.33 + \left(1 - \frac{|1-1|}{3}\right)0.33 = 1 \tag{10.38}$$

Table 10.10 Position of each alternative in the rankings for the group without an expert, at the first cycle

	X_1	X_2	X_3	X_4
Group-E_1	2	4	1	3
Group-E_2	3	4	1	2
Group-E_3	2	4	1	3

$$CW(X_4) = \left(1 - \frac{|2-2|}{3}\right)0.33 + \left(1 - \frac{|4-2|}{3}\right)0.33 + \left(1 - \frac{|2-2|}{3}\right)0.33 = 0.55 \tag{10.39}$$

$$\overline{CW} = \frac{0.77 + 0.77 + 1 + 0.55}{4} = 0.77 \tag{10.40}$$

In Step 3, as the current mean level of consensus is lower than 0.9, the process is directed to Step 4.

In Step 4, the collective results are recalculated for three different formations of the group. In each formation, a particular expert is excluded from the group. Table 10.10 shows the final rankings of the alternatives, obtained for the group without considering a different expert each time.

In Step 5, the levels of consensus for the group, without considering a particular expert, are calculated. The obtained results are shown in Table 10.11.

In Step 6, the contribution of each expert is calculated by applying (10.19) and (10.20), by taking into account the data from Table 10.11. The obtained results are shown in Table 10.12.

In Step 7, by applying (10.21) and (10.22) with $b = 2$, the weights are updated as

$$w_1 = \frac{0.58}{1.06} = 0.55 \tag{10.41}$$

$$w_2 = \frac{0.15}{1.06} = 0.14 \tag{10.42}$$

$$w_3 = \frac{0.33}{1.06} = 0.31 \tag{10.43}$$

Table 10.11 Levels of consensus for the group without considering a particular expert, at the first cycle

	X_1	X_2	X_3	X_4
Group-E_1	0.67	0.67	1	0.67
Group-E_2	0.83	0.83	1	1
Group-E_3	0.83	0.83	1	0.67

Table 10.12 Contribution of each expert to the consensus, at the first cycle

	X_1	X_2	X_3	X_4	$\bar{D}^y$
$D^1(X_k)$	0.11	0.11	0	0.11	0.33
$D^2(X_k)$	−0.06	−0.06	0	−0.22	−0.33
$D^3(X_k)$	−0.06	−0.06	0	0.11	0

In Step 8, the new collective results are obtained, using WAM with the weights calculated in Step 6:

$$\mathbf{ND}^C = [0.63\ 0.39\ 1\ 0.78] \tag{10.44}$$

In Step 2, the current level of consensus per alternative is calculated with the use of (10.15) and (10.16), as follows:

$$CW(X_1) = \left(1 - \frac{|3-3|}{3}\right)0.55 + \left(1 - \frac{|2-3|}{3}\right)0.14 + \left(1 - \frac{|4-3|}{3}\right)0.31 = 0.85 \tag{10.45}$$

$$CW(X_2) = \left(1 - \frac{|4-4|}{3}\right)0.55 + \left(1 - \frac{|3-4|}{3}\right)0.14 + \left(1 - \frac{|3-4|}{3}\right)0.31 = 0.85 \tag{10.46}$$

$$CW(X_3) = \left(1 - \frac{|1-1|}{3}\right)0.55 + \left(1 - \frac{|1-1|}{3}\right)0.14 + \left(1 - \frac{|1-1|}{3}\right)0.31 = 1 \tag{10.47}$$

$$CW(X_4) = \left(1 - \frac{|2-2|}{3}\right)0.55 + \left(1 - \frac{|4-2|}{3}\right)0.14 + \left(1 - \frac{|2-2|}{3}\right)0.31 = 0.91 \tag{10.48}$$

$$\overline{CW} = \frac{0.85 + 0.85 + 1 + 0.91}{4} = 0.901 \tag{10.49}$$

In Step 3, as the current mean level of consensus is considered satisfactory, the process is interrupted. In this example, it is interesting to observe that the level of the weighted consensus increases but the individual results, as well as the collective result, are kept constant. Thus, the achieved consensus may be considered artificial, as it is based on the negligence of some experts when calculating the level of weighted consensus. But it is important to make clear that such a result does not follow a rule. In some cases, it may be possible to modify the collective result with the use of this procedure (and, obviously, to improve the level of consensus).

10.6 Consensus Schemes in Fuzzy Environment

This section presents three consensus schemes which can be utilized with the different techniques for the multicriteria analysis of $\langle \mathbf{X}, \boldsymbol{R} \rangle$ models described in Chapter 7. The first consensus scheme to be presented here requires the preferences of the experts to be expressed as fuzzy estimates. This consensus scheme is named the consensus scheme based on fuzzy estimates (CSFE). The second consensus scheme to be described requires the preferences of the experts to be expressed in terms of nonreciprocal fuzzy preference relations. This means that, if each expert utilizes a different preference format to express their opinions, then it is necessary to make use of adequate transformation functions to translate all information to nonreciprocal fuzzy preference relations, which is taken as a common format for associating and comparing opinions. Here, this consensus scheme is named the consensus scheme based on fuzzy preference relations (CSFPR). The third consensus scheme to be described here requires the preferences to be expressed as the ordering from the most important alternative to the least important alternative. Here, it is named the consensus scheme based on the ranking of alternatives (CSRA).

It should be mentioned that the three consensus schemes admit computational components for executing supervision functions that are usually delegated to a human moderator. In their description, it is assumed that the variable *cycle* indicates the current iteration; and the variable *elast* is a vector utilized to store the index of the expert requested to update his/her opinion, at each cycle. Further, it is also considered that they require a human moderator to specify three input parameters, namely:

- *minconsensus*: this defines the minimum acceptable level of consensus;
- *maxcycles*: this defines for how many cycles, at most, the discussion should persist;
- *maxreviews*: this stores the maximum number of times each expert can be successively invited by the moderator to review his/her opinion.

These input parameters should be specified by considering some important aspects of the process. In order to prevent the discussion from taking a longer time than expected (or allocated), it may be necessary to consider more than one stopping condition to interrupt the discussion. As indicated in Ekel *et al.* (2009), in general, the overall discussion process is interrupted when any of the following conditions is fulfilled:

- an acceptable consensus level among the specialists has been achieved;
- the previously specified maximum number of iterations has been achieved;
- the same expert remains as the most discordant one after a specific number of subsequent iterations and the moderator cannot persuade this expert to change his/her opinion;
- in real-time applications, the allotted time has expired and the process is then interrupted.

With the consideration of a stopping condition besides the minimum acceptable level of consensus, such as, for example, the maximum number of iterations (*maxcycles*), the parameter *minconsensus* can have a high value, such as $0.8 \leq minconsensus \leq 1$. In this way, when the group takes a long time to achieve an acceptable level of consensus, other stopping conditions guarantee a better distribution of time over all the subjects under discussion.

The parameter *maxreviews* is utilized in situations when an expert remains as the least concordant member of the group for several subsequent cycles. Under these circumstances, it is interesting to invite this expert to justify his/her current position. If he/she has the persuasive ability, maybe the other experts can be convinced to change their opinions. However, in such a case, it becomes necessary to give other members of the group the chance to update their respective opinions. Indeed, whenever possible, it is important to give all experts an opportunity to review their respective opinions, at least once. Otherwise, the discussion may be polarized by just a small group of experts. Depending on the proportion between the values of the parameters *maxcycles* and *maxreviews*, some experts may have no opportunity to review their opinions. In this way, a priori, the value of *maxcycles* should be higher than or equal to the product of the value of the parameter *maxreviews* and the size of the group.

Next, CSFE is summarized as a stepwise guidance procedure by assuming that all experts provide their opinions in terms of fuzzy estimates: $F_p^y(X_k)$, $k = 1, 2, \ldots, n$; $y = 1, 2, \ldots, v$; $p = 1, 2, \ldots, q$. In this way, the following steps must be repeated for each alternative and each criterion.

10.6.1 Guidance Procedure of CSFE

Step 1. Initialize $cycle = 1$ and ask the moderator to specify the weight w_z for each expert, as well as the input parameters *minconsensus*, *maxcycles*, and *maxreviews*.

Step 2. Collect the opinion of each expert concerning only the pth criterion and the kth alternative.

Step 3. Aggregate the individual estimates in a temporary collective opinion $F_p^C(X_k)$ with the use of the aggregation operator given by (9.1).

Step 4. Calculate the consensus level within group members by means of (10.5).

Step 5. If the maximum number of cycles or a minimum level of consensus is achieved, then go to Step 10. If no stopping condition is satisfied, then go to Step 6.

Step 6. Calculate the concordance level for each specialist with the use of (10.4).

Step 7. Identify the least concordant expert and verify, in vector *elast*, if he/she has been the least concordant expert for the last *maxreviews* cycles. If this is true, repeat Step 7 for the second least concordant expert and so on (in order to avoid the same expert being excessively requested).

Step 8. Add 1 to the value of variable *cycle*, store the index of the expert selected in Step 7 in *elast*(*cycle*), and invite this expert to update his/her opinion.

Step 9. Collect the opinion of the selected expert and go to Step 3.

Step 10. Interrupt the procedure. The output data is the current collective fuzzy estimate.

Example 10.7. Consider the use of CSFE to construct a collective evaluation of an alternative X_1 for a group of four experts.

In Step 1, the variable *cycle* is initialized as 1; the input parameters are fixed as $minconsensus = 0.6$, $maxcycles = 10$, $maxreviews = 1$ and the weights of the three experts are set as $w_1 = w_2 = w_3 = w_4 = 0.25$.

In Step 2, the experts express their evaluations as $F^1(X_1) = \{0.25, 0.4, 0.6, 0.75\}$, $F^2(X_1) = \{0.5, 0.65, 0.85, 1\}$, $F^3(X_1) = \{0.5, 0.65, 0.85, 1\}$, $F^4(X_1) = \{0, 0.15, 0.35, 0.5\}$.

Table 10.13 Level of concordance and of consensus, reported at the first iteration

	$S_{FE}^{1,C}$	$S_{FE}^{2,C}$	$S_{FE}^{3,C}$	$S_{FE}^{4,C}$	Consensus
Concordance	0.80	0.52	0.52	0.36	0.55

In Step 3, the collective estimate, obtained using (9.1), is given by $F^C(X_1) = \{0.31, 0.46, 0.66, 0.81\}$.

In Step 4, by applying (10.5), we obtain the level of consensus as 0.55 and, in Step 5, as none of the stopping conditions is satisfied, we go to Step 6.

In Step 6, the level of concordance between the collective and each individual opinion is calculated using (10.4), with $\beta = 0.5$, as given in Table 10.13.

In Step 7, E_4 is identified as the least concordant expert in accordance with the data shown in Table 10.13.

In Step 8, the variable *cycle* is incremented (*cycle* = 2); the index of E_4 is stored in the vector *elast*(2) and this particular expert is invited to review his/her opinion.

In Step 9, E_4 provides a new evaluation $F^4(X_1) = \{0.25, 0.4, 0.6, 0.75\}$ and the process goes directly to Step 3.

In Step 3, the collective fuzzy estimate is updated to $F^C(X_1) = \{0.38, 0.53, 0.73, 0.88\}$.

In Step 4, by applying (10.5), we obtain the level of consensus as 0.64 (refer to Table 10.14 for the level of concordance and of consensus reported at the second iteration), which the moderator considered an acceptable degree of consensus. In this way, the session is terminated.

Next, the guidance procedure of CSFPR is described, considering that the discussion is to be divided into independent sessions focused on each criterion separately.

10.6.2 Guidance Procedure of CSFPR

Step 1. Initialize *cycle* = 1 and ask the moderator to specify the weight w_z for each expert, as well as the input parameters *minconsensus*, *maxcycles*, *maxreviews*.

Step 2. Collect the opinion of each specialist. All experts are supposed to evaluate or compare all alternatives using any preference format among utility values, ordering of all alternatives, fuzzy or linguistic estimates, multiplicative preference relations, fuzzy preference relations.

Step 3. If any expert expresses his/her preferences in terms of a multiplicative preference relation or of a fuzzy preference relation, then verify whether the supplied relations are consistent. The process should not continue until a satisfactory

Table 10.14 Level of concordance and of consensus, reported at the second iteration

	E_1	E_2	E_3	E_4	Consensus
Concordance level	0.63	0.65	0.65	0.63	0.64

level of consistency has been achieved. As discussed in Chapter 9, a priori, weak transitivity is the minimum acceptable level of consistency.

Step 4. Make the required data transformation to the fuzzy preference relation, with the use of proper transformation functions, as discussed in Chapter 6. The supplied information must be converted to fuzzy preference relations, which is accepted as the generic form for comparing and associating individual opinions, as well as for performing the multicriteria analysis later on.

Step 5. Calculate the collective opinion. The individual opinions are aggregated into a temporary collective opinion, expressed as a fuzzy preference relation, using WAM, WGM, min, or OWA. Note that the selection of the most appropriate aggregation operator should be made by the group, before the beginning of the discussion.

Step 6. Verify whether the collective opinion satisfies weak transitivity. If not, an analyst can modify the collective fuzzy preference relation in order to improve its consistency using an algorithm such as the one described in Chapter 6.

Step 7. Calculate the consensus level achieved per relation given by (10.11).

Step 8. If the maximum number of cycles or a minimum level of consensus has been achieved, then go to Step 13. If none of the stopping conditions are satisfied, then go to Step 9.

Step 9. Calculate the concordance level for each expert with the use of (10.7) and (10.8), in such a way that a table like Table 10.15 can be fulfilled.

Step 10. Identify the least concordant expert and verify, in vector *elast*, if he/she has been the least concordant expert for the last *maxreviews* cycle. If this is true, repeat Step 10 for the second least concordant expert and so on (in order to prevent any expert from being excessively demanded).

Step 11. Add 1 to the value of variable *cycle*, store the index of the expert selected in Step 10 in *elast*(*cycle*), and invite this expert to update his/her opinion.

Step 12. Collect the opinion of the selected expert. If this expert expresses his/her opinion in terms of a fuzzy preference relation or a multiplicative preference relation, then go to Step 3. Otherwise, go directly to Step 4.

Step 13. Interrupt the procedure. The output data are the current collective fuzzy preference relation.

Table 10.15 Levels of concordance per alternative and per relation for each expert

Concordance level per alternative	E_1	E_2	...	E_v
X_1	$SX_1^{1,C}$	$SX_1^{2,C}$	...	$SX_1^{v,C}$
X_2	$SX_2^{1,C}$	$SX_2^{2,C}$	...	$SX_2^{v,C}$
...	...	...	...	...
X_n	$SX_n^{1,C}$	$SX_n^{2,C}$	...	$SX_n^{v,C}$
Concordance level per relation	$SR^{1,C}$	$SR^{2,C}$	...	$SR^{v,C}$

Example 10.8. Consider the use of CSFPR to construct a collective preference for a group of three experts. Here, three alternatives are to be compared by taking into account the criterion F_1.

In Step 1, the variable *cycle* is initialized as 1; the input parameters are fixed as $minconsensus = 0.75$, $maxcycles = 5$, $maxreviews = 1$ and the weights of the three experts are set as $w_1 = w_2 = w_3 = 0.333$.

In Step 2, the experts express their preferences, each one using a different preference format:

- Preferences of E_1, expressed in terms of fuzzy estimates

$$F_1^1(X_1) = \{0.85, 0.95, 1, 1\},\ F_1^1(X_2) = \{0.6, 0.7, 0.8, 0.9\},$$
$$F_1^1(X_3) = \{0.35, 0.45, 0.55, 0.65\} \tag{10.50}$$

- Preferences of E_2, expressed in terms of the utility values (defined on an interval scale) of each alternative:

$$\boldsymbol{U}_1^2 = \begin{bmatrix} 0.8 & 0.7 & 0.7 \end{bmatrix} \tag{10.51}$$

- Preferences of E_3, expressed in terms of the order of each alternative:

$$\boldsymbol{O}_1^3 = \begin{bmatrix} 3 & 1 & 2 \end{bmatrix} \tag{10.52}$$

In Step 3, as no experts have expressed their respective opinions in terms of a fuzzy preference relation or a multiplicative preference relation, it is not necessary to check the consistency of the provided preferences.

In Step 4, with the use of adequate transformation functions, all preferences are transformed to the following nonreciprocal fuzzy preference relations:

- Preferences of E_1:

$$\boldsymbol{R}_1^1 = \begin{bmatrix} 1 & 1 & 1 \\ 0.25 & 1 & 1 \\ 0 & 0.25 & 1 \end{bmatrix} \tag{10.53}$$

- Preferences of E_2:

$$\boldsymbol{R}_1^2 = \begin{bmatrix} 1 & 1 & 1 \\ 0.9 & 1 & 1 \\ 0.9 & 1 & 1 \end{bmatrix} \tag{10.54}$$

- Preferences of E_3:

$$\boldsymbol{R}_1^3 = \begin{bmatrix} 1 & 0 & 0.25 \\ 1 & 1 & 1 \\ 1 & 0.25 & 1 \end{bmatrix} \tag{10.55}$$

Table 10.16 Levels of consensus per alternative and per relation, reported during the first cycle

	X_1	X_2	X_3
Consensus level per alternative	0.65	0.74	0.75
Consensus level per relation		0.71	

In Step 5, a temporary collective fuzzy preference relation is constructed with the use of WAM:

$$\boldsymbol{R}_1^C = \begin{bmatrix} 1 & 0.67 & 0.75 \\ 0.71 & 1 & 1 \\ 0.63 & 0.5 & 1 \end{bmatrix} \tag{10.56}$$

In Step 6, it is verified that $\boldsymbol{R}_1^C$ satisfies the weak-transitivity condition.

In Step 7, the consensus level per relation is calculated. Table 10.16 presents the consensus level per alternative, obtained with (10.10), as well as the consensus level per relation, given by (10.11).

In Step 8, as none of the stopping conditions have been satisfied, the process moves to Step 9.

In Step 9, the concordance level between the individual opinions and the collective opinion is calculated for the individual alternative and the relation, with the use of (10.7) and (10.8). However, here we applied (10.7) and (10.8) to the version of the collective fuzzy preference relation given by (10.57), which was derived from (10.56), in accordance with the procedure described in Section 10.3:

$$\boldsymbol{R}^{MC} = \begin{bmatrix} 1 & 0.67 & 1 \\ 1 & 1 & 1 \\ 0.63 & 0.5 & 1 \end{bmatrix} \tag{10.57}$$

Table 10.17 contains the levels of concordance per alternative and per relation.

In Step 10, E_3 is selected as the least concordant expert, based on the concordance level per relation of each expert given in Table 10.17.

In Step 11, the variable *cycle* is incremented (*cycle* $= 2$); the index of E_3 is stored in the vector *elast*(2) and this particular expert is invited to review his/her opinion.

Table 10.17 Levels of concordance per alternative and per relation for each expert, reported during the first cycle

Concordance level per alternative	E_1	E_2	E_3
X_1	0.57	0.82	0.55
X_2	0.67	0.77	0.77
X_3	0.78	0.81	0.66
Concordance level per relation	0.67	0.80	0.66

Table 10.18 Levels of consensus per alternative and per relation, at the second cycle

	X1	X_2	X3
Consensus level per alternative	0.77	0.76	0.82
Consensus level per relation		0.78	

In Step 12, the new preferences of E_3 are collected and the process goes directly to Step 4:

$$\boldsymbol{O}_1^3 = [2 \quad 1 \quad 3] \tag{10.58}$$

In Step 4, the fuzzy preference relation $\boldsymbol{R}_1^3$ is updated yielding

$$\boldsymbol{R}_1^3 = \begin{bmatrix} 1 & 0.25 & 1 \\ 1 & 1 & 1 \\ 0.25 & 0 & 1 \end{bmatrix} \tag{10.59}$$

In Step 5, the temporary collective fuzzy preference relation is updated to

$$\boldsymbol{R}_1^C = \begin{bmatrix} 1 & 0.75 & 1 \\ 0.72 & 1 & 1 \\ 0.38 & 0.42 & 1 \end{bmatrix} \tag{10.60}$$

In Step 6, it is verified that $\boldsymbol{R}_1^C$ satisfies the weak transitivity condition.

In Step 7, the new consensus level per relation is calculated with the use of (10.11). Table 10.18 shows the consensus level per alternative and the consensus level per relation.

In Step 8, as the level of consensus is considered satisfactory, the process moves on to Step 13.

In Step 13, the procedure is terminated and the output data is the collective fuzzy preference relation given by (10.60).

It should be indicated that an important aspect to be considered in the choice between CSFE and CSFPR is associated with the fact that CSFE is less efficient than CSFPR when the experts can have and do have very different levels of exigency. This aspect is analyzed in the following example.

Example 10.9. Let us consider that two experts, namely E_1 and E_2, the former being more exigent than the latter, evaluated the alternatives X_1, X_2, and X_3 using the set of five normalized fuzzy values *S*(*F*)={*very low*, *low*, *middle*, *high*, *very high*} shown in Figure 7.31. The collected evaluations are as follows:

- Preferences of E_1: $F^1(X_1)$ =*very low*, $F^1(X_2)$ =*low*, $F^1(X_3)$ =*middle*.
- Preferences of E_2: $F^2(X_1)$ =*middle*, $F^2(X_2)$ = *high*, $F^2(X_3)$ =*very high*.

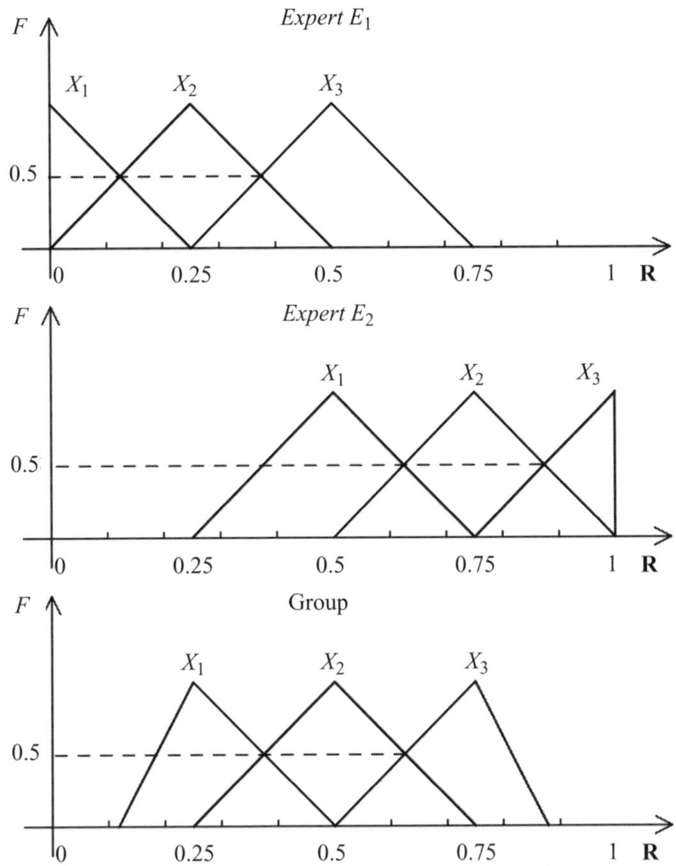

Figure 10.1 Example of evaluations through fuzzy estimates that result in similar fuzzy preference relations.

If we consider $w_1 = w_2$, (9.1) yields the collective fuzzy estimates shown in Figure 10.1. As can be seen in Figure 10.1, there is a significant difference between the collective opinion and each individual opinion.

On the other hand, if we use (6.14) and (6.15), we obtain the fuzzy preference relations associated with the preferences of each expert, that is,

$$\boldsymbol{R}^1 = \begin{bmatrix} 1 & 0.5 & 0 \\ 1 & 1 & 0.5 \\ 1 & 1 & 1 \end{bmatrix} \tag{10.61}$$

and

$$\boldsymbol{R}^2 = \begin{bmatrix} 1 & 0.5 & 0 \\ 1 & 1 & 0.5 \\ 1 & 1 & 1 \end{bmatrix} \tag{10.62}$$

which are identical to each other. Therefore, aggregation of (10.61) and (10.62) with the use of any aggregation operator among WAM, WGM, OWA, or min results in a fuzzy preference relation which is equal to (10.61) and (10.62). In this situation, the use of CSFPR allows the group to interrupt the discussion process sooner under a higher level of consensus on the collective fuzzy preference relation. On the other hand, although the use of CSFE may make difficult for the group to achieve an acceptable level of consensus, its use may be helpful, when it is desirable to achieve a consensus on the evaluation of each alternative individually (this sort of application usually requires the definition of a standardized level of exigency for every criterion).

Now let us consider the guidance procedure of CSRA. In practice, the most substantial difference between CSFPR and CSRA lies in the fact that whereas CSFPR aims at constructing a consensus on the preferences per criterion, CSRA aims at constructing a consensus on the final ranking of the alternatives. Actually, it should be noted that, when CSFPR is utilized, an acceptable level of consensus may be reached on the preferences per criterion, but the level of consensus on the final ranking of the alternatives is not necessarily satisfactory. CSFPR is particularly useful when the group is formed by experts coming from different areas, who are invited to participate in the decision process by comparing or evaluating the alternatives only for the criteria related to their respective areas of expertise. On the other hand, CSRA is particularly useful when all experts can solve the overall problem individually and it is important to guarantee a consensus on the final results rather than on the preferences of the experts. It is interesting also to note that CSRA may admit a considerable level of discordance among the experts for a criterion with a low level of importance for the final decision.

10.6.3 Guidance Procedure of CSRA

Step 1. Initialize $cycle = 1$ and ask the moderator to specify the weight w_z of each expert as well as the input parameters *minconsensus*, *maxcycles*, *maxreviews*.

Step 2. Collect the opinion of each specialist. All experts are supposed to evaluate or compare all alternatives using any preference format among utility values, ordering of all alternatives, fuzzy or linguistic estimates, multiplicative preference relations, fuzzy preference relations. It should be indicated that the same expert is also allowed to express his/her opinions using a different preference format for each criterion.

Step 3. If any expert expresses his/her preferences in terms of a multiplicative preference relation or a fuzzy preference relation, then verify whether the supplied relations are consistent. The process should not continue until a satisfactory level of consistency has been achieved.

Step 4. Make the required data transformations to the fuzzy preference relation, with the use of proper transformation functions, as discussed in Chapter 6. The supplied information must be translated to fuzzy preference relations, which is regarded as the generic format for carrying out the multicriteria analysis for each expert.

Step 5. Run the multicriteria analysis for each expert individually. It is important to ensure that all members of the group select the same method for the multicriteria analysis. As discussed in Chapter 9, the use of different methods makes it difficult to meet a satisfactory level of consensus on the results, even if all experts have achieved a perfect consensus on their opinions about the alternatives.

Step 6. Calculate the collective results. The individual results, which correspond to the fuzzy nondominance level of each alternative, are aggregated into a temporary collective result, using WAM, WGM, min, or OWA.

Step 7. Calculate the mean level of concordance for each expert by using (10.17) and (10.18) subsequently.

Step 8. Calculate the mean level of consensus achieved by the group using (10.19) followed by (10.20).

Step 9. If the maximum number of cycles or the minimum consensus level has been reached, then move on to Step 13. If none of the conditions are met, then go to Step 10.

Step 10. Identify the least concordant expert and verify, in vector *elast*, if he/she has been the least concordant expert for the last *maxreviews* cycles. If this holds, repeat Step 10 for the second least concordant expert and so on (in order to prevent any expert from being excessively demanded).

Step 11. Increment the variable *cycle*, store the index of the expert selected in Step 10 in *elast*(*cycle*), and invite this expert to update his/her opinion.

Step 12. Collect the opinion of the selected expert. If this expert expresses his/her opinion in terms of a fuzzy preference relation or a multiplicative preference relation, then go to Step 3. Otherwise, go directly to Step 4.

Step 13. Interrupt the discussion. The output data of the consensus scheme is the current collective result. But, if the current level of consensus is still unacceptable, then the moderator should consider the possibility of activating the procedure for constructing an optimized consensus.

Example 10.10. Let us consider that a multicriteria decision problem, which involves ranking four alternatives considering three criteria, is to be solved by three experts using the *first technique* for multicriteria analysis, described in Chapter 7. All experts are supposed to solve the problem individually, following the guidance procedure of CSRA.

In Step 1, the moderator sets the values of the input parameters as $minconsensus = 0.9$, $maxcycles = 10$, $maxreviews = 2$ and the weights of the three experts are set as $w_1 = w_2 = w_3 = 0.333$.

In Step 2, each expert expresses their preferences, taking into account all criteria and using different preference formats:

- Preferences of E_1, expressed in terms of the following fuzzy estimates:

$$F_1^1(X_1) = \{0.85, 0.95, 1, 1\},\ F_1^1(X_2) = \{0.35, 0.45, 0.55, 0.65\},$$
$$F_1^1(X_3) = \{0.35, 0.45, 0.55, 0.65\},\ F_1^1(X_4) = \{0.1, 0.2, 0.3, 0.4\}$$

$$
\begin{aligned}
&F_2^1(X_1) = \{0.1, 0.2, 0.3, 0.4\},\ F_2^1(X_2) = \{0.35, 0.45, 0.55, 0.65\} \\
&F_2^1(X_3) = \{0.6, 0.7, 0.8, 0.9\},\ F_2^1(X_4) = \{0.35, 0.45, 0.55, 0.65\} \\
&F_3^1(X_1) = \{0.35, 0.45, 0.55, 0.65\},\ F_3^1(X_2) = \{0.6, 0.7, 0.8, 0.9\} \\
&F_3^1(X_3) = \{0.6, 0.7, 0.8, 0.9\},\ F_3^1(X_4) = \{0.85, 0.95, 1, 1\}
\end{aligned} \tag{10.63}
$$

- Preferences of E_2, expressed in terms of the interval-scale utility values of each alternative:

$$
\boldsymbol{U}_1^2 = [0.9 \quad 0.6 \quad 0.5 \quad 0.3],\ \boldsymbol{U}_2^2 = [0.5 \quad 0.7 \quad 0.9 \quad 0.6],\ \boldsymbol{U}_3^2 = [0.4 \quad 0.3 \quad 0.9 \quad 0.6] \tag{10.64}
$$

- Preferences of E_3, expressed in terms of the order of each alternative by taking into account each criterion separately:

$$
\boldsymbol{O}_1^3 = [1 \quad 2 \quad 3 \quad 4],\ \boldsymbol{O}_2^3 = [4 \quad 3 \quad 2 \quad 1],\ \boldsymbol{O}_3^3 = [4 \quad 3 \quad 1 \quad 2] \tag{10.65}
$$

In Step 3, as none of the experts have expressed their opinion in terms of a fuzzy preference relation or a multiplicative preference relation, it is not necessary to check the consistency of the provided information.

In Step 4, with the use of adequate transformation functions, all preferences are transformed to the nonreciprocal fuzzy preference relations as shown below:

- Preferences of E_1:

$$
\boldsymbol{R}_1^1 = \begin{bmatrix} 1 & 1 & 1 & 1 \\ 0 & 1 & 1 & 1 \\ 0 & 1 & 1 & 1 \\ 0 & 0.25 & 0.25 & 1 \end{bmatrix},\ \boldsymbol{R}_2^1 = \begin{bmatrix} 1 & 0.25 & 0 & 0.25 \\ 1 & 1 & 0.25 & 1 \\ 1 & 1 & 1 & 1 \\ 1 & 1 & 0.25 & 1 \end{bmatrix},
$$
$$
\boldsymbol{R}_3^1 = \begin{bmatrix} 1 & 0.25 & 0.25 & 0 \\ 1 & 1 & 1 & 0.25 \\ 1 & 1 & 1 & 0.25 \\ 1 & 1 & 1 & 1 \end{bmatrix} \tag{10.66}
$$

- Preferences of E_2:

$$
\boldsymbol{R}_1^2 = \begin{bmatrix} 1 & 1 & 1 & 1 \\ 0.54 & 1 & 1 & 1 \\ 0.43 & 0.82 & 1 & 1 \\ 0.25 & 0.54 & 0.67 & 1 \end{bmatrix},\ \boldsymbol{R}_2^2 = \begin{bmatrix} 1 & 0.67 & 0.43 & 0.82 \\ 1 & 1 & 0.67 & 1 \\ 1 & 1 & 1 & 1 \\ 1 & 0.82 & 0.54 & 1 \end{bmatrix},
$$
$$
\boldsymbol{R}_3^2 = \begin{bmatrix} 1 & 1 & 0.33 & 0.67 \\ 0.82 & 1 & 0.25 & 0.54 \\ 1 & 1 & 1 & 1 \\ 1 & 1 & 0.54 & 1 \end{bmatrix} \tag{10.67}
$$

- Preferences of E_3:

$$\boldsymbol{R}_1^3 = \begin{bmatrix} 1 & 1 & 1 & 1 \\ 0.33 & 1 & 1 & 1 \\ 0.17 & 0.33 & 1 & 1 \\ 0 & 0.17 & 0.33 & 1 \end{bmatrix}, \quad \boldsymbol{R}_2^3 = \begin{bmatrix} 1 & 0.33 & 0.17 & 0 \\ 1 & 1 & 0.33 & 0.17 \\ 1 & 1 & 1 & 0.33 \\ 1 & 1 & 1 & 1 \end{bmatrix},$$

$$\boldsymbol{R}_3^3 = \begin{bmatrix} 1 & 0.33 & 0 & 0.17 \\ 1 & 1 & 0.17 & 0.33 \\ 1 & 1 & 1 & 1 \\ 1 & 1 & 0.33 & 1 \end{bmatrix} \tag{10.68}$$

In Step 5, the multicriteria analysis is carried out for each expert following the *first technique*. It should be mentioned that when two or more alternatives are considered nondistinguishable, the weight of each criterion is differentiated in a subsequent analysis as $\lambda_1 = 0.25$, $\lambda_2 = 0.35$, and $\lambda_3 = 0.4$. The individual results are:

- Expert E_1:

$$\mathbf{ND}^1 = [0.59 \quad 0.25 \quad 1 \quad 0.85] \tag{10.69}$$

- Expert E_2:

$$\mathbf{ND}^2 = [0.90 \quad 0.43 \quad 1 \quad 0.54] \tag{10.70}$$

- Expert E_3:

$$\mathbf{ND}^3 = [0.52 \quad 0.60 \quad 1 \quad 0.80] \tag{10.71}$$

In Step 6, by calculating the collective result with the use of WAM, we obtain:

$$\mathbf{ND}^c = [0.67 \quad 0.43 \quad 1 \quad 0.73] \tag{10.72}$$

Table 10.19 presents the rankings of the alternatives, which are derived from each individual result and from the collective results.

Table 10.19 Position of each alternative in the rankings for each expert and for the group, at the first cycle

	X_1	X_2	X_3	X_4
E_1	3	4	1	2
E_2	2	4	1	3
E_3	4	3	1	2
Group	3	4	1	2

Table 10.20 Levels of concordance for each expert, at the first cycle

	$SO_1^{y,C}$	$SO_2^{y,C}$	$SO_3^{y,C}$	$SO_4^{y,C}$	$SO^{y,C}$
E_1	1	1	1	1	1
E_2	0.42	1	1	0.42	0.71
E_3	0.4	0.42	1	1	0.71

In Step 7, the level of concordance for each individual alternative and for the group of alternatives is calculated with the use of (10.17) with $b = 0.5$ and (10.18). Table 10.20 shows the values of $SO_k^{y,C}$ for $y = 1, 2, 3$ and $k = 1, 2, \ldots, 4$, and the average level of concordance $SO^{y,C}$for $y = 1, 2, 3$.

In Step 8, the mean level of consensus achieved by the group is calculated using (10.19) and (10.20), as can be seen in Table 10.21.

In Step 9, as none of the stopping conditions are satisfied, the discussion process is not interrupted.

In Step 10, although E_2 and E_3 are identified to be the least concordant experts (which results from the data shown in Table 10.20), only E_2 is invited to review his/her opinion.

In Step 11, the value of the variable *cycle* is updated to 2, the index of E_2 is stored in *elast*(2), and E_2 is invited to review his/her opinion.

Considering the fact that $O_1^C - O_1^2 > 0$, $O_2^C - O_2^2 = 0$, $O_3^C - O_3^2 = 0$, and $O_4^C - O_4^2 < 0$, the advice mechanism generates the following recommendations: E_2 should increase the evaluations of X_4, decrease the evaluations of X_1, and maintain the evaluations of X_2 and X_3.

In Step 12, the modified preferences of E_2 are collected:

$$\boldsymbol{U}_1^2 = [0.9 \quad 0.6 \quad 0.5 \quad 0.75] \tag{10.73}$$

As can be seen, only the preferences related to the criterion F_1 are modified. The unique change made by the expert is related to X_4 and is in accordance with the information supplied by the advice mechanism.

In Step 4, the fuzzy preference relation $\boldsymbol{R}_1^2$ is updated to (all other fuzzy preference relations remain the same)

$$\boldsymbol{R}_1^2 = \begin{bmatrix} 1 & 1 & 1 & 1 \\ 0.54 & 1 & 1 & 0.74 \\ 0.43 & 0.82 & 1 & 0.6 \\ 0.74 & 1 & 1 & 1 \end{bmatrix} \tag{10.74}$$

Table 10.21 Levels of consensus for each individual alternative for the set of alternatives

CO_1	CO_2	CO_3	CO_4	CO
0.61	0.81	1	0.81	0.81

Table 10.22 Position of each alternative in the rankings for each expert and for the group, reported at the second cycle

	X_1	X_2	X_3	X_4
E_1	3	4	1	2
E_2	3	4	1	2
E_3	4	3	1	2
Group	3	4	1	2

In Step 5, the multicriteria analysis is performed again for E_2 and the following results are obtained:

$$\mathbf{ND}^2 = [0.90 \quad 0.43 \quad 1 \quad 0.94] \tag{10.75}$$

In Step 6, the collective results are updated to

$$\mathbf{ND}^C = [0.67 \quad 0.43 \quad 1 \quad 0.86] \tag{10.76}$$

The new rankings of the alternatives are presented in Table 10.22.

In Step 7, the mean level of concordance is obtained for each specialist with the use of (10.17), where $b = 0.5$. Table 10.23 shows the values of $SO_k^{y,C}$ for $y = 1, 2, 3$ and $k = 1,2,\ldots,$ 4, and the average level of concordance $SO^{y,C}$ for $y = 1, 2, 3$.

In Step 8, the mean level of consensus achieved by the group is calculated using (10.19) and (10.20). The results are shown in Table 10.24.

As the current level of consensus is higher than *minconsensus*, in Step 9, the procedure goes to Step 13 and the discussion is interrupted.

It is important to emphasize that, although it is possible to achieve a satisfactory level of consensus in the given examples, it is not always possible or easy to achieve a satisfactory level of consensus for the collective results. In real-world applications, the most common obstacles for achieving consensus are as follows:

- The discordant experts may not want to modify their opinions and may not present convincing arguments in favor of their positions.
- The experts may modify their opinions in the opposite direction recommended by the advice mechanism. Indeed, as discussed in Section 10.4, the moderator should cautiously utilize the feedback mechanism, as it may produce undesirable effects on the discussion process.

Table 10.23 Levels of concordance for each expert, at the second cycle

	$SO_1^{y,C}$	$SO_2^{y,C}$	$SO_3^{y,C}$	$SO_4^{y,C}$	$SO^{y,C}$
E_1	1	1	1	1	1
E_2	1	1	1	1	1
E_3	0.42	0.42	1	1	0.71

Table 10.24 Levels of consensus for each individual alternative for the set of alternatives

CO_1	CO_2	CO_3	CO_4	CO
0.81	0.81	1	1	0.905

10.7 An Application Related to the Balanced Scorecard Methodology

The balanced scorecard methodology (BSC) is a tool for strategic management, developed by Kaplan and Norton (1992), which can help organizations to translate strategic objectives into relevant performance measures, that is, to put strategy into action. It does so by integrating four perspectives, each one addressing a main question (Bremser and White, 2000):

- **Financial perspective:** In order to succeed financially, how should the organization perform for its shareholders?
- **Customer perspective:** In order to achieve its vision, how should the organization be manifested to its customers?
- **Internal business perspective:** In order to satisfy customers as well as shareholders, in which business processes should the organization stand out?
- **Learning and growth perspective:** How can the organization sustain its ability of changing and improving?

In this context, let us consider that the enterprise's board of directors, which includes four members (E_1, E_2, E_3, E_4), is to plan the development of strategy initiatives for the following five years. Five possible strategies have been marked:

- Strategy X_1: to outsource certain activities traditionally done in-house.
- Strategy X_2: to design products based on customer requirements.
- Strategy X_3: to adopt new technologies to be used in the production phase, in order to increase product quality.
- Strategy X_4: to improve after-sales service quality by widening the service network.
- Strategy X_5: to adopt new technologies to be used in the production phase, in order to improve the production process by solving some of the existing operational problems.

It is necessary to compare these projects to select the most important ones, as well as order them from the point of view of their importance, taking into account the four criteria (categories) suggested by the BSC (note that all of them are of the maximization type):

- F_1: financial perspective,
- F_2: the customer satisfaction,
- F_3: internal business process perspective,
- F_4: learning and growth perspective.

All experts are invited to make a complete analysis of the problem. It is assumed that all experts selected the *first technique* for multicriteria analysis described in Chapter 7. When two or more alternatives are considered indistinguishable, the weight of each criterion is differentiated in a subsequent analysis as $\lambda_1 = 0.4$, $\lambda_2 = 0.25$, $\lambda_3 = 0.2$ and $\lambda_4 = 0.15$. The CSRA consensus scheme is utilized to construct a consensual decision for the group, as described next.

In Step 1, the moderator sets *minconsensus* $= 0.80$, *maxcycles* $= 10$, *maxreviews* $= 2$. The professionals are considered of the same importance, except for E_1, whose opinions are judged to be more important. Therefore, the parameters w_y are set as $w_1 = 0.2$, $w_2 = 0.3$, $w_3 = 0.3$, and $w_4 = 0.2$.

In Step 2, each expert expresses their preferences using a different preference format:

- Preferences of E_1, expressed in terms of the fuzzy preference relations:

$$\boldsymbol{R}_1^1 = \begin{bmatrix} 0.5 & 0.9 & 0.7 & 0.8 & 0.9 \\ 0.1 & 0.5 & 0.25 & 0.4 & 0.7 \\ 0.3 & 0.75 & 0.5 & 0.6 & 0.8 \\ 0.2 & 0.6 & 0.4 & 0.5 & 0.8 \\ 0.1 & 0.3 & 0.2 & 0.2 & 0.5 \end{bmatrix}, \quad \boldsymbol{R}_2^1 = \begin{bmatrix} 0.5 & 0.2 & 0.15 & 0.3 & 0.8 \\ 0.8 & 0.5 & 0.4 & 0.6 & 0.9 \\ 0.85 & 0.6 & 0.5 & 0.7 & 0.9 \\ 0.7 & 0.4 & 0.3 & 0.5 & 0.9 \\ 0.2 & 0.1 & 0.1 & 0.1 & 0.5 \end{bmatrix}$$
$$\boldsymbol{R}_3^1 = \begin{bmatrix} 0.5 & 0.2 & 0.1 & 0.1 & 0.1 \\ 0.8 & 0.5 & 0.3 & 0.2 & 0.1 \\ 0.9 & 0.7 & 0.5 & 0.4 & 0.3 \\ 0.9 & 0.8 & 0.6 & 0.5 & 0.4 \\ 0.9 & 0.9 & 0.7 & 0.6 & 0.5 \end{bmatrix}, \quad \boldsymbol{R}_4^1 = \begin{bmatrix} 0.5 & 0.9 & 0.95 & 0.8 & 0.85 \\ 0.1 & 0.5 & 0.7 & 0.3 & 0.4 \\ 0.05 & 0.3 & 0.5 & 0.15 & 0.2 \\ 0.2 & 0.7 & 0.85 & 0.5 & 0.6 \\ 0.15 & 0.6 & 0.8 & 0.4 & 0.5 \end{bmatrix} \tag{10.77}$$

- Preferences of E_2, expressed in terms of the ranking of all alternatives by taking into account each criterion separately:

$$\boldsymbol{O}_1^2 = [1 \quad 4 \quad 2 \quad 3 \quad 5], \ \boldsymbol{O}_2^2 = [4 \quad 2 \quad 1 \quad 3 \quad 5],$$
$$\boldsymbol{O}_3^2 = [5 \quad 4 \quad 3 \quad 2 \quad 1], \ \boldsymbol{O}_4^2 = [1 \quad 4 \quad 5 \quad 2 \quad 3] \tag{10.78}$$

- Preferences of E_3, expressed in terms of the fuzzy estimates, selected from the set of linguistic estimates shown in Figure 10.2:

$F_1^3(X_1) = \textit{very large}$, $F_1^3(X_2) = \textit{small}$, $F_1^3(X_3) = \textit{large}$, $F_1^3(X_4) = \textit{middle}$,
$F_1^3(X_5) = \textit{very small}$

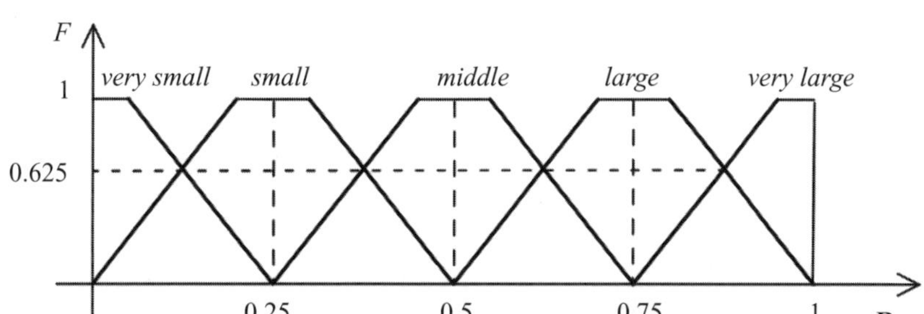

Figure 10.2 Membership functions of normalized fuzzy values.

$F_2^3(X_1) = small,\ F_2^3(X_2) = large,\ F_2^3(X_3) = very\ large,\ F_2^3(X_4) = middle,$
$F_2^3(X_5) = small$

$F_3^3(X_1) = very\ small,\ F_3^3(X_2) = small,\ F_3^3(X_3) = middle,\ F_3^3(X_4) = large,$
$F_4^3(X_5) = very\ large$

$F_4^3(X_1) = very\ large,\ F_4^3(X_2) = small,\ F_4^3(X_3) = very\ small,\ F_4^3(X_4) = middle,$
$F_4^3(X_5) = large$ (10.79)

- Preferences of E_4, expressed in terms of the ratio-scale utility values of all alternatives:

$$\boldsymbol{U}_1^4 = [0.8 \quad 0.4 \quad 0.7 \quad 0.6 \quad 0.1],\ \boldsymbol{U}_2^4 = [0.3 \quad 0.8 \quad 0.9 \quad 0.75 \quad 0.25]$$
$$\boldsymbol{U}_3^4 = [0.15 \quad 0.2 \quad 0.4 \quad 0.7 \quad 0.95],\ \boldsymbol{U}_4^4 = [0.65 \quad 0.2 \quad 0.1 \quad 0.2 \quad 0.3]$$

In Step 3, it is verified that the fuzzy preference relations $\boldsymbol{R}_1^1$, $\boldsymbol{R}_2^1$, $\boldsymbol{R}_3^1$, and $\boldsymbol{R}_4^1$ satisfy weak transitivity.

In Step 4, with the use of adequate transformation functions, all preferences are transformed to nonreciprocal fuzzy preference relations:

- Preferences of E_1:

$$\boldsymbol{R}_1^1 = \begin{bmatrix} 1 & 1 & 1 & 1 & 1 \\ 0.11 & 1 & 0.33 & 0.67 & 1 \\ 0.43 & 1 & 1 & 1 & 1 \\ 0.25 & 1 & 0.67 & 1 & 1 \\ 0.11 & 0.43 & 0.25 & 0.25 & 1 \end{bmatrix},\ \boldsymbol{R}_2^1 = \begin{bmatrix} 1 & 0.25 & 0.18 & 0.43 & 1 \\ 1 & 1 & 0.67 & 1 & 1 \\ 1 & 1 & 1 & 1 & 1 \\ 1 & 0.67 & 0.42 & 1 & 1 \\ 0.25 & 0.11 & 0.11 & 0.11 & 1 \end{bmatrix}$$
$$\boldsymbol{R}_3^1 = \begin{bmatrix} 1 & 0.25 & 0.11 & 0.11 & 0.11 \\ 1 & 1 & 0.43 & 0.25 & 0.11 \\ 1 & 1 & 1 & 0.67 & 0.42 \\ 1 & 1 & 1 & 1 & 0.67 \\ 1 & 1 & 1 & 1 & 1 \end{bmatrix},\ \boldsymbol{R}_4^1 = \begin{bmatrix} 1 & 1 & 1 & 1 & 1 \\ 0.11 & 1 & 1 & 0.43 & 0.67 \\ 0.05 & 0.43 & 1 & 0.18 & 0.25 \\ 0.25 & 1 & 1 & 1 & 1 \\ 0.18 & 1 & 1 & 0.67 & 1 \end{bmatrix} \tag{10.80}$$

- Preferences of E_2:

$$\boldsymbol{R}_1^2 = \begin{bmatrix} 1 & 1 & 1 & 1 & 1 \\ 0.13 & 1 & 0.25 & 0.38 & 1 \\ 0.38 & 1 & 1 & 1 & 1 \\ 0.25 & 1 & 0.38 & 1 & 1 \\ 0 & 0.38 & 0.13 & 0.25 & 1 \end{bmatrix},\ \boldsymbol{R}_2^2 = \begin{bmatrix} 1 & 0.25 & 0.13 & 0.38 & 1 \\ 1 & 1 & 0.38 & 1 & 1 \\ 1 & 1 & 1 & 1 & 1 \\ 1 & 0.38 & 0.25 & 1 & 1 \\ 0.38 & 0.13 & 0 & 0.25 & 1 \end{bmatrix}$$
$$\boldsymbol{R}_3^2 = \begin{bmatrix} 1 & 0.38 & 0.25 & 0.13 & 0 \\ 1 & 1 & 0.38 & 0.25 & 0.13 \\ 1 & 1 & 1 & 0.38 & 0.25 \\ 1 & 1 & 1 & 1 & 0.38 \\ 1 & 1 & 1 & 1 & 1 \end{bmatrix},\ \boldsymbol{R}_4^2 = \begin{bmatrix} 1 & 1 & 1 & 1 & 1 \\ 0.13 & 1 & 1 & 0.25 & 0.38 \\ 0.25 & 0.38 & 1 & 0.13 & 0.25 \\ 0.38 & 1 & 1 & 1 & 1 \\ 0.25 & 1 & 1 & 0.38 & 1 \end{bmatrix} \tag{10.81}$$

- Preferences of E_3:

$$\boldsymbol{R}_1^3 = \begin{bmatrix} 1 & 1 & 1 & 1 & 1 \\ 0 & 1 & 0 & 0.63 & 1 \\ 0.63 & 1 & 1 & 1 & 1 \\ 0 & 1 & 0.63 & 1 & 1 \\ 0 & 0.63 & 0 & 0 & 1 \end{bmatrix}, \quad \boldsymbol{R}_2^3 = \begin{bmatrix} 1 & 0 & 0 & 0.63 & 1 \\ 1 & 1 & 0.63 & 1 & 1 \\ 1 & 1 & 1 & 1 & 1 \\ 1 & 0.63 & 0 & 1 & 1 \\ 1 & 0 & 0 & 0.63 & 1 \end{bmatrix}$$
$$\boldsymbol{R}_3^3 = \begin{bmatrix} 1 & 0.63 & 0 & 0 & 0 \\ 1 & 1 & 0.63 & 0 & 0 \\ 1 & 1 & 1 & 0.63 & 0 \\ 1 & 1 & 1 & 1 & 0.63 \\ 1 & 1 & 1 & 1 & 1 \end{bmatrix}, \quad \boldsymbol{R}_4^3 = \begin{bmatrix} 1 & 1 & 1 & 1 & 1 \\ 0 & 1 & 1 & 0.63 & 0 \\ 0 & 0.63 & 1 & 0 & 0 \\ 0 & 1 & 1 & 1 & 0.63 \\ 0.63 & 1 & 1 & 1 & 1 \end{bmatrix} \tag{10.82}$$

- Preferences of E_4:

$$\boldsymbol{R}_1^4 = \begin{bmatrix} 1 & 1 & 1 & 1 & 1 \\ 0.25 & 1 & 0.32 & 0.44 & 1 \\ 0.76 & 1 & 1 & 1 & 1 \\ 0.56 & 1 & 0.73 & 1 & 1 \\ 0.01 & 0.06 & 0.02 & 0.03 & 1 \end{bmatrix}, \quad \boldsymbol{R}_2^4 = \begin{bmatrix} 1 & 0.14 & 0.11 & 0.16 & 1 \\ 1 & 1 & 0.79 & 1 & 1 \\ 1 & 1 & 1 & 1 & 1 \\ 1 & 0.88 & 0.69 & 1 & 1 \\ 0.69 & 0.10 & 0.08 & 0.11 & 1 \end{bmatrix}$$
$$\boldsymbol{R}_3^4 = \begin{bmatrix} 1 & 0.56 & 0.14 & 0.04 & 0.02 \\ 1 & 1 & 0.25 & 0.08 & 0.04 \\ 1 & 1 & 1 & 0.32 & 0.18 \\ 1 & 1 & 1 & 1 & 0.54 \\ 1 & 1 & 1 & 1 & 1 \end{bmatrix}, \quad \boldsymbol{R}_4^4 = \begin{bmatrix} 1 & 1 & 1 & 1 & 1 \\ 0.09 & 1 & 1 & 1 & 0.44 \\ 0.02 & 0.25 & 1 & 0.25 & 0.11 \\ 0.09 & 1 & 1 & 1 & 0.44 \\ 0.21 & 1 & 1 & 1 & 1 \end{bmatrix} \tag{10.83}$$

In Step 5, the problem is solved for each expert individually, using the *first technique* for multicriteria analysis, described in Chapter 7. The degrees of fuzzy nondominance of each alternative, obtained by considering separately the preferences of each expert, are as follows:

- Expert E_1:

$$\mathbf{ND}^1 = [0.86 \quad 0.58 \quad 0.75 \quad 1 \quad 0.44] \tag{10.84}$$

- Expert E_2:

$$\mathbf{ND}^2 = [0.88 \quad 0.51 \quad 0.88 \quad 0.82 \quad 0.53] \tag{10.85}$$

- Expert E_3:

$$\mathbf{ND}^3 = [0.85 \quad 0.38 \quad 1 \quad 0.74 \quad 0.38] \tag{10.86}$$

Table 10.25 Positions of the alternatives for each expert and for the group, at the first cycle

	X_1	X_2	X_3	X_4	X_5
E_1	2	4	3	1	5
E_2	1.5	5	1.5	3	4
E_3	2	4.5	1	3	4.5
E_4	2	5	4	1	3
Group	1	5	3	2	4

- Expert E_4:

$$\mathbf{ND}^4 = [0.95 \quad 0.2 \quad 0.56 \quad 1 \quad 0.58] \tag{10.87}$$

In Step 6, the collective result is calculated with the use of WAM:

$$\mathbf{ND}^C = [0.88 \quad 0.42 \quad 0.83 \quad 0.87 \quad 0.48] \tag{10.88}$$

Table 10.25 shows the ranking of the alternatives for each expert and for the group, in accordance with the results shown above.

In Step 7, the mean level of concordance is calculated for each specialist using (10.17) with $b = 1$ and (10.18) (refer to Table 10.26).

In Step 8, the level of consensus achieved by the group is calculated using (10.19) and (10.20). Refer to Table 10.27 for the level of consensus for each alternative and the mean level of consensus for the set of alternatives.

In Step 9, as none of the stopping conditions have been satisfied, the procedure goes on to Step 10.

In Step 10, the least concordant expert is E_3, as confirmed in Table 10.26.

In Step 11, the value of variable *cycle* is updated to 2, E_3 is invited to review his/her opinion, and his/her index is stored in $elast(2)$.

Considering the fact that $O_1^C - O_1^3 < 0$, $O_2^C - O_2^3 > 0$, $O_3^C - O_3^3 > 0$, $O_4^C - O_4^3 < 0$ and $O_5^C - O_5^3 < 0$, the advice mechanism generates the following recommendations: E_3 should increase the evaluations of X_1, X_4 and X_5, and decrease the evaluations of X_2 and X_3.

In Step 12, the modified preferences of E_3 are collected:

$$\boldsymbol{F}_3^3(X_3) = \textit{very small} \tag{10.89}$$

Table 10.26 Levels of concordance for each expert at the first cycle

	$SO_1^{y,C}$	$SO_2^{y,C}$	$SO_3^{y,C}$	$SO_4^{y,C}$	$SO_5^{y,C}$	$SO^{y,C}$
E_1	0.75	0.75	1	0.75	0.75	0.8
E_2	0.88	1	0.63	0.75	1	0.85
E_3	0.75	0.88	0.5	0.75	0.88	0.75
E_4	0.75	1	0.75	0.75	0.75	0.8

Table 10.27 Levels of consensus for each individual alternative for the set of alternatives

CO_1	CO_2	CO_3	CO_4	CO_5	CO
0.76	0.91	0.72	0.75	0.78	0.78

In Step 4, the fuzzy preference relation $\boldsymbol{R}_3^3$ is updated to

$$\boldsymbol{R}_2^2 = \begin{bmatrix} 1 & 0.63 & 1 & 0 & 0 \\ 1 & 1 & 1 & 0 & 0 \\ 1 & 0.63 & 1 & 0 & 0 \\ 1 & 1 & 1 & 1 & 0.63 \\ 1 & 1 & 1 & 1 & 1 \end{bmatrix} \tag{10.90}$$

In Step 5, the multicriteria analysis is performed again for E_3 and the following result is obtained:

$$\mathbf{ND}^2 = [1 \quad 0.38 \quad 0.95 \quad 0.74 \quad 0.38] \tag{10.91}$$

In Step 6, the collective result is updated to

$$\mathbf{ND}^C = [0.93 \quad 0.42 \quad 0.81 \quad 0.87 \quad 0.48] \tag{10.92}$$

Table 10.28 shows the ranking of the alternatives for each expert and for the group, in accordance with the results shown above.

In Step 7, the levels of concordance are calculated using (10.17) with $b = 1$ and (10.18). Table 10.29 shows the values obtained for $SO^{y,C}$ and for $SO_k^{y,C}$ where $y = 1, 2, 3, 4$ and $k = 1, 2, \ldots, 5$.

In Step 8, the mean level of consensus achieved by the group is calculated using (10.19) and (10.20), as can be seen in Table 10.30.

In Step 9, as the current level of consensus is already viewed as satisfactory, the procedure is transferred to Step 13 and then terminated. The output is the collective result given by (10.92).

Table 10.28 Rankings of the alternatives for each expert and for the group, at the second cycle

	X_1	X_2	X_3	X_4	X_5
E_1	2	4	3	1	5
E_2	1.5	5	1.5	3	4
E_3	1	4.5	2	3	4.5
E_4	2	5	4	1	3
Group	1	5	3	2	4

Table 10.29 Levels of concordance for each expert, at the second cycle

	$SO_1^{y,C}$	$SO_2^{y,C}$	$SO_3^{y,C}$	$SO_4^{y,C}$	$SO_5^{y,C}$	$SO^{y,C}$
E_1	0.75	0.75	1	0.75	0.75	0.8
E_2	0.88	1	0.63	0.75	1	0.85
E_3	1	0.88	0.75	0.75	0.88	0.85
E_4	0.75	1	0.75	0.75	0.75	0.8

10.8 Conclusions

This chapter considered two different approaches which can be utilized for forming a consensual result for discrete multicriteria decision-making problems in a group environment. They are the consensus schemes and the procedures for constructing an optimized consensus.

Three different consensus schemes were presented. These are the consensus scheme based on fuzzy estimates (CSFE), consensus scheme based on fuzzy preference relations (CSFPR), and consensus scheme based on the ranking of the alternatives (CSRA). As discussed in the chapter, they find different applications in practice.

CSRA aims at constructing a consensus on the final ranking of the alternatives. It is particularly useful when all experts can solve the entire problem individually and it is important to guarantee a consensus on the final results, rather than a consensus on the preferences of the experts.

On the other hand, CSFPR aims at constructing a consensus on the preferences per criterion. When CSFPR is used, it may be that an acceptable level of consensus is reached on the preferences per criterion, but the level of consensus on the final ranking of the alternatives is not necessarily satisfactory. CSFPR is particularly useful when the group is formed of experts from different areas, who are invited to contribute their opinions over the alternatives, considering only the criteria related to their respective areas of expertise.

Finally, CSFE allows the discussion to be organized in such a way that each alternative is considered in an independent session, which may involve different groups of experts. However, as indicated in the text, CSFE can be utilized only when the experts involved in the decision-making process have similar levels of exigency or are supposed to follow certain standards of exigency. Otherwise, it may be impossible to obtain a consensus on the evaluation of each alternative.

Regardless of the type of consensus scheme being utilized, it is not always possible or easy to achieve a satisfactory level of consensus on the collective results. In real-world applications, some discordant experts may be reluctant to modify their respective opinions. To deal with these cases, we consider the use of a procedure for obtaining an optimized consensus. It consists of an algorithm for modifying the weight of each opinion in the construction of the aggregated opinion, in such a way that the level of consensus, reflected by a weighted

Table 10.30 Levels of consensus for each individual alternative for the set of alternatives

CO_1	CO_2	CO_3	CO_4	CO_5	CO
0.85	0.91	0.78	0.75	0.85	0.83

index of consensus, is increased until an acceptable level has been reached. However, it is important to mention that, in some cases (as can be confirmed by the numerical example included in Section 10.5), the collective result remains unmodified even after running this procedure.

In the last section, the use of CSRA was demonstrated through an application example, which consists of solving a multicriteria problem generated with the use of the balanced scorecard methodology for strategic planning in an organization.

Exercises

Problem 10.1. Consider that four experts are invited to compare four alternatives. Given the individual opinions expressed in terms of fuzzy preferences, verify which expert is the least concordant member of the group. Utilize WAM to obtain the collective fuzzy preference relation and assume that all professionals are of the same importance, that is, $w_y = 0.2$, $y = 1, 2, \ldots, 5$:

$$\boldsymbol{R}_1^1 = \begin{bmatrix} 1 & 0 & 0.5 & 1 \\ 1 & 1 & 1 & 1 \\ 1 & 0.44 & 1 & 1 \\ 0.53 & 0 & 0 & 1 \end{bmatrix}, \quad \boldsymbol{R}_1^2 = \begin{bmatrix} 1 & 0.53 & 0.54 & 1 \\ 1 & 1 & 1 & 1 \\ 1 & 1 & 1 & 1 \\ 1 & 0.61 & 0.51 & 1 \end{bmatrix}, \quad \boldsymbol{R}_1^3 = \begin{bmatrix} 1 & 0 & 0.64 & 1 \\ 1 & 1 & 1 & 1 \\ 1 & 0.72 & 1 & 1 \\ 1 & 0 & 0.53 & 1 \end{bmatrix},$$

$$\boldsymbol{R}_1^4 = \begin{bmatrix} 1 & 0.63 & 0.63 & 1 \\ 1 & 1 & 1 & 1 \\ 1 & 1 & 1 & 1 \\ 0.63 & 0 & 0 & 1 \end{bmatrix}, \quad \boldsymbol{R}_1^5 = \begin{bmatrix} 1 & 0.41 & 1 & 1 \\ 1 & 1 & 1 & 1 \\ 1 & 0.32 & 1 & 1 \\ 0.6 & 0 & 0.64 & 1 \end{bmatrix}$$

Problem 10.2. Calculate the current level of consensus per alternative and per relation among the experts discussed in Problem 10.1.

Problem 10.3. Consider that four experts are invited to solve a multicriteria problem. It consists of ranking four alternatives in accordance with their importance. The individual results obtained by the experts are as follows:

- Expert E_1:

$$\mathbf{ND}^1 = [0.5 \quad 0.6 \quad 0.45 \quad 0.7]$$

- Expert E_2:

$$\mathbf{ND}^2 = [0.45 \quad 0.75 \quad 0.5 \quad 0.6]$$

- Expert E_3:

$$\mathbf{ND}^3 = [0.5 \quad 0.45 \quad 0.3 \quad 0.6]$$

- Expert E_4:

$$\mathbf{ND}^4 = [0.6 \quad 0.5 \quad 0.4 \quad 0.55]$$

- Expert E_5:

$$\mathbf{ND}^5 = [0.5 \quad 0.4 \quad 0.65 \quad 0.5]$$

Verify which expert is the least concordant in the group is. Utilize WAM to obtain the collective fuzzy preference relation and assume that all professionals are of the same importance, that is, $w_y = 0.2$, $y = 1, 2, \ldots, 5$.

Problem 10.4. Calculate the current level of consensus per alternative and the mean level of consensus, considering the preferences given by the experts in Problem 10.3.

Problem 10.5. Execute each step of the procedure to construct an optimized consensus described in Section 10.5, in order to obtain an optimized consensus. Consider the preferences given by the five experts discussed in Problem 10.3.

References

Ben-Arieh, D. and Chen, Z. (2006) Linguistic-labels aggregation and consensus measure for autocratic decision-making using group recommendations. *IEEE Transactions on Systems, Man, and Cybernetics – Part A: Systems and Humans*, **36** (3), 558–568.

Bernardes, P., Ekel, P., Kotlarewski, J., and Parreiras, R. (2009) Fuzzy set based multicriteria decision-making and its applications, in *Progress on Nonlinear Analysis* (ed. E. T. Hoffmann), Nova Science, Hauppauge, pp. 247–272.

Bremser, W.G. and White, L.F. (2000) An experimental approach to learning about the balance scorecard. *Journal of Accounting Education*, **18** (3), 241–255.

Bui, T.X. and Jarke, M. (1986) Communications design for Co-oP: a group decision support system. *ACM Transactions on Office Information Systems*, **4** (2), 81–103.

Ekel, P., Queiroz, J., Parreiras, R., and Palhares, R. (2009) Fuzzy set based models and methods of multicriteria group decision-making. *Nonlinear Analysis: Theory, Methods & Applications*, **71** (12), e409–e419.

Eklund, P., Rusinowska, A., and De Swart, H. (2007) Consensus reaching in committees. *European Journal of Operational Research*, **178** (1), 185–193.

Griffith, T.L., Fuller, M.A., and Northcraft, G.B. (1998) Facilitator influence in group support systems: intended and unintended effects. *Information Systems Research*, **9** (1), 20–36.

García-Lapresta, J.L. (2008) Favoring consensus and penalizing disagreement in group decision-making. *Journal of Advanced Computational Intelligence and Intelligent Informatics*, **12** (5), 416–421.

Herrera-Viedma, E., Alonso, S., Chiclana, F., and Herrera, F. (2007) A consensus model for group decision-making with incomplete fuzzy preference relations. *IEEE Transactions on Fuzzy Systems*, **15** (5), 863–877.

Herrera-Viedma, E., Herrera, F., and Chiclana, F. (2002) A consensus model for multiperson decision-making with different preference structures. *IEEE Transactions on Systems, Man, and Cybernetics – Part A: Systems and Humans*, **32** (3), 394–402.

Hsu, H.M. and Chen, C.T. (1996) Aggregation of fuzzy opinions under group decision-making. *Fuzzy Sets and Systems*, **79** (3), 279–285.

Jiang, J.J. and Klein, G. (2000) Side effects of decision guidance in decision support systems. *Interacting with Computers*, **12**, 469–481.

Kaplan, R.S. and Norton, D.P. (1992) The balanced scorecard—-measures that drive performance. *Harvard Business Review*, **70** (1), 71–79.

Lu, C., Lan, J., and Wang, Z. (2006) Aggregation of fuzzy opinions under group decision-making based on similarity and distance. *Journal of Systems Science and Complexity*, **19** (1), 63–71.

Madu, C.N. and Kuei, C.H. (1995) Stability analyses of group decision-making. *Computers and Industrial Engineering*, **28** (4), 881–892.

Salo, A.A. (1995) Interactive decision aiding for group decision support. *European Journal of Operational Research*, **84** (1), 134–149.

Shanteau, J. (2001) What does it mean when experts disagree?, in *Linking Expertise and Naturalistic Decision-Making* (eds E. Salas and G. Klein), Lawrence Erlbaum, Mahwah, NJ.

Index

Analytic Hierarchy Process (AHP) 70
A posteriori approach 105
A priori approach 105
Adaptive approach 106
Adaptive interactive decision-making system 116, 118
Additional information representation 106
Additive reciprocal fuzzy preference relation 163
Aggregated payoff matrix 251, 258
Aggregated risks 252, 253
Aggregation operator 82, 114, 222, 269
Alternative 1, 4, 9–10, 278
Alpha (α) cut 33, 38, 194
Analytical model 5
Assignment principle 147
Attribute 9–10
Automorphism 139
Axiom of independence of irrelevant alternatives 146
Axiom of independence of rejected alternatives 205
Axiom of symmetry of indifference and incomparability relations 147

Bellman-Zadeh approach to decision-making 112–13, 120
Binary fuzzy relation
  definition 56, 137, 138
  properties 55, 56, 140
Boldur's method 109
Brainstorming 278

Characteristic estimates 250, 252
Characteristic function 23, 42
Choice criteria 250, 252
Choice of alternatives 4, 202–3, 207
Classic approach to dealing with uncertainty of information 247
Clustering 76
Complement of a fuzzy set 46, 138
Concordance index 296–8, 302
Consensus 13, 293
Consensus index 296–8, 302, 305
Consensus scheme
  Definition 295, 309–10
  Guidance procedures 311–12, 318
Consistency
  of fuzzy preference relation 148, 163
  of multiplicative preference relation 55, 137
  under group settings 275
Continuous decision space 10
Continuous models of multicriteria decision-making 103
Cooperative decision-making 12, 264, 293
Criteria
  Hurwicz 248, 250
  Laplace 248, 250
  Savage 248, 250
  Wald 248, 250
Criterion 10

Decision-maker (DM) 2–4
Decision-making 2, 4, 6, 9
Decision-making problem 2, 4, 6, 9
Decision-making process 2–3, 13
Decision space 7, 10, 278
Decision support 4, 6, 12
Decision support system 4, 6, 12
Decision uncertainty region 6–8
Decision variable 10
Dictatorial consensus 305
Discrete decision space 10, 278

Efficient solution 104–5
Extension principle 84

Feasible solutions 6
Fuzzy arithmetic 87
Fuzzy C-Means (FCM) 76
Fuzzy estimate 66, 160
Fuzzy number 28
Fuzzy objective function 76
Fuzzy preference relation 11, 141, 163, 202
Fuzzy quantifier 222, 224
Fuzzy relation 47, 49
  Cartesian product 49, 52
  Cylindrical extension 54
  Equivalence 57
  Operations 51
  Projection 53
  Reconstruction 55
  Transposition 52
Fuzzy set
  Cardinality 34
  Convexity 33
  Core 32
  Equality 35
  Equalization 80
  Interpretation 22, 24
  Normality 31
  Normalization 32
  Operations 41
  Support 32
  Specificity 37
Fuzzy set of non-dominated alternatives 204
Fuzzy solution 113

General scheme of multicriteria decision making under uncertainty 253
Generalized preference relation 167
Goal programming 111
Goals 10, 111
Group decision environment 264
Group decision process 278
Group decision-making 11
Group decision-making problem 263

Harmonious solutions 112
Human participation 6, 9
Human-oriented interfaces 17

Ill-structured problem 4
Importance factor 110, 116, 207, 228, 268
Importance of objective function 110, 124
Importance weight of experts 268, 294,305
Incommensurable unit 9
Incomparability relation 142–3, 145–6, 150
Indifference relation 142–3, 145–6, 150
Individual decision-making 6, 12
Information granularity 74
Information of qualitative character 16
Interactive approach 106
Intersection of fuzzy sets 44
Interval coefficients 254
Interval scale 158

Large preference relation 142
Lexicographic goal programming 112
Lexicographic character 207
Linguistic variable 29, 160
LP_{τ}-sequence 253

Measure of fuzziness
  Energy 36
  Entropy 36
Membership function 26
  Definition 23
  Elicitation 63
  Gaussian 27
  Triangular 26
  Trapezoidal 27
Method of global criterion 112

Method of successive concessions 111
Moderator 8, 294–5, 304
Modified criterion
  Hurwicz 253
  Laplace 252
  Savage 252
  Wald 252
Modified payoff matrix 251
Multiattribute decision-making 10–11, 202
Multicriteria decision-making 3, 9–10, 103, 193
Multicriteria power and energy shortage allocation 120, 122
Multicriteria Resource Allocation 115
Multiobjective decision making 10, 103
Multiperson and Multiattribute Aggregation Modes 265
Multiple criteria 2, 9
Multiplicative preference relation 70, 161

Negation 46
Non-compensatory behavior 124, 221, 272
Non-cooperative decision making 12, 264
Non-dominated solutions 104–5
Non-local search 116
Nonstrict preference relation 142
Nonreciprocal fuzzy preference relation 163–4
Normalization of objective functions 107–8

Objective function 6–7, 103
Objective space 104–5, 278
Objectives 9–10, 112
Operational research 3, 5
Optimal solution 3, 6–7
Optimization 6
Optimized (dictatorial) consensus 305
Ordered weighted average 203, 221, 271, 272, 274
Orlovsky choice function 203, 205
Outranking relation 203, 216

Pareto-optimal front 105
Pareto-optimal solution set 105
Pareto-optimal solutions 105
Particular risks 252–3
Payoff matrices 247–8
Positive association principle 146
Preference elicitation
  of fuzzy estimates 160
  of multiplicative preference relations 161–2
  of utility functions 157
  Process 155
Preference formats 156
  Fuzzy preference relation 163
  Fuzzy estimates 160
  Multiplicative preference relation 70, 161
  Ordering of alternatives 156
  Utility values 157
Preference function 226
Preference structures
  Definition 145
  Construction 146
Principle of justifiable granularity 73
Principle of optimality 107
Principle of just compromise 110
Principle of uniform optimality 110
Priorities of objective functions 107
Probabilistic methods15
Promethee 216
Proximity relation 58

Quantifier guided dominance degree 224
Quantifier guided non-dominance degree 224

Ranking of the alternatives 203, 230, 266
Ratio scale 158, 161
Real-time decision support 5
Reciprocity (reciprocality)
  multiplicative 70,
  additive 140
Regret 249–50
Representation theorem 39
Representative combinations of initial data 247–8
Risk 249–50, 252
Risk matrix 250

Scenario 247, 253
Semi-structured problem 4–5
State of nature 249
Strict preference relation 141–3, 146, 203
Structured problem 4

Triangular norm 43
T-conorm 45
T-norm 44
Transformation function
  Definition 172
  For additive reciprocal fuzzy preference relation 172
  For nonreciprocal fuzzy preference relation 180
Transitivity
  Additive transitivity 163, 173, 176–8, 186, 273, 275,
  Multiplicative transitivity 70, 162, 164, 174, 176, 178, 185–6
  Min-transitivity 56, 141, 150
  T-transitivity 140–1
  Weak-transitivity 141, 149–50, 205, 271, 275

Uncertainty 2, 5–6
Uncertainty factor 2
Uncertainty of goals 7, 9
Uncertainty of information 6, 14
Union of fuzzy sets 45
Unstructured problems 4, 5

Weak domination 104
Weak Pareto optimal solutions 105
Weak preference relation 142
Weighted arithmetic mean 270–2, 274
Weighted geometric mean 271–2, 274

< **X**, **M** > models 11, 17, 103, 107
< **X**, ***R*** > models 11, 17, 202–3